HINTS & HACKS

for the Radio Amateur

Edited by Steve Ford, WB8IMY

ARRL The National Association for Amateur Radio®
225 Main Street, Newington, CT 06111-1400 USA
www.arrl.org

Printed in the USA

ISBN: 978-1-62595-197-7

First Edition

Contents

Preface

Welcome to the first edition of *Hints & Hacks for the Radio Amateur*. This book is a compilation of material published in *QST* magazine's popular "Hints & Hacks" column between January 2017 and December 2023.

Between these covers you'll find handy advice on a broad range of topics from antennas to vintage radio restoration. The great thing about "Hints & Hacks" is that all these tips come from your fellow amateurs, people just like you who've discovered creative solutions to everyday problems. That's why "Hints & Hacks" has held an honored position for decades as one of the most-read sections of *QST*.

If you have hint or hack to contribute, by all means send it to our Hints & Hacks editor. You can reach the editor by e-mail at **hh@arrl.org**.

Please note: The information in this book was current when it was originally published in *QST*. Call signs, postal addresses, e-mail addresses and websites cited in articles may now be out of date.

US Customary to Metric Conversions

International System of Units (SI)—Metric Prefixes

Prefix	*Symbol*			*Multiplication Factor*
exa	E	10^{18}	=	1 000 000 000 000 000 000
peta	P	10^{15}	=	1 000 000 000 000 000
tera	T	10^{12}	=	1 000 000 000 000
giga	G	10^{9}	=	1 000 000 000
mega	M	10^{6}	=	1 000 000
kilo	k	10^{3}	=	1 000
hecto	h	10^{2}	=	100
deca	da	10^{1}	=	10
(unit)		10^{0}	=	1
deci	d	10^{-1}	=	0.1
centi	c	10^{-2}	=	0.01
milli	m	10^{-3}	=	0.001
micro	μ	10^{-6}	=	0.000001
nano	n	10^{-9}	=	0.000000001
pico	p	10^{-12}	=	0.000000000001
femto	f	10^{-15}	=	0.000000000000001
atto	a	10^{-18}	=	0.000000000000000001

Linear
1 metre (m) = 100 centimetres (cm) = 1000 millimetres (mm)

Area
1 $m^2 = 1 \times 10^4\ cm^2 = 1 \times 10^6\ mm^2$

Volume
1 $m^3 = 1 \times 10^6\ cm^3 = 1 \times 10^9\ mm^3$
1 litre (l) = 1000 $cm^3 = 1 \times 10^6\ mm^3$

Mass
1 kilogram (kg) = 1 000 grams (g)
(Approximately the mass of 1 litre of water)
1 metric ton (or tonne) = 1 000 kg

US Customary Units

Linear Units
12 inches (in) = 1 foot (ft)
36 inches = 3 feet = 1 yard (yd)
1 rod = 5½ yards = 16½ feet
1 statute mile = 1 760 yards = 5 280 feet
1 nautical mile = 6 076.11549 feet

Area
1 ft^2 = 144 in^2
1 yd^2 = 9 ft^2 = 1 296 in^2
1 rod^2 = 30¼ yd^2
1 acre = 4840 yd^2 = 43 560 ft^2
1 acre = 160 rod^2
1 $mile^2$ = 640 acres

Volume
1 ft^3 = 1 728 in^3
1 yd^3 = 27 ft^3

Liquid Volume Measure
1 fluid ounce (fl oz) = 8 fluidrams = 1.804 in^3
1 pint (pt) = 16 fl oz
1 quart (qt) = 2 pt = 32 fl oz = 57¾ in^3
1 gallon (gal) = 4 qt = 231 in^3
1 barrel = 31½ gal

Dry Volume Measure
1 quart (qt) = 2 pints (pt) = 67.2 in^3
1 peck = 8 qt
1 bushel = 4 pecks = 2 150.42 in^3

Avoirdupois Weight
1 dram (dr) = 27.343 grains (gr) or (gr a)
1 ounce (oz) = 437.5 gr
1 pound (lb) = 16 oz = 7 000 gr
1 short ton = 2 000 lb, 1 long ton = 2 240 lb

Troy Weight
1 grain troy (gr t) = 1 grain avoirdupois
1 pennyweight (dwt) or (pwt) = 24 gr t
1 ounce troy (oz t) = 480 grains
1 lb t = 12 oz t = 5 760 grains

Apothecaries' Weight
1 grain apothecaries' (gr ap) = 1 gr t = 1 gr a
1 dram ap (dr ap) = 60 gr
1 oz ap = 1 oz t = 8 dr ap = 480 fr
1 lb ap = 1 lb t = 12 oz ap = 5 760 gr

Multiply →

Metric Unit = Conversion Factor × US Customary Unit

← Divide

Metric Unit ÷ Conversion Factor = US Customary Unit

Metric Unit =	Conversion Factor ×	US Unit
(Length)		
mm	25.4	inch
cm	2.54	inch
cm	30.48	foot
m	0.3048	foot
m	0.9144	yard
km	1.609	mile
km	1.852	nautical mile
(Area)		
mm^2	645.16	$inch^2$
cm^2	6.4516	in^2
cm^2	929.03	ft^2
m^2	0.0929	ft^2
cm^2	8361.3	yd^2
m^2	0.83613	yd^2
m^2	4047	acre
km^2	2.59	mi^2
(Mass)	(Avoirdupois Weight)	
grams	0.0648	grains
g	28.349	oz
g	453.59	lb
kg	0.45359	lb
tonne	0.907	short ton
tonne	1.016	long ton

Metric Unit =	Conversion Factor ×	US Unit
(Volume)		
mm^3	16387.064	in^3
cm^3	16.387	in^3
m^3	0.028316	ft^3
m^3	0.764555	yd^3
ml	16.387	in^3
ml	29.57	fl oz
ml	473	pint
ml	946.333	quart
l	28.32	ft^3
l	0.9463	quart
l	3.785	gallon
l	1.101	dry quart
l	8.809	peck
l	35.238	bushel
(Mass)	(Troy Weight)	
g	31.103	oz t
g	373.248	lb t
(Mass)	(Apothecaries' Weight)	
g	3.387	dr ap
g	31.103	oz ap
g	373.248	lb ap

Common Schematic Symbols Used in Circuit Diagrams

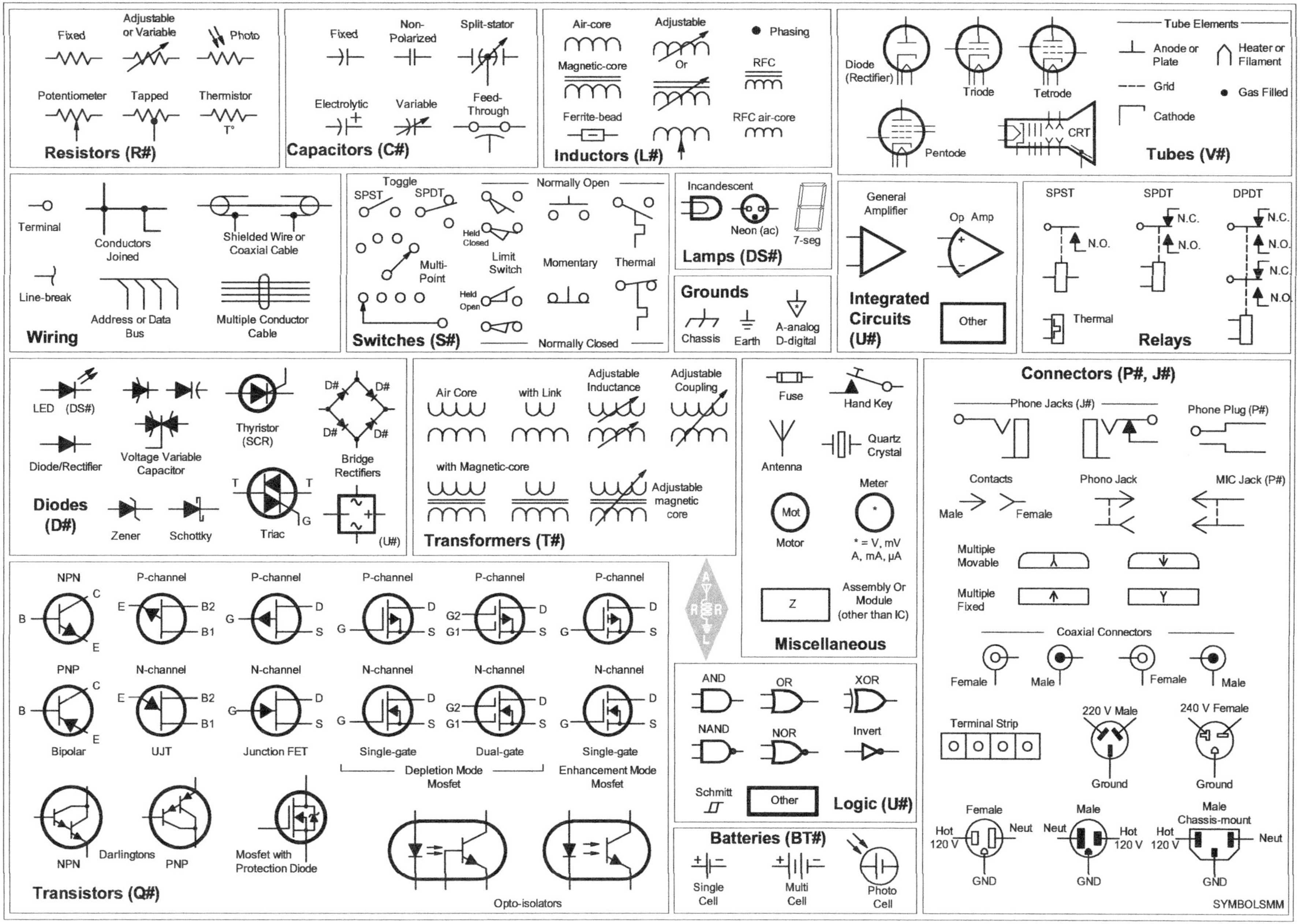

About ARRL

We're the American Radio Relay League, Inc. — better known as ARRL. We're the largest membership association for the amateur radio hobby and service in the US. For over 100 years, we have been the primary source of information about amateur radio, offering a variety of benefits and services to our members, as well as the larger amateur radio community. We publish books on amateur radio, as well as four magazines covering a variety of radio communication interests. In addition, we provide technical advice and assistance to amateur radio enthusiasts, support several education programs, and sponsor a variety of operating events.

One of the primary benefits we offer to the ham radio community is in representing the interests of amateur radio operators before federal regulatory bodies advocating for meaningful access to the radio spectrum. ARRL also serves as the international secretariat of the International Amateur Radio Union, which performs a similar role internationally, advocating for amateur radio interests before the International Telecommunication Union and the World Administrative Radio Conferences.

Today, we proudly serve nearly 150,000 members, both in the US and internationally, through our national headquarters and flagship amateur radio station, W1AW, in Newington, Connecticut. Every year we welcome thousands of new licensees to our membership, and we hope you will join us. Let us be a part of your amateur radio journey. Visit www.arrl.org/join for more information.

ARRL The National Association for Amateur Radio®

225 Main Street
Newington, CT 06111-1400 USA
Tel: 860-594-0200
FAX: 860-594-0259
Email: membership@arrl.org

www.arrl.org

CHAPTER 1

Antennas, Tips, and Tools

Identifying a Balun

When you find an old unlabeled balun in your junk box and can't remember if it's a 1:1, 4:1, or 9:1, there's a simple way to find out. All you need are two 100 Ω resistors and either an antenna analyzer or your rig and an SWR meter.

Any small resistor will work, but it should not be the wire-wound type. [Use an ohmmeter to get an accurate value for the resistors before you begin testing. — Ed.] If you are using an antenna analyzer, you don't have to worry about the power dissipation of the resistors. If you're using a rig and SWR meter, use low power for testing — the lowest your meter will read. You can use 100 Ω, 2 W composition resistors, but don't key the transmitter for an extended period if 5 W is your lowest power level. [Note that if you overheat a resistor, its resistance value can change, perhaps significantly. This is especially true if it becomes too hot to touch. — Ed.]

Connect the two 100 Ω resistors in series across the antenna terminals of the balun. Connect the feed-line side of the balun to your antenna analyzer, or SWR meter and transmitter. Use a relatively low frequency, such as 80 or 40 meters, and measure the SWR. Don't go lower than 80 meters, because most baluns are designed to operate in the 3.5 – 30 MHz range. If you go above 40 meters, you run the risk of lead inductance and internal inductance in the resistors confusing your readings.

Now take an SWR reading. If your balun is 4:1, the impedance at the transmission line side of the balun should be 200 Ω divided by four, or 50 Ω, providing a perfect 1:1 match. But, that's in a perfect world where resistors and their leads have no inductance, and the balun's turn ratio is a perfect 4:1. In the real world, you'll get a low reading — something like 1.5:1 or less. If the balun is 1:1, the 200 to 50 Ω impedance ratio should be four, giving an SWR measurement of 4:1. In reality, it may be something on the order of 3:1 or more, due to internal losses and that pesky stray inductance.

Let's suppose you get that high reading, indicating your balun is a 1:1 type. There's an easy way to verify that. Place the two 100 Ω resistors in parallel across the antenna terminals of the balun. In parallel, they will be 50 Ω and show an excellent match. Again, it's an imperfect world, so you might not get exactly zero reflected power and a perfect 1:1 SWR, but it should be close, say 1.2:1 or 1.5:1.

I once wound a 9:1 toroidal balun for a Beverage receiving antenna. A 450 Ω resistor was used to test it, and what a delight it was to see a 1:1 match on the antenna analyzer. There's something very reassuring, maybe even thrilling, about seeing theory and practical application merge. — *73, Richard Foster, W5TA,* **w5ta@arrl.net**

Coax-Seal Substitute

I find that Coax-Seal works quite well, as evidenced by the condition of the connector when the seal material is removed. However, with the closing of so many radio parts stores, this product is becoming harder to find. In searching for a more readily available substitute at a local home improvement store, I found a product that works equally well. It's called Mortite Weather Strip and Caulking Cord. It appears to be made out of a similar material as Coax-Seal, and will not harden with age. — *73, Ken Slusher, N2DF,* **n2df@arrl.net**

Give Me Some Slack

For years, I used a combination of black electrical tape and rubber splicing tape as a homebrew strain relief for the feed line of my horizontal full-wave loop.

The loop is tied off to five different trees and needs to be able to swing up and down during high winds as the trees sway back and forth. I decided to change out my balun and the outside portion of the feed line. I wanted a stronger strain relief for the feed line that would still allow movement with the wind. A simple piece of 1⁄2-inch ID rubber hose works perfectly to allow the feed line to move with the wind, and at the same time, it keeps the coax nice and solid at the post. [The hose may need to be checked periodically for UV damage. — Ed.] — *73, Mark Burton, KG1Q,* **kg1q@arrl.net**

Wilderness Antenna

I use an aluminum staff when I hike, as an aid to keep balance on a rocky trail. A rubber duck is not a very good long-distance antenna, so I built this antenna bracket that clamps on the end of my staff, providing a better counterpoise than the handheld and allowing me to elevate my antenna for improved coverage (see **Figure 1**).

A right-angle bracket cut from aluminum sheet clamps to the staff and holds a BNC panel jack for my rubber duck or a gain-type antenna. I used about 8 feet of RG-174 coax from the BNC jack to my radio. RG-58 would have very slightly less loss, but the RG-174 makes this adapter as easy as possible to coil up and carry. The antenna measures 1.4:1 standing-wave ratio (SWR) on 146.52, weighs about 3.5 ounces, and fits easily in a pocket. An on-the-air test with two repeaters about 15 miles away showed S-5 on the display of my Icom IC-W32

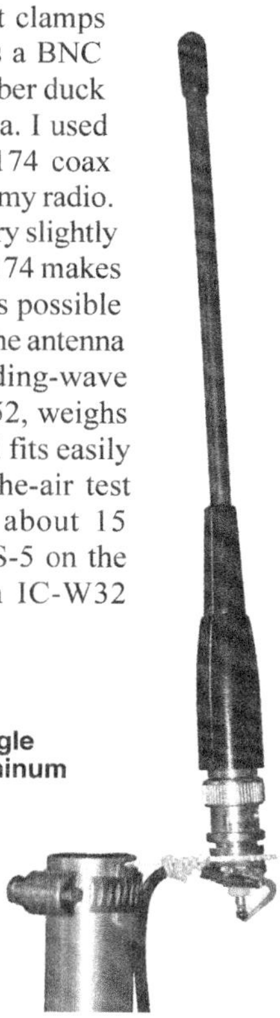

Figure 1 — A right-angle bracket cut from aluminum sheet clamps to the staff and allows me to elevate my antenna for improved coverage. [Scott McCann, W3MEO, photo]

with the rubber ducky on the rig, and S-9 using the wilderness antenna. I'm sure these readings are not equal to a lab-grade receiver, but they do show an appreciable improvement with the very small wilderness antenna. If you use a fiberglass or wooden hiking staff, you can screw or epoxy a 19-inch-long metal strip to the end of your staff and clamp the wilderness antenna to that. — *73, H. Scott McCann, W3MEO,* **achess@juno.com**

Self-Amalgamating Silicone Tape

In addition to the old-fashioned, gummy coax connector sealant, there is a readily available product that seals well and leaves absolutely no residue. Self-amalgamating (or self-fusing) silicone tape is sold under a variety of brand names and is used by the military and aviation industries.

It takes a little practice to apply it correctly, because unlike adhesive tape, silicone tape doesn't adhere to anything other than itself. The tape is applied by stretching and simultaneously wrapping in overlapping turns. As it is stretched and wrapped over itself, it fuses together to form a watertight seal.

One of the drawbacks of silicone tape is that you need to have 360-degree access around the connector to get the tape stretched as it is wrapped. Mounted antenna connectors are difficult to access, so it's best to attach and seal the coax before mounting to the mast. The silicone has UV resistance, but a few wraps of quality electrical tape on top will provide an additional UV barrier. — *73, Benson Wills, NE4W,* **ne4w@arrl.net**

Hose Strain Relief

I finally got my G5RV antenna up, ladder line to the side of my wife's garden barn, and ran a length of RG-11 coax to my shack. I had to go between two buildings and needed a strain relief on each eave. A search of the QST archives didn't show anything quite like what I needed. WB3DDM's solution for a strain relief at the feed point of his inverted long wire (from the April 1987 "Hints & Kinks" column) inspired me to look at old garden hose as an option.

After a little practice with a scrap of coax, angling the galvanized wood screws to miss the cable and penetrate the other side of the hose, I had my strain relief. Start the screws through so the coax slips through the hose. Put the hose/coax up in position, then turn the screws one or two turns into the mounting surface (see **Figure 2**). Your wire should still slip. Adjust as needed, and run the screws down a few turns at a time until the coax doesn't slip without pulling hard. Don't tighten any more than necessary. The inside of the hose will collapse around your coax and push it away from the screw (up to a certain point). — *73, Jon Sumpter, AC7CD,* **jonsmptr@msn.com**

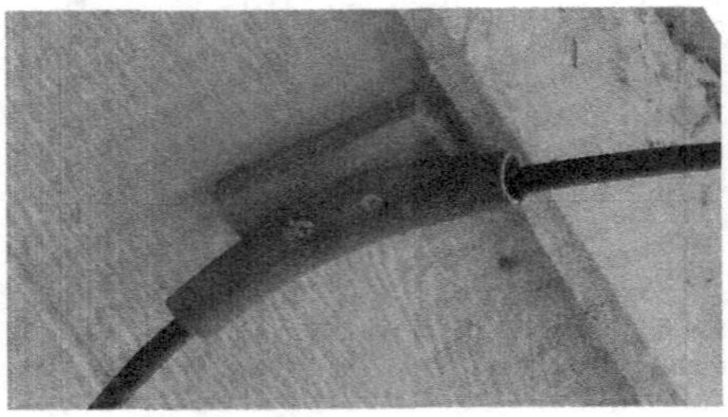

Figure 2 — Old garden hose serves as a strain relief, while still allowing the wire to slip. [Jon Sumpter, AC7CD, photo]

Mounting Modification for Butternut Vertical Doorknob Capacitor

In researching vertical base station antennas, I decided on the Butternut HF6V, a Don Newcomb, WØDN (SK), design, but I learned that the mounting bracket for the 30-meter 67 pF doorknob capacitor may result in its failure due to mechanical stress over time. Some years ago, a "Hints & Kinks" article by ARRL member Phil Salas, AD5X, offered a fix. His, as well as other solutions I've seen, requires either modification of the existing bracket or fabrication of a new one from raw material. I wanted a quicker, ready-made solution, which I found at a local home improvement store. The product I adapted is an Everbilt Mending Plate, Model #14579 (SKU #266920). Advertised as galvanized, rust resistant, and sold as a two-pack, it costs less than $3. At 2 inches long and 5⁄8 inches wide, with countersunk holes 13⁄16 inches apart at their centers, it works perfectly, accommodating #6-32 stainless-steel hardware. Due to the short lengths of the mounting screws I used (5⁄16 and 3⁄8 inches), the purchase of those items was completed at a regular hardware store selling individual stainless-steel screw types and sizes. For stress relief, a mending plate is mounted on one side of the existing bracket, with a #6-32 stainless-steel 3⁄8-inch flat Phillips head screw, flat washer, split lock washer, and securely tightened nut. The top of the plate attaches to one side of the doorknob capacitor, using a #6-32 stainless-steel 5⁄16-inch flat Phillips head screw with a split lock washer. The other side of the assembly is completed with a 13⁄4-inch piece of insulated, stranded #14 copper wire soldered to terminals and attached at the top to the doorknob capacitor with a #6-32 stainless-steel 5⁄16-inch round head Phillips screw with a split lock washer. The other end attaches to the original bracket with a #6-32 stainless-steel 3⁄8-inch round head Phillips screw, flat washer, split lock washer, and securely tightened nut. I found it more practical to remove the 30-meter coil to complete the modification on a table. The two views of the end result, shown in **Figures 3** and **4**, should provide enough clarity. This is a rather quick, easy fix, which I expect to solve the problem long term. — *73, Thomas Houston, K6OKE,* **houston.thomas@gmail.com**

Figure 3 — The mending plate and its two flat screws are visible behind the doorknob capacitor. [Thomas Houston, K6OKE, photo]

Figure 4 — Note the wire between the doorknob capacitor and the original bracket. [Thomas Houston, K6OKE, photo]

Pants Hanger Portable Antenna

A simple antenna for low-power operation may be fabricated from a plastic pants hanger and small-gauge wire. The pants hanger consists of a hanger bar with clamps on either side.

I cut the hanger as shown in **Figure 5**, removing the clamps. I then drilled holes for the antenna wires and tie-wrapped it in the hanger section to hold the coax. The clamps were then drilled to hold the antenna wire ends.

The completed antenna uses the clamps to hold the ends to various supports, and the hanger to a tree limb or mast. — *73, Charles Rankin, WA2HMM,* **wa2hmm013@gmail.com**

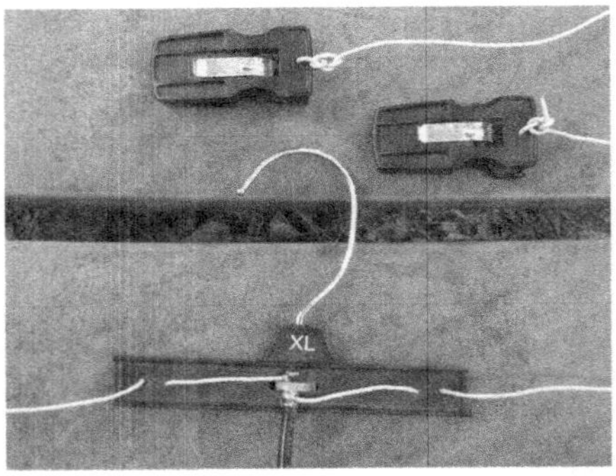

Figure 5 — The finished pants hanger antenna holds the ends to supports using the clamps, and the hanger to a tree limb or mast. [Charles Rankin, WA2HMM, photo]

Removing Flex-Weave Insulation

Flex-Weave is a super-flexible aerial wire, and I've found that removing the insulation from Flex-Weave without cutting individual strands can be tricky. Plus, Flex-Weave seems to be larger in diameter than normal electrical wire of the same gauge. To remove the insulation, score 14-gauge Flex-Weave with wire strippers made for 12-gauge electrical wire, clamping the strippers tightly and revolving them around the score. Next, heat the insulation to be removed with a match or grill lighter, softening the insulation. Then use the 12-gauge wire stripper at the location of the original score to pull the softened insulation from the Flex-Weave. A little practice with a piece of scrap Flex-Weave will improve this skill. — *73, Dean Sheeley, W7DWS,* **deansheeley@gmail.com**

More Easy-Off Connector Seals

I read the procedure for easy removal of connector seals by Evan Rolek, K9SQG, in the September 2016 edition of "Hints & Kinks." I've been using a similar method for over 30 years. I am a retired RF/microwave engineer, formerly employed by Telco.

Instead of painter's tape, I used ordinary, black electrical tape in a somewhat unconventional manner. When weatherproofing a connector, the first wrap of the tape was to wrap the connector with the black electrical tape "sticky side" out. I ensured that the tape was stretched enough to fill the low spots on the connector, then I wrapped the connector in the usual manner with the weather-seal product.

If the connection must be opened, everything peels off with a careful knife cut and no effort, and nothing sticks to the parts involved. Using black tape in this manner requires nothing new to put in your toolbox.

I've opened connectors and splices after many years using this method, and they looked as good as the day they were originally wrapped. — *73, Terry White, VE5TLW,* **twhite@sasktel.net**

Radial Plate Reference

For the vertical antenna at my previous station, I purchased the DX Engineering stainless-steel radial plate. The plate has 60 holes for bolts to hold the radials. Using the wire at my disposal, I attached a few radials and began to use the antenna. As time passed, the radials disappeared into the turf of the lawn. I had no way of telling how long each of the radials was, or in which direction it was laid down. When more wire became available, I could not tell in which directions additional radials should best be oriented.

Upon moving to a new station location, I solved the problem. Using a protractor and a sharp marker, I drew lines through each bolt hole, starting at the corner holes at 45 degrees, with each subsequent hole six degrees different. Each time I laid down a new radial, I laid it out in the direction indicated by the marker line, and I marked the length of the radial along the marker line, as you can see in **Figure 6**. Since the photo was taken, I have added many radials, and it has been easy to tell in which directions radials are few or short.

If I had to do it again, I would draw the lines before the plate is mounted on the mast; it's hard to draw good lines while kneeling in the mud. — *73, Mike Conlon, N2EM,* **mpconlon@zoominternet.net**

Figure 6 — I laid new radials out in the direction indicated by the marker line, and marked the length of the radial along the marker line. "80 m" indicates that the radial is ¼ wavelength long for the 80-meter band. [Mike Conlon, N2EM, photo]

Figure 7 — An old wet/dry vac's tubing holds the window line and allows it to slide freely. [Jeff Louden, K7ZSA, photo]

Window Line Strain Relief

My house has a metal roof and gutters, so keeping the balanced window feed line clear of these obstacles while turning 90 degrees from the balun to the center feed point has gone through various evolutions. My latest method provides a smooth sweep through the turn and does not alter the physical characteristics of the feed line, nor does it create any pinching that could eventually lead to breakage as the line gets whipped in all directions by the dipole "dancing" up and down in the wind. The flexible tubing came from a discarded wet/dry vac, and the line is free to slide inside (see **Figure 7**). I'm sure similar alternatives found in home improvement stores would work just as well. — *73, Jeff Louden, K7ZSA,* **k7zsa@yahoo.com**

Coax Cable-Stripping Tip

I recently acquired some LMR-240-UF coax to make short jumper leads between my transmitter and tuner, and the tuner and the balun on the other side of the shack wall. I wanted to install PL-259 connectors and UG-76 reducers. I had not worked with this coax before, and I found it very difficult to strip the ends. The hard part was removing the multi-ply bonded aluminum foil outer conductor between the braided shield and the hard foam dielectric. The foil tape is securely attached to the dielectric and very difficult to cut with a razor blade. I have a good selection of various wire strippers and none of these would do the job either. I ended up using a small file with a good "bite." I found it would pull the foil off the dielectric foam easily if I raked the file across the foil tape and twisted the cable as I went. When doing this, be sure to go slowly so you won't abrade the foam. — *73, Skip Burroughs, KS5KIP,* **ks5kip@arrl.net**

Threading PL-259 Connectors

I have a simple (though effective) solution for threading PL-259 connectors onto cable. I simply cannibalize another PL-259 connector

— that is, I just turn the "coupler" around and screw it onto the center pin connector, using it as a tool to twist the connector onto the cable. If additional torque is required, I use a pair of pliers on the cannibalized coupler. After the connector is installed on the cable, I unscrew the cannibalized coupler and keep it in my toolbox for future use. — *73, William Couloufacos, KC2RIM, 4233 192nd Street, Flushing, NY, 11358*

Radial Ring for the Claw

The Claw Mount for antennas (sold by Quicksilver Radio, www.qsradio.com) quickly mounts a monopole to a fence, picnic table, or other sturdy object. However, the size and position of the single, nut-bolt arrangement for attaching radials/RF ground are far from ideal. The bolt is small and very close to the 3/8-inch antenna mount, leaving virtually no room for connectors and wires.

My solution was to attach a "radial ring" made of a single 13-centimeter (just over 5-inch) piece of 12-gauge copper wire, rigid enough to hold its shape. Onto one end of the wire, I soldered a ring connector; onto the other end, an alligator clip. I fit the ring connector onto the bottom side of the small grounding bolt (next to the Claw's SO-239 connector). From there, the wire was bent to wrap around and clip onto the other side of the mount, presenting a strong, well-connected copper loop to which many radials could be clipped (four are shown in **Figure 8**, with lots of room for more). — *73, Rick Cincotta, KI4FW,* **ki4fw@arrl.net**

Figure 8 — A "radial ring" copper loop to which four radials are clipped. [Rick Cincotta, KI4FW, photo]

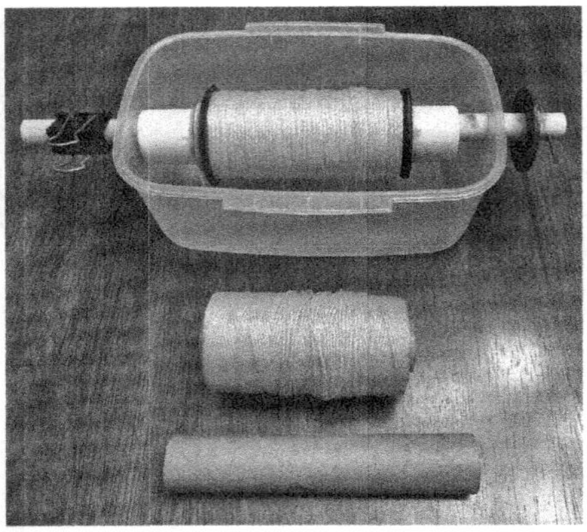

Figure 9 — The author's plastic enclosure mount for construction string prevents tangling and slippage when reeling the string back in. [Bob Haynes, WB4AKA, photo]

Pulling Antenna Wire into Trees Tangle-Free

Harbor Freight sells 40-pound construction string for $6, which is ideal for pulling antenna wire into trees while camping, and for ARRL Field Day. But coiling it up again can be messy because the slippery strands ooze off the ends of the cardboard roller while being reeled in, leaving the string tangled and unusable. I considered using kite reels, but they didn't hold enough string for my antenna work.

I decided to use PVC pipe and some washers from my plumbing junk box to make a reel. I glued the plumbing washers in place to keep string from creeping toward the ends. I drilled a hole in the pipe to secure the end, poked the string through, and tied a knot in the end before I rolled up the string.

Then I placed the reel on a wooden dowel and used a plastic food container as a dispenser. I secured the dowel with a large binder clip and a washer, then glued a nail in place through the other end. The string can be neatly unrolled from this container to avoid the twisting that happens when string is pulled off the end of a spool (**see Figure 9**).

To reel the string in, I simply remove the clip and take the roller off the dowel. I rotate the ends of the pipe by hand to rewind. This way, the string doesn't get twisted and I can continue using it over and over. — *73, Bob Haynes, WB4AKA,* **rhaynes5@tampabay.rr.com**

Supporting Antenna Masts with Fence Post Brackets

For the last 20 years, I have used steel fence post brackets to support small antenna masts along my backyard fence. These inexpensive brackets are intended for wooden privacy fences to attach 2 × 3 wood rails to standard 2 3/8-inch round, steel fence posts.

Figure 10 — A single bracket, and two brackets mounted back to back with 1/4-20 hardware. [Tim Brannon, WA5MD, photo]

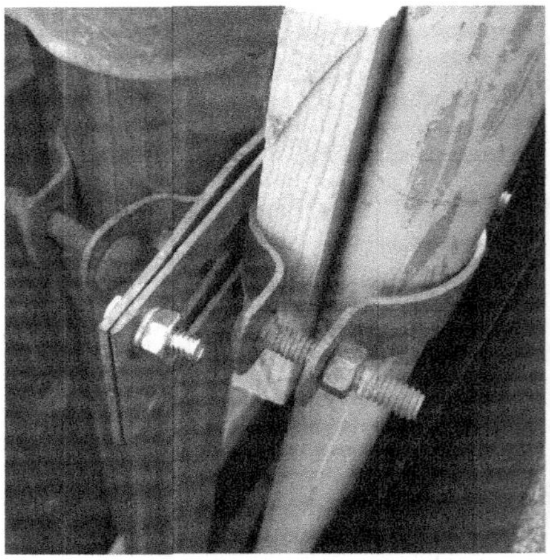

Figure 11 — A dual bracket mounting a 1 1/4-inch, 15-foot TV-type mast to a 2 3/8-inch round fence post. Note the shim, a 1 × 2 wood block. A VHF/UHF vertical antenna is at the top. [Tim Brannon, WA5MD, photo]

They are often sold in the fencing section of home improvement or building supply stores, or online. I have also salvaged these brackets from old fencing after a neighbor built a new fence.

Masts smaller than 1 3/4 inches in diameter may need a shim to fit tightly in the bracket. Lag screws or 1/4-20 bolts can be used to attach the bracket to a wooden fence, or two brackets can be connected back to back with 1/4-20 hardware in a dual arrangement to mount the antenna mast to a steel fence post (see **Figure 10**). It may not be essential, but I usually add a flat washer and a split lock washer to both 5/16-inch strap bolts to prevent the wind vibration of the mast from loosening the nuts.

I have not load-tested these brackets, but they can easily support common 1 1/4-inch TV antenna mast sections (using a shim). Multiple brackets may be needed to support a heavier mast, or to avoid a single point of failure. I have used four of these dual brackets together to support a 2-inch diameter, 10-foot steel mast with a large HF trap vertical antenna at the top, and it was still standing after 5 years (see **Figure 11**). — *73, Tim Brannon, WA5MD,* **wa5md@arrl.net**

Cleaning an SO-239 Coaxial Connector

Cleaning corrosion off a PL-259 connector is relatively easy because the contact surfaces are all external. An SO-239 connector presents more difficulty if corrosion has developed in the center contact. However, a .177-caliber brass/bronze rifle cleaning brush reaches nicely into that center contact. Because a .177/17-caliber brush is a common bore diameter for air rifles, these brushes are readily available from popular stores. To reach completely into the bottom of the center contact, it may be necessary to clip off the tip of the center core of the brush. — *73, Ken Slusher, N2DF,* **n2df@arrl.net**

Repairing Hustler Antenna Resonators

Hustler has been making their BTV series vertical antennas for a long time. They are sturdy and effective when installed properly, as any vertical can be, and are easily concealed and camouflaged in today's stealthy-antenna world. A friend gave me a fairly beaten up 4BTV model that covers 10 – 40 meters after I mentioned wanting to try one on a 20 × 20 foot metal shed roof.

The antenna was dented in places, very corroded, and the weather boots had rotted off. The boots were easily acquired from Hustler. One of the trap covers had separated from its mounting and I was disheartened to learn that the part could not be purchased by itself, so I would have to buy the whole resonator. To try and be frugal, I looked at ways to fix the separated base clamp and sleeve. Heliarc/TIG (tungsten inert gas) welding came to mind first, but the resonator clamp is heavier and probably a different grade of aluminum than the sleeve, which is much thinner. It takes great finesse to weld thin-to-thick without vaporizing the thin, so I looked for another option.

It seemed simple enough that screws would hold the clamp to the sleeve, with the screw heads holding the edge of the clamp (which has a tapered profile where it was originally attached by crimping) to the sleeve. The clamp would not pass through into the sleeve; it just set in the opening like a lid on a pan. I drilled three holes equidistant around the perimeter of the sleeve base, at an angle at the joint line (see **Figure 12**).

I did this twice after discovering two things:

1) Unless you have really short screws, drilling at an angle makes the screws protrude into the opening, which blocks the resonator tubing from passing through.

2) When attaching two things like this together, drilling at an angle into the joint will allow the part to misalign if the screw loosens. Drilling straight down into the resonator cover and letting the screw heads hold the clamp is the way to go.

I then covered three #6 sheet-metal screws with anti-oxidizing compound and, holding the assembly tightly together with a bar clamp, tightened the three screws equally until they were snug. The result was a secure trap cover attached to its base clamp (see **Figure 13**). If you don't have a bar clamp, a helper pressing down on the sleeve clamp with a piece of wood will prevent the screws from "walking" that base clamp out from the sleeve. It is highly unlikely that the small bit of protruding screw threads will have any appreciable effect on the resonator's function. — *73, Charlie Liberto, W4MEC,* **w4mec@arrl.net**

Figure 12 — The author drilled holes equidistant around the perimeter of the sleeve base, which worked best when drilling straight down into the resonator cover and letting the screw heads hold the clamp. [Charlie Liberto, W4MEC, photo]

Figure 13 — The trap cover attached securely to its base clamp. [Charlie Liberto, W4MEC, photo]

Figure 14 — An inexpensive, simple window line support made with rebar, PVC, and small cable ties. [Harvey Serreze, W1HBS, photo]

Inexpensive Window Line Support

I needed a simple way to inexpensively support a 200-foot run of window line from my ham shack to my antenna through some woods, while using non-conductive materials in the vicinity of the line to avoid introducing loss. I first drove 4-foot lengths of steel rebar halfway into the ground at about 20-foot intervals, keeping them as vertical as possible. Atop each of these I slid a 5-foot piece of 1⁄2-inch PVC conduit that had a small slot cut into one end. I then attached the window line to each slot with a small cable tie (see **Figure 14**). Using lightweight cable ties assures that they (and not the window line) will break in the event that a tree or heavy branch falls on the line. So far, this technique has worked well — even after a large tree fell across my line in a recent storm. — *73, Harvey Serreze, W1HBS,* **w1hbs@arrl.net**

Easy Alpha Loop Remote Tuning

When we began planning to relocate to an apartment or condominium, I started exploring my antenna options. I purchased an Alpha Loop package, installed it on the porch, and attached it to an Elecraft KX2 running 10 W. The results were good on digital modes, including JT65, JT9, FT8, and even PSK modes.

However, to change bands, the loop requires tuning a capacitor to maximize receiver noise. (The KX2's ATU takes care of the rest, always achieving 1:1 matching.) This means disconnecting the sound card, turning up the volume, opening the window so receiver noise can be heard at the antenna, and going outside to do the tuning. This makes frequent band changes painful.

I needed a solution to make band changes

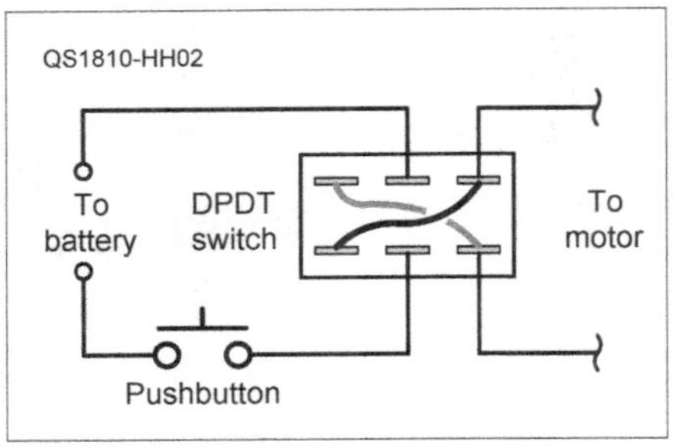

Figure 15 — A schematic of the author's design.

easy. I designed a pushbutton system to power an electric motor (see **Figure 15**). A DPDT (double pole, double throw) switch to change dc polarity would enable forward and reverse motion and the pushbutton would deliver power to the motor, which turns the capacitor shaft. The capacitor does not have hard stops at full-open and full-meshed positions, so high current motor stalls will not be an issue.

I built this project mainly from items in my junk box, including a small, geared 12 V motor; the motor's 5-millimeter shaft (hollowed, and threaded inside the hollowing); a 1⁄4-inch collar; a 1⁄4-inch shaft coupler; a small push-button switch designed for circuit board mounting; 5-centimeter-long bolts and nuts; an old pill bottle, and 4-centimeter-long spacers. I used four-wire telephone station cable to connect the remote motor to a controller in my station (only two wires were needed).

Box Modification

The capacitor's housing is a plastic box, so I removed the cover. Two holes are needed to secure the spacers supporting the motor to the box. I used the motor to locate these holes (mine were offset from the axis of the capacitor shaft).

Caution: I tried unmounting the box from the support pole. One of the clamps is secured with a nut beneath the board holding the capacitor, and I had to disassemble the works to reattach that bolt. A better plan is to remove the support tripod and loop from the pole, leaving the metal connecting piece attached to the box.

Motor Modification

The motor shaft must be enlarged to fit the 1⁄4-inch coupler. The collar may be force-fit to the shaft if its inner diameter is machined to match the shaft, or the collar may be secured with a bolt.

Motor Assembly

I attached the coupler to the motor and secured it with the set screw. I then attached the motor to the box using the 4-centimeter spacers and the 5-centimeter bolts and nuts. I next slid the coupler over the capacitor shaft and secured it with the set screw. I tested the installation with a 9 V battery, making sure to try both directions. The motor has good torque, so minor misalignments should be no problem.

Controller Assembly

I removed any cardboard from the lid of the pill bottle, then mounted the DPDT switch to the plastic lid after attaching wires per the diagram. The pole (center terminal) wires go to the battery and pushbutton, and the throw wires (end terminals cross-wired) go to the motor via the station cable (see **Figure 16**).

My pushbutton is designed for PCB mounting, so I drilled four small holes in the lid for its mounting feet then mounted the switch, bending thefeet over to hold it. Solder blobs on the underside secure the switch. I attached one contact to a pole of the DPDT switch, and attached a 9 V battery to the other pushbutton contact and the other pole of the DPDT switch.

Figure 16 — The control "box," with DPDT switch on top. [Monty Shultes, K2DLJ, photo]

Using the remote is easy. I just flip the DPDT switch, hold the pushbutton, and wait for the received noise to peak. Overshooting the sweet spot is inevitable, but all I need to do is reverse directions and poke the button as often as needed. Success comes quickly after a few trials. Be sure to mark the DPDT positions as **UP** and **DOWN** once you learn which way the motor turns (see the final product in **Figure 17**).

Tuning now is almost fun. It's so easy that I find myself trying new bands. — *73, A LaMont "Monty" Shultes, K2DLJ,* **montys@mindspring.com**

Figure 17 — The completed motor attached to the capacitor box. [Monty Shultes, K2DLJ, photo]

Faux Mobile Mount Portable Antenna

When there are no trees around to hang a dipole and no car to mount an antenna on, I use a portable antenna as my on-the-go HF setup. This antenna is compact, inexpensive, and easy to set up (see **Figure 18**). The trick is to fool a leftover antenna you have lying around into thinking it's mounted on top of a metal bus roof. You'll need three components — the old mobile antenna, a base plate to attach everything to, and four aluminum sectional tent poles to serve as radials.

The base plate could easily be an old aluminum road sign from the local road department scrap heap (or you can buy a blank sign), around 12 × 12 inches (mine was 18 × 18 inches; the dimensions aren't critical).

Depending on the mobile antenna mount type, you may need to drill a hole or mount a small 4 × 8 inch platform to the base. To hold the radials (the tent poles), mount four eye

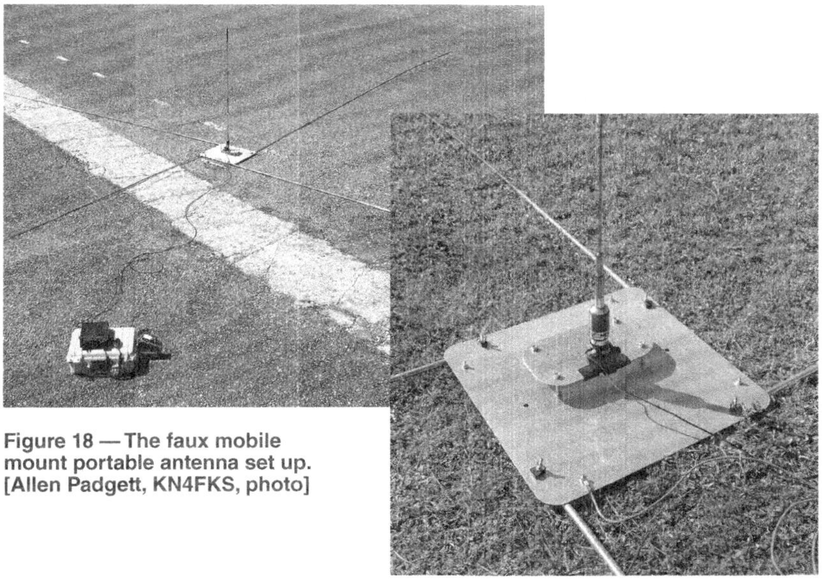

Figure 18 — The faux mobile mount portable antenna set up. [Allen Padgett, KN4FKS, photo]

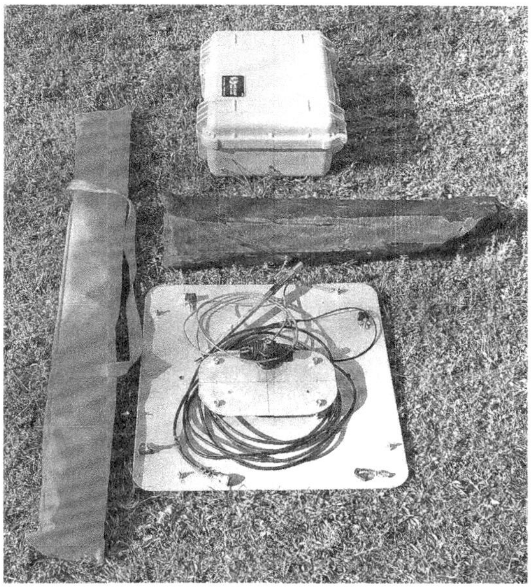

Figure 20 — The antenna bag, the radials bag, the base plate, and the radio/battery box — compact and portable. [Allen Padgett, KN4FKS, photo]

bolts to hold the ends, then four **J**-bolts with wing nuts to clamp the first section of the tent poles to the base plate. Add a short piece of flexible wire attached to a metal tent peg for your ground (see **Figure 19**). If you do use an old road sign, be sure to mount the radials on the bare metal side, so they make contact.

Obtain four shock-corded collapsible flexible tent poles, each about 20 feet long, to serve as radials. This type of tent pole is used on larger, dome-shaped tents. Be sure to check your garage or a store that sells used outdoor gear before looking online. These aluminum poles collapse to about 18 inches and snap together with the internal shock cord.

Find a suitable field, parking lot, beach, or attic and place the base plate down. Next, extend the radials and clamp them to the base plate. Screw in/clamp the mobile antenna to the base plate and nail the tent peg ground into the earth if possible. Once you hook up your coax and radio, you're ready to go. Setup takes less than 10 minutes and can easily be done by just one person (see **Figure 20**). Note that you are putting an RF source on the ground where it's reachable, so make sure kids or curious adults keep away if you're in a public place.

The antenna exhibits the same characteristics, bandwidth, and SWR as it would mounted on a vehicle. The radial length is a few feet short for 40 meters, so there's slightly higher SWR there. You can extend it an additional 13 feet using wire clamped to the end of the tent poles if desired. On 75 meters, the 20-foot radials need to be doubled up using adjustable hose clamps to extend the length. Thanks to John Law, KB4QXI, for the tip on 75-meter operation. — *73, Allen Padgett, KN4FKS,* **on_rope@windstream.net**

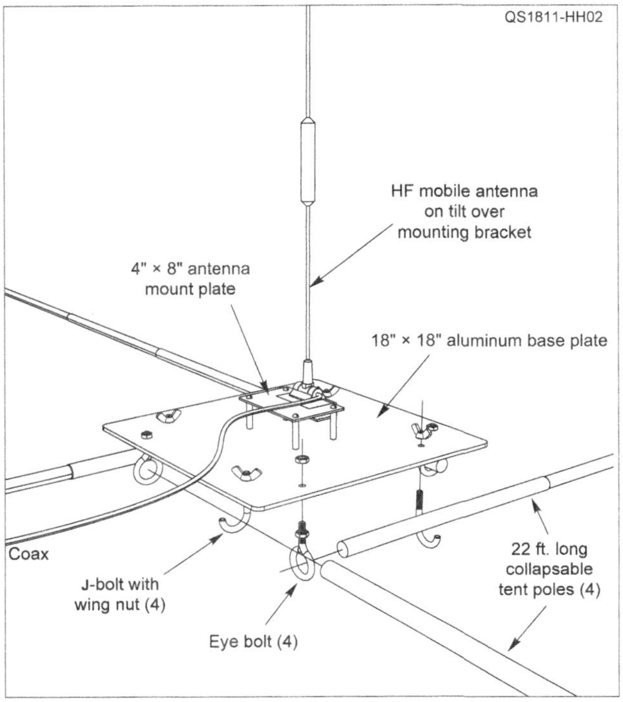

Figure 19 — A diagram of the faux mobile mount antenna's setup.
Base plate: 18-inch square (#215-026000 at store.hallsigns.com/Alodized-Aluminum_c_255.html) or 12-inch square (#215-014000).
Antenna mount plate: 4 × 8 inches
Radials: Four 20-foot-long, 8.5-millimeter flexible aluminum tent poles (17 inches per section, shock-corded for easy assembly). Available on www.amazon.com.
Base plate radial mount hardware:
Four eye bolts, sized to fit pole ends with an extra lock nut per bolt
Four J-bolts, sized to clamp over the pole section
Four wing nuts to fit the J-bolts
Four spacers for J-bolt thread depth
Two bolts with nuts for ground wire
About 2 feet of flexible copper wire eye lugs, crimped on both ends
One blade-type tent peg, 9-inch steel (available from https://www.stansport.com/tent-stake-9-inch-809)
Antenna mount plate hardware:
Four hex head bolts, ¼ × 4 inches
Four lock washers
12 nuts to fit

Cleaning and Protecting RF Connectors

Here are some tips to assist in maintaining low-loss connections. Type N and PL-259/SO-239 connectors will develop oxidation on the center pin connection. You can restore low RF loss to the center of the SO-239 using a cotton swab that has most of the cotton ball removed and has been lightly coated with a cleaning agent, like Brasso. Follow-up with dry cotton swabs until they come out of the center pin hole clean. Use rubbing alcohol to remove any residue from the cleaning agent. The threads of the SO-239 can be cleaned using a cleaning agent and an old toothbrush or brass brush. Finally, you can use a heat gun or can of air to dry out the connector.

For outdoor connectors, consider very lightly coating the contact surfaces with a dielectric grease applied with a cotton swab. Automotive parts stores are a good source for packets of spark plug and dielectric grease. I always keep a packet with a cotton swab ready in my tool bag.

I also cover outdoor connections with a single layer of aluminum duct tape prior to using waterproof tape. Regarding Bob Mudra's, AK9RM, "Breathable Waterproofing" item in the March 2018 "Technical Correspondence" column, I place cotton socks filled with cat litter to absorb some of the humidity that is always present in breathing outdoor enclosures. — *73, Joe Ostrowski, KI5FJ,* **ki5fj@arrl.net**

Thin Cables for Stealth Feed-Throughs

Routing antenna feed lines through windows is a common problem. Even 0.195-inch diameter RG-58 cable makes it difficult to close a window sash, and even if it does close there may be drafts through gaps in the weather strip around the cable. Some manufacturers sell weather seals or cable extensions that allow the window sash to close over cables, but I wanted to be able to fully close and lock the sash. Plus, the only window in my radio room is at the front of the house, so several large coaxial cables hanging from the bottom of the window would not win the approval of my family.

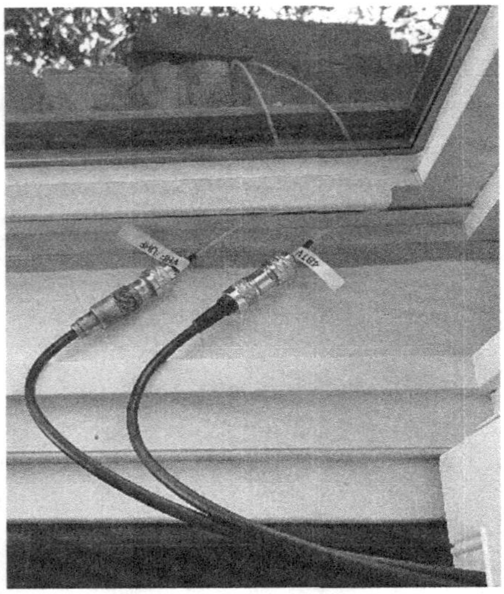

Figure 22 — Two cables inside the window mating to RG-8X cables extending to the radio position, routed at an angle under the sash to minimize any pinch or sharp bends. The sash is fully closed and locked. [Tim Brannon, WA5MD, photo]

While browsing online, I found several vendors selling RG-316 coaxial patch cables intended for interconnecting Wi-Fi devices and RF test equipment. They're available with a variety of crimp-on RF connectors and in various short lengths up to 15 feet. The outer jacket is translucent, so the visible copper braid shield gives the cable a pinkish gold color, which I thought would blend well with the brick veneer of my home.

RG-316 coaxial cable has a low-loss dielectric and an outer jacket diameter of only 0.098 inches (compared to RG-213 cable at 0.400 inches). Though RF loss is higher in small-diameter cables, RG-316 has about half the loss of the similar-sized and more common RG-174 cable. Using the Times Microwave Systems online calculator (www.timesmicrowave.com/calculator), I found that a 3-foot length of RG-316 should have a loss of only 0.1 dB at 28 MHz and was rated for up to 1,100 W — easily acceptable for HF. At 440 MHz, the same short cable would have 0.6 dB loss and was rated to 270 W. I decided this much loss would be tolerable even for my 50 W 440 MHz D-STAR/FM transceiver, so I ordered five of the 3-foot-long cables with regular PL-259 connectors for about $6 each (see **Figure 21**).

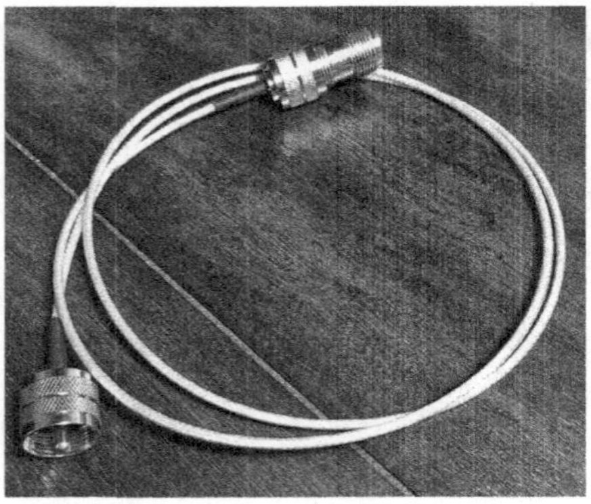

Figure 21 — A new 3-foot RG-316 cable with PL-259 connectors, and a PL-258 barrel connector at the right end. [Tim Brannon, WA5MD, photo]

A plastic cable box with a hinged cover just below my shack window provides weather protection for several coaxial surge protectors. I ran the RG-316 cables from the surge protectors through the window into the shack, and the color helped camouflage them against the brick. The sash could easily close on the thin cables without crushing them, and there was no noticeable draft (see **Figure 22**). Inside the window, I mated a PL-258 UHF barrel con-nector with a 12-foot RG-8X cable extending to the radio desk.

Especially for my UHF antenna, I wanted to verify that the RF loss in the window feed-through cables was not excessive, so I set up a test using my station wattmeters and a dummy load. To get a more accurate value using this simple equipment, I connected all five of my 3-foot RG-316 cables together with barrel connectors as a single string for testing. Assuming the loss in all five was roughly equal, I divided the loss value for the whole string by five to estimate the loss of a single 3-foot cable. I also tested one of the 12-foot RG-8X cables as a control sample. A Daiwa CN-101L wattmeter was used on 10 and 6 meters and a Comet CM-720 wattmeter was used on 2 meters and 440 MHz. Both wattmeters have a 15 W full-scale range, so a test signal power of about 15 W was used to improve accuracy. Because line loss increases with frequency, measurements were made on frequencies 20 kHz below the top of each band to model a worst-case scenario. The wattmeter was first placed at the transmitter end of the cable under test, a power measurement was made, and then the meter was moved to the dummy load end of the cable under test and another measurement was made. Loss in decibels was then calculated by the formula

Figure 23 — A view of the shack window from about 10 feet away. Two RG-316 cables are barely visible entering the lower left corner of the window, just above the tan-painted cable box (containing well-grounded surge protectors). They cannot be seen from the street. All cables enter and exit the cable box through grommets in the bottom. [Tim Brannon, WA5MD, photo]

Loss dB = 10log(P load/P input)

with P input and P load representing the power readings at the input end and the dummy load end of the cable under test, respectively. SWR remained 1.2:1 or less for all tests, except for 440 MHz at the load end of the RG-316 test string, where the SWR was 1.5:1. The MFJ-260C dummy load is only rated up to 650 MHz, so this was not unexpected.

For line losses, my measurements showed a loss of 0.99 dB at 440 MHz for a single 3-foot RG-316 cable, close to the Times Microwave estimate of 0.6 dB, and small compared to the 2.17 dB loss in the 12 feet of RG-8X (which I replaced with a shorter cable). Because I only operate FM and D-STAR modes at 440 MHz (not weak-signal modes), this loss should not be an issue. Loss at frequencies below 10 meters was considered insignificant. Regarding power handling, I only run up to 100 W on HF. Though the specifications indicate that RG-316 is rated for 1,100 W at 10 meters, caution should be used before running higher power.

These thin patch cables provide a convenient and stealthy way of routing antenna feed lines through a window into the shack (see **Figure 23**). They allow my window to fully close and lock, and the RF loss is not prohibitive up through the 440 MHz band. They would also be useful for temporary locations such as hotels or vacation rentals. Remember to weather-seal or otherwise protect the cable connectors outside the window. The cables have been in place for about a year with no significant signs of weathering. — *73, Tim Brannon, WA5MD,* **wa5md@arrl.net**

A Foldable Wooden Antenna Support

As soon as my son earned his General-class license, I wanted to get him an antenna system outside for HF, VHF, and UHF that he could easily maintain himself. Fortunately, he owns a home in the country, so antenna installation is not an issue. My two antennas of choice to start with were an end-fed wire for HF and a dual-band J-pole antenna for 2 meters and 70 centimeters.

I set out to design a structure from pressure-treated two-by-fours that would be simple to build, inexpensive, and would support both of these antennas (see **Figure 24**). The chosen location behind the house dictated that the bottom two "tower" two-by-fours would need to be 12 feet in length to clear the edge of the roof. In addition, I wanted a tip-up/down mast added to the top of this tower that would use one additional 12-foot two-by-four. This would give me a total height of about 19 feet. If one person was to do any antenna work at

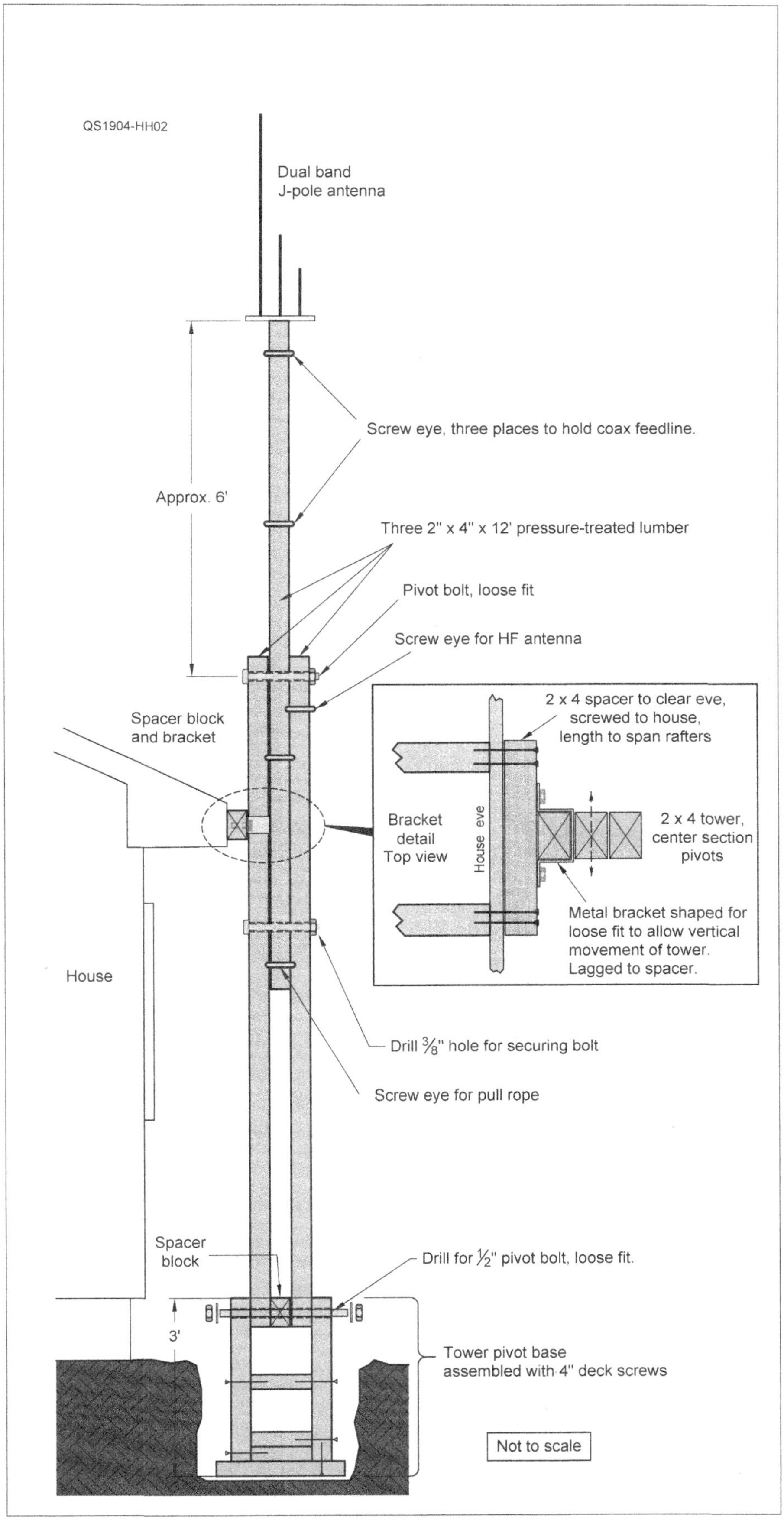

Figure 24 — A schematic of the author's wooden, foldable antenna support.

Figure 25 — This wooden fold-over antenna support allows for easy antenna work. [Karl Schwab, KO8S, photo]

Figure 26 — The completed wooden antenna support. [Karl Schwab, KO8S, photo]

this height, the tower and mast would have to be designed to fold down to the ground for safety reasons (see **Figure 25**).

The materials list for this build was short. It required:

- Four pressure-treated two-by-fours, at 12 feet long
- One ½ × 12 inch threaded steel rod with flat washers and nuts
- Two ⅜ × 5 inch bolts with flat washers and nuts
- Four large eye screws (sized to pass a PL-259 connector through)
- Two ¼ × 1½ inch lag screws and a small box of 4-inch long #10 deck screws

To hold the tower against the house, we needed a metal **U**-shaped bracket that fit loosely around the three sides of one of the two-by-fours. It was made from a 10-inch length of 3⁄32-inch-thick steel that was 11⁄2 inches wide. This bracket would allow the completed assembly to settle into the dug hole in the ground while not pulling down on the attachment point at the house.

For tools, I used a powered miter saw to make all the necessary cuts and a battery-powered hand drill to drill the holes. Accurate saw cuts and perpendicularly drilled holes for the bolts were essential for proper fit and alignment. A drill press would have been perfect here, but none was available. One two-by-four was cut into all of the required pieces, and the other three required no cutting at all.

With all materials procured, I started the build early in the morning with my sons — *Mark, N8KPV,* and *Alan, KN4IJR*. I measured out the wood pieces and made the cuts while the boys drilled holes for the bolts. We then started the assembly on the ground, screwing and bolting the various pieces together. Three screw eyes were spaced down from the top of the mast to hold the coax and end-fed antenna wire. The last screw eye was attached at the bottom for a pull-down rope. At this point, all the building was complete.

We dug a hole in the ground about 2 feet deep that would except the 3-foot-long fold-over base assembly. Directly above this hole, we attached a 24-inch length of two-by-four at the fascia using 4-inch-long deck screws. This two-by-four would space the tower out from the house in order to clear the edge of the roof shingles.

We placed the fold-over base in the hole and got the tower assembly (minus the mast) raised and held to its vertical position against the house. We made adjustments by moving the base in the hole and tweaking its position against the house to make sure everything was plumb.

Next, we attached the metal bracket to the house with two lag screws securing the towers inside the two-by-four to the house. The hole we dug was then filled with dirt and the mast was bolted in its center position in the tower and pivoted down for attaching the two antennas. We easily raised the mast into its vertical position by pulling down on the rope, and secured it using a single ⅜ × 5 inch long bolt and then removed the rope. At this location, the pivoting mast could be pivoted down in either of two directions if needed. The far end of the end-fed wire antenna was then pulled up high in a nearby tree, thus completing all this antenna work in a single day (see **Figure 26**).

If this assembled unit ever needs to be brought down to the ground for any reason, only the two lag screws that secure the metal bracket to the house need to be removed and then it will fold down. Careful planning made this a quick, inexpensive ham radio project.

After this antenna work was completed, it was time to try it out. For the first time, good signal reports were given on the local repeater. On 20-meters SSB, our first contact was with Ecuador, and the second with an aircraft flying at 38,000 feet over Ecuador! It was a nice reward for a good day's work. — *73, Karl Schwab, KO8S,* **ko8s@arrl.net**

A Handy Tool for Ground Radials

While putting up a vertical antenna, I needed to install ground radials. I decided to build a tool that would help me, so I wouldn't have to get down on my hands and knees a lot. To make this ground radial tool, I used some 1 × 1 inch square aluminum solid stock, purchased from a metal scrap dealer. Wood with the same dimensions would also work. The model measurements were 26 inches long, with a 5-inch-long handle (see **Figure 27**).

First, I marked an "X" on the end of the aluminum stock. This gave me the center. I also used a strong, round ½ × 2 inch magnet. I drilled a hole 4 inches deep — the same size as the magnet — so I could get a snug fit. You could also use some glue to secure it in place if needed.

To make the end piece, I used some flat aluminum strips, 1 × ⅛ inches. I used a ruler to make marks at 2 × 1 × 2 inches. Then I took the 1-inch square aluminum and bent the flat aluminum around it using a small hammer. I put this piece on the end where the hole was and made a mark to get the center of the hole. I measured ¼ inch from both sides and marked ½ inch long. I used a ³⁄₁₆-inch drill bit and drilled across, then used a file to finish rounding out the hole. The end product should be an oval slot ³⁄₁₆ inches wide and ½ inch long.

Next, I secured the end piece with epoxy glue. For the 5-inch-long handle, I measured 1 inch from the edge and made a mark to define where the square was located. Then I measured ½ inch, marked the center,

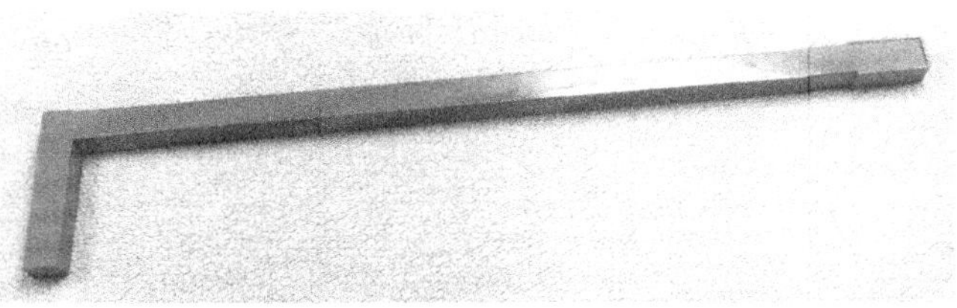

Figure 27 — The author's tool, which allows for easy installation of ground radials. [David Doonan, KIØNN, photo]

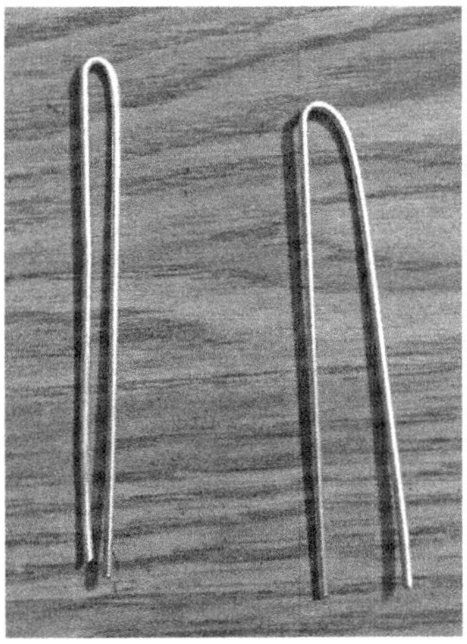

Figure 28 — Staples made from metal wire hangers. [David Doonan, KIØNN, photo]

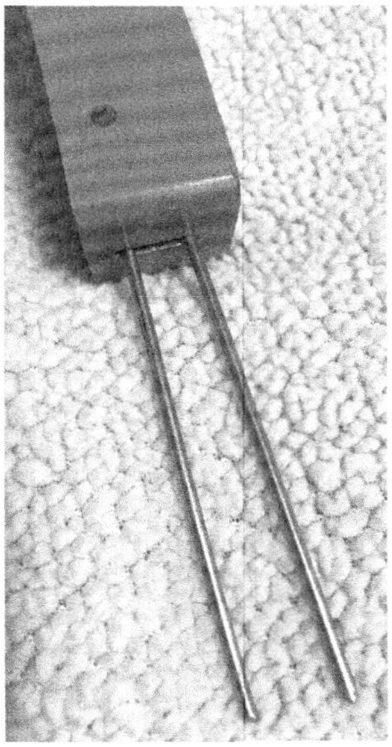

Figure 29 — The homebrewed staples in the ground radial tool. [David Doonan, KIØNN, photo]

and measured 1/4 inch in from both sides and marked that using a center punch. I drilled two 3/16-inch holes. I used a 12-24 × 2 inch countersunk screw with a countersunk bit so the screw was flush with the surface. I looked at the end of the long piece to make sure the slot was in a horizontal position with the handle. This way, when I put the staple in the ground, the handle would be in line with the wire.

I put the long, square piece in the vise so the end was flush with the top of the vise. I then took the handle and squared it up. I put the 3/16-inch drill bit in the hole and tapped it with a hammer to make an indentation, which helped tell me where to drill a 3/16-inch hole. Once I drilled the hole, I used a 12-24 tap to tap the hole. The only thing left to do was put on the handle. The screws went right in. For the staples, I got a bunch of metal clothes hangers and cut the wire just below where it twists. After straightening out the kinks, I wound up with around 32 inches of wire. I measured 4 inches, made a **U** bend, and cut it off square (see **Figure 28**).

Once I put the staple in and the tool touched the ground, I lifted the tool, turned the handle 1/4 turn, and pushed the staple in the ground (see **Figure 29**). This tool should not cost more than $10 to make. — *73, David Doonan KIØNN,* **d.doonan@ruraltel.net**

Protecting Antenna Support Rope from Wildlife

Like many hams, I use trees as antenna supports. After three unexplained line (rope) failures, I discovered that squirrels had been chewing through the line where it passed over the tree limb.

I solved this problem by using a 6-inch loop of small-diameter stainless-steel aircraft cable screwed to the tree holding a pulley for the antenna support rope. Alternatively, you could form the loop using ordinary stranded copper wire (such as #14 AWG) and either crimping or soldering the ends together.

While setting this up, it occurred to me that instead of using a single piece of rope through the pulley, I made it a loop. A rope through a pulley means that the bottom end is susceptible to running free out of the pulley unless it is always held. The rope loop ensures a bit more security.

The insulator is fastened with a heavy tie-wrap where the ends of the loop through the pulley are tied together. This allows me to lower the antenna from ground level for repairs or changes without climbing. Plus, the antenna support rope is now out of reach of the squirrels' sharp little teeth. — *73, Doug McCray, K2QWQ,* **k2qwq@comcast.net**

Straightforward Coax Pulling Tool

Installing a coax cable with the ends already terminated in a wall or conduit can be difficult, so I constructed a simple tool to make that job easier. This works for coax with PL-259s on the ends.

All you'll need is a PL-259 connector, a PL-259 coax adapter (the size doesn't matter), an SO-239 barrel connector, and a short piece of parachute cord or similar.

Construction

Remove the center pin from the PL-259. I did this by heating it with a large soldering iron. Next, push the parachute cord through the PL-259. Tie a jam knot on the inside end and tie loop (a bowline knot) on the outside end (see Figure 1). Assemble the PL-259. Screw the SO-239 barrel connector to the PL-259.

To use this tool, insert your pull cord (or rope) in the hole or conduit you want to install the coax in. Tie one end of your pull cord to the loop of the tool. Screw the PL-259 on the coax to the SO-239 barrel connector, then gently pull the coax through the hole or conduit.

By using this tool, you will be pulling the coax by a connector instead of to the side, so there is less chance of it getting hung up. — *73, Joe Birsa, N3TTE,* **jjbirsa@yahoo.com**

Protecting Your Feed Line

After reading John Portune's, W6NBC, article, "A Novel Approach to Using Window Line," in the August 2018 issue of *QST*, I thought I might add my own experience, as I had done something similar. My Windom runs from an antenna mount on my chimney to the Anderson Powerpole in front of the house. It is fed from the rear of the house and the feed line runs across the roof. Over time, the 300 Ω feed line became worn and the conductors were chewed up by the shingles on which it ran.

My simple solution was to pull the feed line through a common garden hose (see **Figure 30**). This quick, inexpensive fix has been in place for almost 10 years now without any issues.

I also used the vent for a different antenna feed line. I used a different vent pipe to run coax to another antenna by drilling the hole in the pipe where it ran through the attic. — *73, George Keller, W2DM,* **georgenann@aol.com**

Figure 30 — The author's feed line protected by a garden hose. [George Keller, W2DM, photo]

Fishing Net Cable Catcher

I've attached a very large fishing net to the base of my crank-up tower to gently catch the large coils of coax and cables that would otherwise fall to the ground or hang loose from the tower. This way, all the cables fall into the net and usually coil themselves into big loops, though a little help is needed when the cables are cold and less flexible. No cables pass through the net (see **Figure 31**).

The largest net I could find was the Ranger Nets model 987T "Big Game" landing net, available at **www.rangernets.com/big-game-landing-nets**. Four feet above the net, I mounted a small loop with an opening, made from UV-resistant PVC parts, to help guide the cables into place. To hold everything in place, I used two stainless-steel hose clamps, positioned not to interfere with the motion of the crank-up tower sections. I've found that the net lasts for many years and replacement nets (measuring 40 × 37 × 48 inches) are inexpensive and easy to install. — *73, Bob Wilson, N6TV,* **n6tv@arrl.net**

Figure 31 — A large fishing net catches any cables that fall from the crank-up tower. [Bob Wilson, N6TV, photo]

Gravity Fuse

Anyone who uses trees to support wire antennas knows the importance of having some strain relief. My end-fed half-wave antenna has a pulley and weight at one end to keep constant tension on the wire while allowing for any motion of the trees. However, the large oak tree my antenna passes under drops limbs large enough to break the wire. The length of line going through the pulley at the end of the antenna isn't enough to allow the antenna to go all the way to the ground. If a limb falls on the wire, something is going to get broken or the mast is going to be damaged.

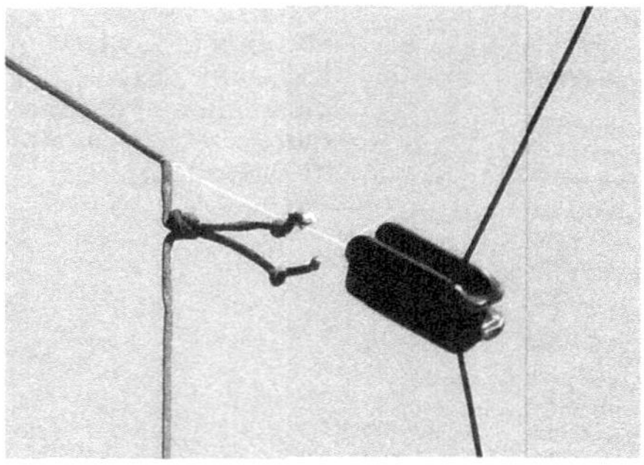

Figure 32 — The author's "gravity fuse" ensures the monofilament line will be broken by falling tree limbs, preventing damage to the antenna. [Allen White, WB4MIO, photo]

My solution is to use a short piece of monofilament fishing line at the end of the antenna and at the fiberglass pole that elevates the center of the antenna (see **Figure 32**). If a limb falls on the antenna, it will just break the monofilament line. Once the limb is removed, it takes just a few minutes to tie in a new piece of line and hoist it back up.

I use 15-pound test line. The weight at the end of the antenna is 5 pounds, so that gives me a good margin for normal stresses. I would suggest using line rated at least three times the normal tension on the line.

This gravity fuse was put to the test during a thunderstorm with 53 MPH winds. Several limbs came down from a neighbor's tree and one of the larger ones landed right across the antenna. The monofilament line snapped and let the antenna wire go to the ground. There was no damage at all to the antenna. During cleanup, it took me less than 10 minutes to tie in a new piece of monofilament and hoist it back up. I use a continuous loop like a flag halyard for the center support hoist, so it is very easy to let it up and down.

Unless you add head-shrink tubing, sunlight will deteriorate the monofilament line, so it may need to be replaced every 2 or 3 years if a limb has not taken it down. — *73, Allen White, WB4MIO,* **wb4mio@arrl.net**

A Homebrew Antenna Launcher

One challenge that most hams are faced with is figuring out how to get their wire antenna up in the air.

Here's a homebrewed slingshot that uses a fishing reel (see **Figure 33**). If you want to try this yourself, please note that some states and municipalities regulate the ownership and sale of slingshots. It would be a good idea to check ahead of time.

The three major components to this device are the slingshot, the fishing reel, and the method to connect the two. I drilled out the bottom of the slingshot handle to accommodate a ¼ × 20 press-in nut (see **Figure 34**). These nuts have a shank that is meant to be inserted into sheet metal or other sheet materials. I ordered these from an industrial supply house (McMaster-Carr at **www.mcmaster.com**), but you may be able to find them at a local hardware store. I used epoxy adhesive to hold the nut in place. I found the ¼ × 20 knob screw at my nearby hardware store.

The reel was sourced at a local discount department store and came with 10-pound test fishing line already spooled. I taped the foot of the reel to a piece of tubing that I found in my junk box (see **Figure 35**).

To complete the assembly, I used a piece of **C** channel aluminum (ordered from the same industrial supply house) that was the same width as the base of the slingshot. I used two **U**-bolts to hold the tubing to the **C** channel aluminum. To finish the project, I purchased some ¾-ounce and 1-ounce sinkers and painted them fluorescent orange and green to denote the weights.

If you choose to use this method, place the weight in the slingshot cup and press the button on the reel to release the line, take aim, and let it fly. Make sure the fishing line is free to unspool from the reel. It's best to do a dozen or so dry runs with small rocks, so you get a feel for how far to pull back the slingshot and how to aim. I have found that the ¾-ounce weight is enough to have the line freely fall back to earth and is relatively easy to find due to the bright color. I've been able to shoot lines up as high as 60 feet.

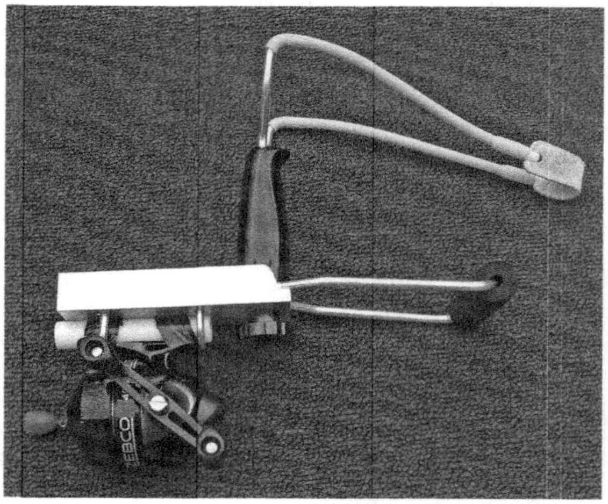

Figure 33 — The finished slingshot launcher. [Steve Allen, KZ4TN, photo]

Figure 34 — The press-in nut and knob screw. [Steve Allen, KZ4TN, photo]

Figure 35 — The attachment of the reel to the aluminum channel. [Steve Allen, KZ4TN, photo]

Figure A — Rex Lester, G8UBJ, outdoors with his portable antenna mount.

Once you have the line over your target branch and the sinker on the ground, remove the sinker and attach a long piece of chalk line. This step requires attention to detail because you must keep the knot small so it doesn't hang up in the crotch of a branch, yet it must be strong enough to not come untied while pulling it up and over the branch. It helps to tape over the knot to smooth the surface of the knot junction.

Once the chalk line is over the tree, you can attach the antenna halyard to the chalk line and pull it back over the branch, again paying close attention to the knotting method. — *Steve Allen, KZ4TN*, **kk1vt@arrl.net**

A Portable Antenna Support

Erecting or supporting portable antennas has always been a challenge for field operators, so I decided to build my own. If you follow the illustrations with the accompanying text you should find this support to be easy to construct.

You will need the following:

- A single section of lumber (¾ × 5¾ × 70 inches)
- Wood glue
- 16 2-inch wood screws
- Two 1¼-inch wood screws
- Primer/undercoat and gloss paint (optional)
- A 3/16-inch drill bit
- A ⅜-inch drill bit

A short length of 1½-inch-diameter PVC or aluminum mast (select a diameter close to the size of the mast you wish to support)

Start by marking up the plank to outline the various pieces you'll need to cut (see **Figure 36**). It helps to number them as well. You'll draw a total of 10 sections on the board.

- Sections 1, 2, 3, and 4: 5½ × 8 inches each
- Sections 5 and 6: 2¾ × 29 inches each
- Section 7: 2¾ × 9 inches
- Sections 8, 9, and 10: 2¾ × 3 each

Carefully cut out the various sections. Assemble the base using sections 5, 6, 7, and 8. This mount is intended for use with a vehicle for stability, so section 7 is raised and sits behind the inner tire wall when in use (see **Figure 37**). Section 8 sits at the far end between sections 5 and 6.

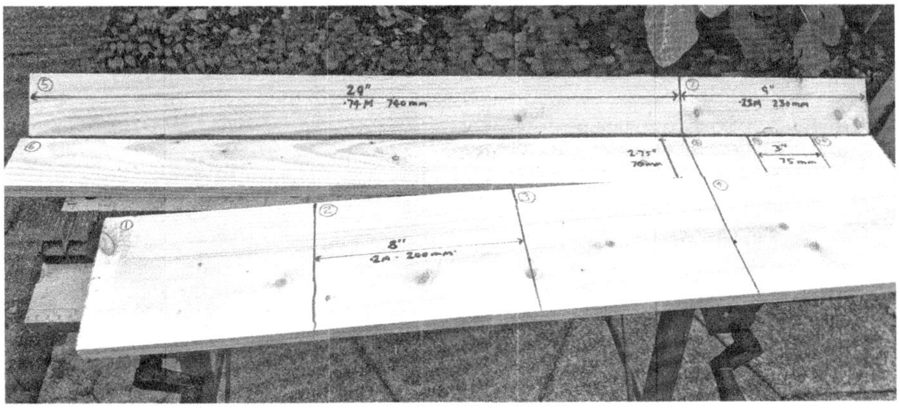

Figure 36 — All the wood parts are made by cutting the plank into numbered sections. Notice how the author marked the sections onto the board for easier cutting. [Rex Lester, G8UBJ, photo]

Figure 37 — Assemble the base using sections 5, 6, 7, and 8. Section 7 is raised and sits behind the inner tire wall when in use. Section 8 sits at the far end between sections 5 and 6. Prepare each screw by first drilling 3/16-inch pilot holes. [Rex Lester, G8UBJ, photo]

Figure 38 — Cut out the four sections that will create the stand. Take length of pipe and use this as a template for the center of the stand. [Rex Lester, G8UBJ, photo]

Prepare each screw by drilling a 3⁄16-inch pilot hole. This allows for easy assembly. Once tight, the screws should be flush. You'll notice that the screws used for securing sections 5 and 6 to section 8 are too short. Use the 3⁄8-inch drill bit to create a countersink for these screws. It needs to be about 1¾ inches deep so the screw threads protrude by an inch. The support won't be strong enough if it's assembled only with screws, so use wood glue on all the joints.

Now cut out the four sections that will comprise the stand and take a length of PVC pipe (or whatever mast section you want to support) and use this as a template for the center (see **Figure 38**). Use a flat surface to mark up where the four sections touch and then drill two holes in each section. Screw the four pieces of wood together in a square formation, which will leave enough room for the pipe.

Figure 39 — Complete the assembly by attaching sections 9 and 10 to the protruding support sections. You can prime and paint everything if you wish. [Rex Lester, G8UBJ, photo]

Once this is complete, secure the support to the base by attaching sections 9 and 10 to the protruding support sections (see **Figure 39**). Wait about a day before attempting to use the stand. This allows enough time for the glue to set. If you need it to last, give it a coat of paint in whatever color you prefer.

Using the support is very simple. Place it behind or in front of the wheel of your car that you choose to use. Position the stand so the raised section 7 is just behind the inner tire wall of the vehicle. Ask a friend to guide you as you drive your vehicle onto the support (see **Figure 40** and **Figure A**). — *73, Rex Lester, G8UBJ,* **g8ubj.io91uj@gmail.com**. *All photos by Rex Lester, G8UBJ.*

Figure 40 — To secure the support in the field, simply roll one of your vehicle tires onto the flat portion. [Rex Lester, G8UBJ, photo]

A Reinforced Speaker-Wire Dipole

When purchasing wire for my dipole antennas, I opted for speaker wire because it is inexpensive, comes in pairs, and can be split for providing a lot more wire for antenna projects. Stranded electrical wiring is probably stronger and comparable in price.

However, I needed to solve the strength problem of stringing the speaker wire up in the air without breakage. My idea came from a worker at the telephone company, who commented about supporting copper wires with steel cable. Mason's cord seemed to be readily available, and would serve a similar purpose, but I worried that it might not hold up well when wet and sun-bleached over time.

During another visit to the store to purchase replacement string for my weed trimmer, it occurred to me that I may have chanced upon a viable alternative for supporting my speaker-wire antennas. I decided to try it with my next project: a speaker-wire inverted-**V** dipole.

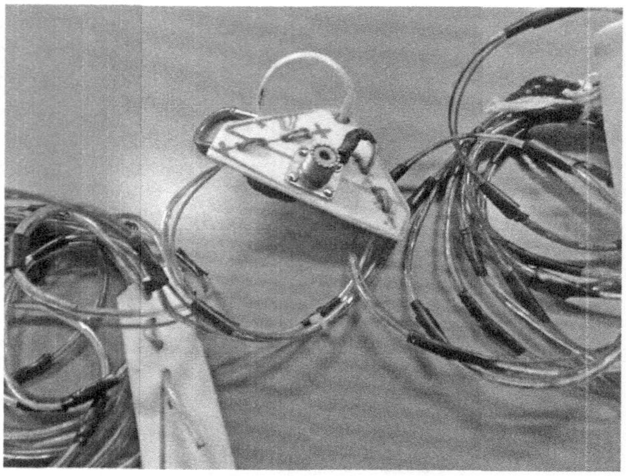

Figure 41 — The finished antenna, made from inexpensive plastic cutting boards and speaker wire. The speaker wire is supported by blue weed-trimmer string. [Manuell Alvarez, KB8ITC, photo]

That same day, I purchased several plastic cutting boards at a local discount store for making the end insulators and center connector. I cut the speaker wire into two lengths that I had calculated for a half-wavelength at 14.250 MHz.

I used electrical tape to secure the speaker wire to the string line. I drilled holes in both the center connector and insulators for weaving the wire and trim line through the holes to relieve the strain and prevent stretching (see **Figure 41**). The result is a lightweight HF antenna that's also great on a budget. — *73, Manuell Alvarez, KB8ITC,* **alvarm@citlink.net**

Figure 42 — Dig a hole at least 4 inches in diameter and bury a short length of 2-inch inside diameter PVC tubing. [Robert Bastone, WC3O, photo]

A Mast Holder That's Always Ready

If you often build and experiment with antennas, you know it can be a challenge to hold the antenna at an easy working level. I've seen sawhorses and custom-made fixtures being used, but here's an easy solution that will be ready whenever you need it.

First, find an area in your yard that you'll use as your antenna "workshop." Dig a hole and bury a short length of 2-inch inside diameter PVC tubing (see **Figure 42**). Make the hole at least 4 inches in diameter and deep enough to extend the tubing below the frost line (don't worry about the excess length of tubing sticking out of the hole yet).

Mix concrete and pour it into the hole so it surrounds the tube. Check to make sure the tube is perfectly plumb. Once the concrete sets, cut the PVC tube flush with the ground and attach a cap or just stuff an old rag into it. Now you can mow right over the tube with no harm done (see **Figure 43**).

Figure 43 — Whenever you need to set up a temporary mast to adjust an antenna, simply remove the cap and slide the mast into the tube. [Robert Bastone, WC3O, photo]

Screwdriver Antenna Solution

Screwdriver antennas are neat multiband mobile radiators, but because of their length, they can be problematic when it comes to using your garage. In my case, I had to unmount the antenna when entering and then remount when leaving (see **Figure 44**). Even if I put the upper whip on an antenna spring or a fold-over detent, it wasn't going to fit.

Because my screwdriver antenna mounts to an SO-239 connector, I realized that I could make a short vertical antenna using a mating PL-259 connector. It would be short enough to make it into the garage, yet long enough for me to check conditions on the bands while driving around town. If I discovered decent conditions, I could stop and mount the screwdriver antenna. Otherwise, I could just continue listening.

I had a stainless-steel element that I picked up at a hamfest, so I cut it to fit. I crimped it to the center pin of a PL-259 and filled the back shell with RTV silicone rubber (see **Figure 45**). However, while gently pulling on it after some use, the element came right out of the PL-259. The crimp hadn't survived all the road vibration and the RTV silicone did not stick to the element.

I decided to solder the stainless-steel antenna to the PL-259 shell and fill the shell with epoxy. The epoxy would dampen the vibrations to center pin to prevent it from gradually separating over time. The problem is that stainless steel won't take solder. I'm a determined ham and I wasn't about to let this stop me. My solution for my solution, so to speak, was to copper plate the end of the antenna.

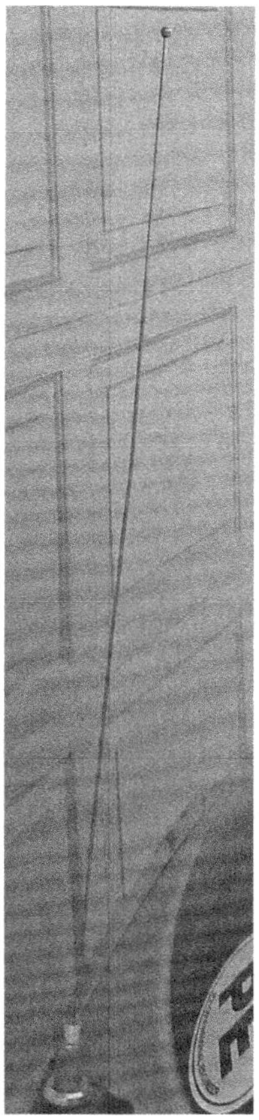

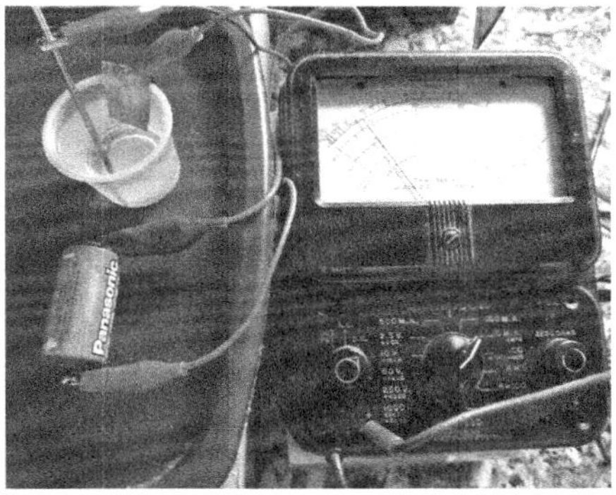

Figure 44 — (Left) The author couldn't drive into his garage without first removing the long whip element from his screwdriver HF mobile antenna. [Richard F. Gillette, W9PE, photo]

Figure 45 — (Right) A shortened element would at least allow the author to drive in and out of the garage, as well as around town, while checking receive conditions to determine whether installing the full-sized element would be worthwhile. However, securing the stainless-steel element to the PL-259 connector presented a problem. [Richard F. Gillette, W9PE, photo]

Figure 46 — (Bottom) The solution to attaching the stainless-steel element was a "solution" indeed — an electroplating solution! The positive battery lead is attached to the copper. [Richard F. Gillette, W9PE, photo]

I gathered a shot glass, vinegar, a NiCad battery, and a small piece of copper foil. I cleaned the end of the stainless-steel element and placed it in the glass of vinegar with the foil hanging over the side (see **Figure 46**).

After a few minutes, I could see copper deposition on the stainless element. When I had a nice copper coating, I tinned it with solder and easily attached it to the PL-259 center pin. This easy technique may be the answer for other situations where you need to securely solder to stainless steel. — *73, Richard F. Gillette, W9PE,* **rf.gillette@ieee.org**

An Easy Strain Relief for Wire Antennas

There are a number of ideas to help alleviate the stress effects of high winds on wire antennas that are strung in trees. Many of these involve a pulley with weights and ropes; some are quite complicated and involve a bit of work to get them right. But many wire antennas can get away with much simpler strain reliefs.

One solution is throttle return springs, which are available at auto parts stores. You can find light ones, about 5 inches in length, that are strong enough to do the job. Attach them to the ends of the antenna between the insulators and the tree branches. You'll also find these springs, such as the Dorman Help! Assortment, for less than $10 on Amazon. — *73, Joe Falletta, W6UDO,* **w6udo@arrl.net**

Figure 47 — A mechanical ascender is normally used for climbing, but it can also keep an antenna support rope nice and snug. [Joe Gore, WA4RTE, photo]

A Snug Support

I have a dipole antenna supported at both ends by Dacron rope. One end is anchored to a tree, but I had trouble keeping the rope snug. I had to use a ladder to reach the anchor point, and it was hard to keep the rope snug while I tied it off.

Searches for an appropriate tool to secure the rope led me to a mechanical ascender designed for climbing. It employs a cam that allows the device to slide freely in the intended direction of movement, but it provides a firm grip on the rope when pulled in the opposite direction. To prevent an ascender from accidentally coming off the rope, a locking mechanism or trigger is deployed.

I found an imported ascender at an online retailer. I fastened my antenna rope to the ascender, and then clipped it to the support rope from the tree (see **Figure 47**). Now, I can just slide the ascender along the support rope to tighten the antenna. It is easily removed if needed. — *73, Joe Gore, WA4RTE*, **joegore1965@bellsouth.net**

A Helping Hand

My wire antennas are installed in trees that are in or on the edge of a forest. Inevitably, when putting a wire antenna into a tree, the pull string used to haul up the rope becomes tangled in leaves and sticks. I have to stop, walk to the other side of the tree, free the string, then walk back and resume pulling the rope into the tree. Of course, the string gets tangled again, so I walk to the other side and repeat.

The best way to put an antenna into a tree is to enlist another ham or a friend. During the many times I've performed this exercise solo, however, frustration was too often the result. I created a solution with some scrap wood, a saw, a wooden dowel, and about 30 minutes of assembly time (see **Figure 48**).

The result is a spooling jig that maintains the rope at the proper upward angle and away from the soil. It also helps keep tension on the line to avoid tangles. I've used this "helping hand" on many occasions and have put several antennas into trees with minimal time and frustration. — *73, Lionel Booth, N5LB*, **n5lb@outlook.com**

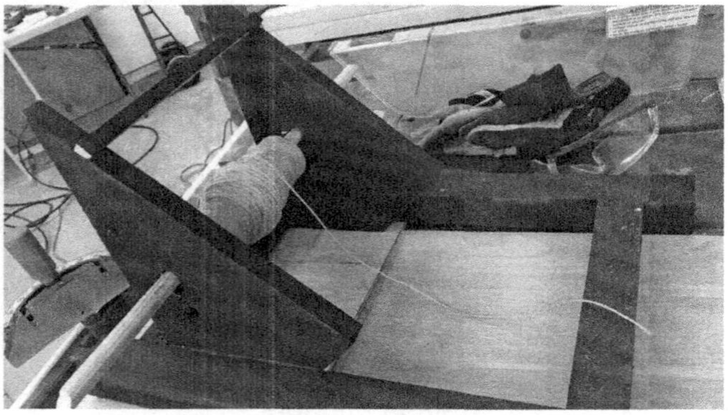

Figure 48 — You can use a dowel and some scraps of wood to create a "helping hand" to smooth the process of getting ropes and strings aloft. [Lionel Booth, N5LB, photo]

Sliding Insulators

I recently had to add extra insulators to my 144-foot wire loop antenna. It was impractical to take down the entire antenna, disconnect the wire, and add the necessary insulators, so I found an easier way.

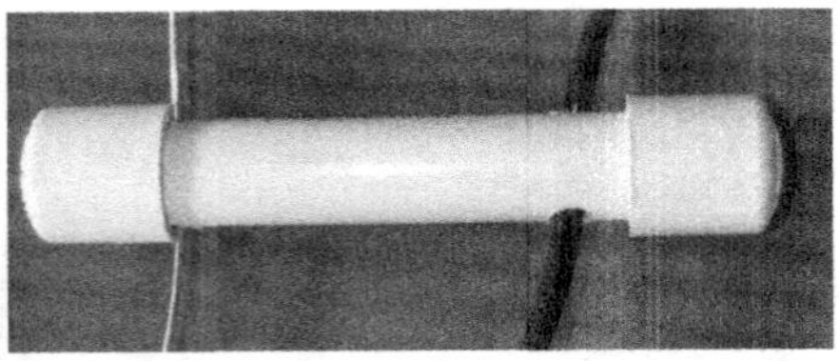

Figure 49 — A movable insulator. The antenna wire is at the left and the guy rope appears at the right. [Jeff Rahmel, KA8ZAW, photo]

I started with a 5½-inch piece of PVC. In one end, I drilled the hole for the rope. On the other end, I drilled a hole for the antenna wire. I then cut two slots 180 degrees apart, connecting the antenna hole to the end of the pipe (see **Figure 49**).

To install this insulator, I fed the antenna wire through the slots and into the antenna wire hole. I then used PVC pipe cement to attach the cap. Next, I cemented a cap on the guy rope end of the insulator and ran the guy rope through that hole. Using this movable insulator allows you to slide it along the antenna wire until you reach the desired connection point and then the tie off the rope. — *73, Jeff Rahmel, KA8ZAW*, **jrahmel@roadrunner.com**

Yagi Boom-to-Element Alignment

While assembling Yagi antennas for various bands over the years, I've always struggled with aligning the elements and squaring them with the boom, whether by eyeball or with a level.

Recently, I purchased a Cushcraft LFA-6M4EL 6-meter Yagi and set about assembling it in my family room on sawhorses, saving the long element parts for later so I could get it out the back door. The familiar problem of aligning the elements with the boom quickly arose. After I marked the boom exactly where each element was to go, and loosely putting it all together, I decided there had to be a better way.

From the pantry I grabbed six cans of soup, making sure that all were the same size. I placed one pair of cans under the reflector assembly, another pair at the last director, and the third pair at a mid-element (see **Figure 50**). After double checking the reflector's position on the boom, I tightened all three assemblies. This became the reference.

Keeping the first two pairs of cans in place at the reflector and last director, I then moved the other two cans from one element to another while I aligned each one and tightened its bolts. All the element-to-boom assemblies were perfectly aligned.

With the antenna still resting on the soup cans, I mounted the boom to mast bracket. To align the bracket, I put a three-foot piece of two-inch aluminum tubing in the clamps and used a small level to get it perfectly vertical, while being careful not to disturb the soup cans.

Figure 50 — Identically sized soup cans become tools to ensure proper boom-to-element alignment. [Glenn Kurzenknabe, K3SWZ, photo]

The Yagi will go up on the tower soon, and I have no doubt that it will look as good as the pictures. For antennas that are too big for indoor assembly, simply apply this technique on a larger surface, like a good, flat cement garage or carport floor. — *73, Glenn Kurzenknabe, K3SWZ*, **k3swz@arrl.net**

Standoffs for 450 Ω Balanced Line

While relocating the feed line for my doublet antenna, I needed standoffs for the 1-inch balanced line, so that I could route the line under the eaves of my house and up a wall.

It's always a challenge to find reasonably priced standoffs for balanced line. I usually shop for antenna insulators in the electric fence section of a local farm supply store, Blain's Farm and Fleet (**www.farmandfleet.com/electric-fence-insulators/**), and on their website I found a selection of Dare post insulators intended to hold electric fence wire. Two possibilities stood out: a 5-inch yellow post insulator and a 3.5-inch black one. Opting for the 3.5-inch insulators, I ordered two bags of 10, and in a few days, I was mounting my feed line. (Note: the availability by size and color tends to vary from time to time.)

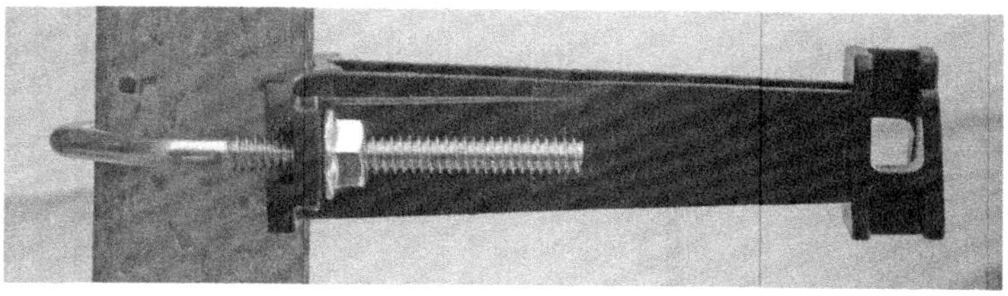

Figure 51 — The 3.5-inch insulator comes with a U-bolt for fastening to a vertical or horizontal pipe. [John Dewey, KA9CAR, photo]

Figure 52 — The insulators can also be nailed into a wood base as shown on this mockup. [John Dewey, KA9CAR, photo]

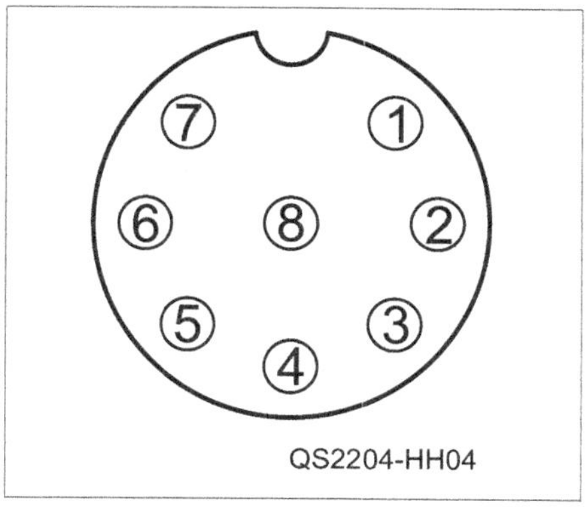

Figure 53 — The pinout of the DIN jack on the rear panel of the Yaesu G-5400B control box. [José Manuel Freixo Nunes, CT1BRM, photo]

This type of insulator includes a **U**-bolt to fasten it to a vertical or horizontal pipe, so it can work on a tower leg about 1 to 2 inches in diameter (**Figure 51**). I did not need the **U**-bolts because I used wood screws to fasten the insulators in place. The base can be oriented with its long axis either parallel or perpendicular to the feed line (see **Figure 52**). The 5-inch insulators can attach to a wood surface with nails or screws. — *73, John Dewey, KA9CAR,* **ka9car@arrl.net**

A Solution for Weed Control

At the Skyview Radio Society, K3MJW, we are forever struggling to keep Mother Nature under control when it comes to our antennas. Weeds and vines are always trying to make their way onto our tower bases, and they frequently climb the towers and guy wires. One could use a string trimmer to keep them at bay, but this runs the risk of severing wires. And, yes, you could resort to chemicals, but we prefer a greener solution.

Our answer is to use carpet remnants for weed control. Many carpet outlets will be more than happy to give these to you. Simply cut up the carpet sections to the appropriate sizes and place them over or around the areas you are trying to protect. The carpeting acts as a highly effective weed block and it lasts for years. You can even paint it to blend with the surroundings. — *73, Bob Bastone, WC3O,* **radiofreebob@gmail.com**

Rotator Mouse Controller

Repurposing a surplus computer mouse to control a Yaesu satellite antenna rotor G-5400B is an extremely easy project that should take only a few minutes. The only thing you'll typically need to buy is an eight-pin DIN plug. I'd had this idea for a while, and it was time to finally make it happen.

With slight wiring modifications, it is also possible to adapt this idea to standard horizontal-only rotator control units that offer external switch connections. Just check your rotator manual for information about the wiring of the external switching jack.

The Yaesu G-5400B control box has an eight-pin female DIN jack on the rear panel (see **Figure 53**) with this pinout:

Pin	*Function*
1	0 to 12 V dc. Indicates the azimuth position.
2	0 to 12 V dc. Indicates the elevation position.
3	Grounding this pin turns the rotator to the left.
4	Grounding this pin turns the rotator to the right.
5	Grounding this pin points the elevation downward.
6	Grounding this pin points the elevation upward.
7	7 – 15 V dc
8	Ground

A computer mouse usually has two, three, or four switches. Four switches would be optimal for a satellite rotator, because they can provide the necessary **UP**, **DOWN**, **LEFT**, and **RIGHT** commands.

The old mouse I used in this experiment had three switches on the top, plus a fourth one on the bottom. The bottom switch can be used to change the mouse's central switch to move the rotator up or to the right. **Figure 54** is a diagram of the mouse switches and their corresponding connections to the DIN plug. Open the mouse and solder the mouse cable wires directly to the mini switches and to the wight-pin male DIN plug. I was able to use the original mouse cable, and after I made all the connections, I checked them with a multimeter. Finally, I labeled each mouse button, as shown in **Figure 55**. — *73, José Manuel Freixo Nunes, CT1BRM,* **brm@netcabo.pt**

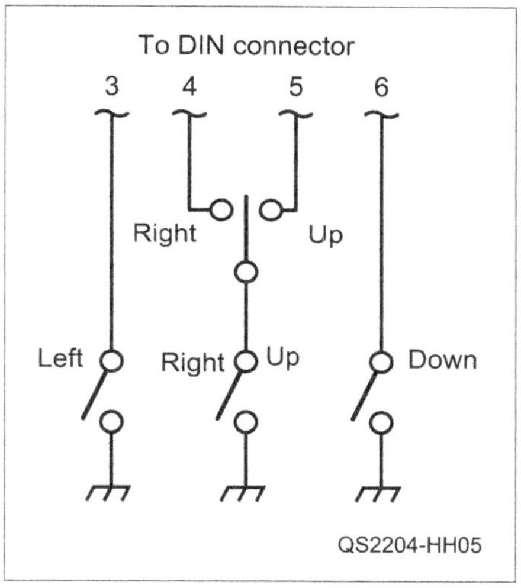

Figure 54 — A schematic diagram of the mouse switches and their corresponding connections to the DIN plug pins. [José Manuel Freixo Nunes, CT1BRM, photo]

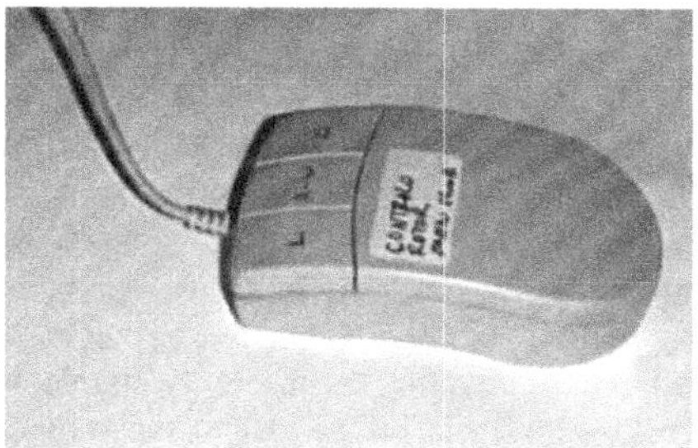

Figure 55 — The mouse switches are labeled. [José Manuel Freixo Nunes, CT1BRM, photo]

Figure 56 — A 1-inch and a ½-inch boom. The ½-inch boom uses hook-end springs, which make it easier to attach the elements. The 1-inch boom uses closed-end springs, which require a bit more dexterity to fasten. Both elements are ⅛ inch in diameter. [Bill Gerhold, K2WH, photo]

Mounting Elements Using Spring Tension

I've investigated different ways to attach and secure antenna elements to a boom. I've used pressure screws, clamps, and even rubber bands — these methods were okay, but it wasn't easy to adjust, assemble, or disassemble the beam.

Eventually, I thought of using spring tension (see **Figure 56**). The boom can be any material. Once it is properly pre-drilled, the elements are held in position by springs. Insert the element into the spring on one side of the boom, and partially slide the element into the boom. As it starts to protrude from the other side, insert it into the other end of the spring. This holds the elements in position and allows for quick adjustments.

You will have to experiment to find the right type of spring, which is dependent on the boom and element diameters. Small spring assortments are available from websites such as Amazon. — *73, Bill Gerhold, K2WH,* **k2wh@optonline.net**

Making a Great Antenna Better

The Hy-Gain AV-680 is a great an-tenna. After assembling three of these, I'd like to pass on a few helpful tips.

The 30- through 80-meter loading coils are near the top of the antenna, which makes it top-heavy and prone to wind damage. One solution is to reinforce the connections between tubes BB and BC, and tubes BC and BD. The manual states that only stainless-steel hose clamps are needed at those junctions. But they are weak points, and if repeatedly subjected to high winds, they tend to fail. To make these connections secure, I first in-stalled the clamps as directed. Then, I drilled holes all the way through the junctions above and below the clamps. Through these holes, I in-stalled #10-32 × 1¼-inch stainless-

Figure 57 — A close-up of one of the tube joints. Notice the stainless-steel bolts above and below the hose clamp. [Al Wasielewski, WA2VJL, photo]

steel bolts, nuts, and lock washers. Now the two joints are extremely rigid (see **Figure 57**).

I live in an area that is often hit by hurricanes, so I also followed Hy-Gain's advice by installing three guy lines at a height of 14 feet. I purchased three thimbles and three guy clamps from Cushcraft, and mounted and secured them with stainless-steel clamps. I attached the lines with ¼ × 6-inch plastic wire ties.

The screws and nuts for the radial mounting brackets are exceedingly small and easily lost. My solution was to put a cloth under the work area to catch anything that drops.

Lastly, I found that it is easier to mount the antenna and install the counterpoise at the bottom *after* installation. — *73, Al Wasielewski, WA2VJL,* **wa2vjl@yahoo.com**

An Easy-to-Construct Coax Puller

Routing coaxial cable through conduit and other tight spaces can be a frustrating exercise, especially when a PL-259 connector is attached. I devised a tool to make the job much easier.

My tool uses easy-to-obtain parts that may be in your junk box. You will need a PL-259 connector with an adapter, an SO-239 barrel connector, and one $\frac{3}{16}$ × 2-inch stainless-steel eyebolt with an extra nut (see **Figure 58**). Begin construction by using a hacksaw to cut the PL-259 adapter just below the spot where the threads end at the bottom of the adapter. Insert the eyebolt (with the extra nut on the eyebolt placed all the way to the top of the threads) through the shortened adapter. Put the remaining nut on the eyebolt and securely tighten it. Use threadlocker if desired.

When the assembly is complete, use a grinder to carefully remove the sharp edges from the bottom nut of the adapter. This will make it fit through the top of the PL-259. Be careful when grinding to prevent damage to the threads. Tighten the assembly to the PL-259 plug. The assembled version is also shown in Figure 1. I used the parts I had on hand, but if you use a smaller stainless-steel eyebolt, you can eliminate the need to use the grinder on the bottom nut.

To use, attach a pull rope to the eyebolt, and then attach the PL-259 connector on the coaxial cable to the barrel connector. Now, you can easily pull the cable to its destination while reducing the odds of its PL-259 connector getting snagged. — *73, Robert Spears, Sr., AE4B,* **ae4b@yahoo.com**

Figure 58 — At the top of this image, you can see the "exploded" version of the coaxial cable puller tool. The assembled version is directly beneath it. [Robert Spears, Sr., AE4B, photo]

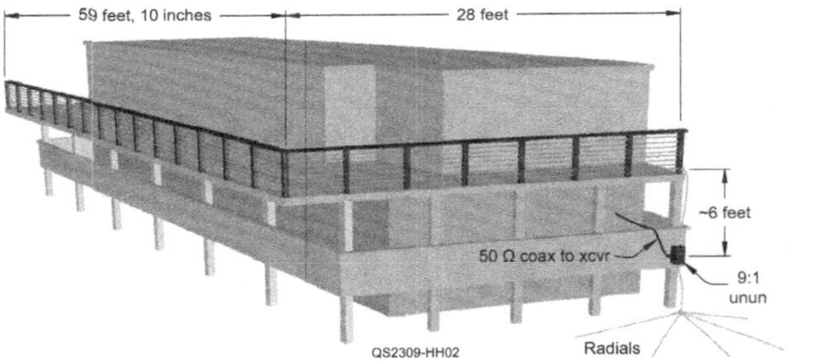

Figure 59 — A diagram of the author's balcony railing antenna. The radial wires are a few feet long. They are not required, but they will likely improve the antenna's efficiency. [Hiroki Kato, AH6CY]

Railing Antenna

I live in an antenna-restricted homeowner's association community. A large metal railing surrounds my second-floor balcony, so I decided to try using the railing as an end-fed antenna. To my surprise, it worked — not only on one band, but on several bands!

I connected a wire at the end of the railing and inserted a 9:1 unun near the spot where the coaxial cable enters my station on the first floor (see **Figure 59**). By using an Icom IC-7610 transceiver with 100 W and an MFJ-994B automatic tuner, I can load the balcony railing on all bands between 160 and 6 meters.

If you have a similar railing or other long metal structure, it may serve as an equally stealthy antenna. Of course, be sure to visit the ARRL RF Exposure web page at **www.arrl.org/rf-exposure** to make sure the antenna complies with the latest regulations. — *73, Hiroki Kato, AH6CY,* **ah6cy@arrl.net**

The Sky Tube: A High-Q Short Vertical Antenna for 40 – 10 Meters

I have long been impressed by the performance of shortened vertical antennas with high-Q loading coils. So, when the contractor who remodeled my house left behind a cardboard concrete form, I decided to build an HF vertical antenna with a large-diameter loading coil.

I named the resulting antenna the "Sky Tube." Its diameter is 12 inches, and its height is 40 inches. I began construction by wrapping the concrete form with 10 ½ turns of braided cable — the kind used for grounding — starting 12 inches from the bottom. I left approximately 2 inches of vertical space between each turn. Next, I placed a telescopic whip, adjustable to a maximum 6-foot length, on top of the form (see **Figure 60**). A wire attached to the base of the whip connects to the top of the coil. To adjust tuning on 30 – 10 meters, I constructed a 24-inch-long braided cable with alligator clips at either end (see **Figure 61**); the bottom of the braid clips onto the bottom of the coil. I found that I didn't need a tuning cable for 40 meters.

The SO-239 connector at the feed point is shown in **Figure 62**. Inside the form, I attached a wire to one of the connector's mounting nuts. This wire connects to the Sky Tube's radials, which consist of three sets of 15-wire computer cables. Each set is about 12 feet long and rests on the soil around the antenna. Another wire is attached between the center conductor of the connector at the bottom of the coil.

If you'd like to build a Sky Tube of your own, you'll find concrete forms at home improvement stores such as Home Depot. A telescopic whip, such as a 6-foot-long MFJ-1963, can be used for the radiator, though it will require a 10-32 stud mount at the top of the form. — *73, Hiroki Kato, AH6CY,* **ah6cy@arrl.net**

Figure 60 — The completed Sky Tube, with the telescopic whip at the top. It is resting on a tripod that stands a few feet above the ground. [Hiroki Kato, AH6CY, photo]

Figure 61 — A braided cable with alligator clips at either end allows me to adjust tuning. The lower end of the cable attaches to the bottom of the coil. [Hiroki Kato, AH6CY, photo]

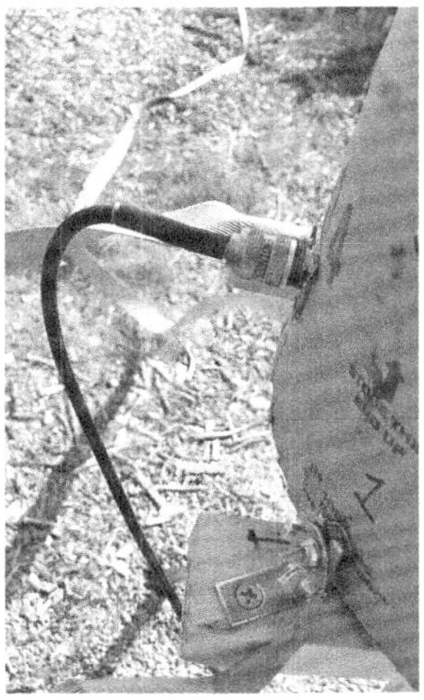

Figure 62 — A view of the SO-239 feed-point connection at the bottom of the coil. A wire from one of the SO-239 mounting nuts inside the tube attaches to the wires, which serve as radials. The connector's center conductor is soldered to a wire that attaches to the screw that secures the coil's end to the tube. [Hiroki Kato, AH6CY, photo]

Turn a Multi-Bit Screwdriver into an Antenna Tool

After assembling an M2 Antenna Systems UHF antenna that used slip-on spring element retainers, I realized I needed a better tool for installing retainers. I found a low-cost, multi-bit screwdriver at a local home improvement store; its ¼-inch internal hex end was the perfect size for pressing retainers down the lengths of the elements. I drilled a ¼-inch hole through the handle and the shank so that the element could fit through it (see **Figure 63**). Now, I have an easy-to-grip tool that allows me to simply slide the retainers down the elements (see **Figure 64**).

Centering the elements in the boom is a bit challenging because once the retainers are in place, they can't be moved backward. However, by measuring the width of the insulators installed in the boom, you can calculate how far each retainer is from a given element tip. After this step, hold the element in a bench vise with the required length exposed. Slip the retainer down until it is against the vise jaw. Install the element in the boom, and slip the other retainer in place to retain the element. — *73, Phil Lefever, KBØNES,* **kb0nes@yahoo.com**

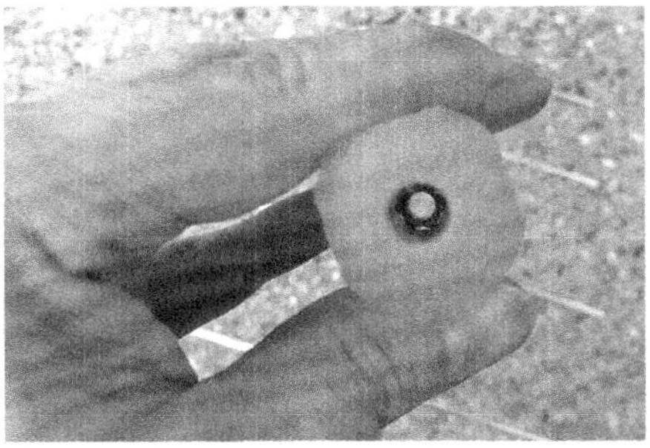

Figure 63 — Modify the screwdriver by drilling a ¼-inch hole through the handle and the shank. In this image, you can see one of the antenna elements inside the screwdriver. [Phil Lefever, KBØNES, photo]

Figure 64 — This modified tool makes it easy to slide element retainers into place. [Phil Lefever, KBØNES, photo]

CHAPTER 2

Around the Shack

IR Thermometer Accuracy

The infrared (IR) thermometer is becoming a popular piece of test equipment. Before buying mine, I reviewed several products and noted that they were calibrated to an emissivity of 0.95. That means that the instrument assumes, when calculating the displayed temperature, that the surface being measured has an emissivity of 0.95. Thus, computed results can be radically in error for transistor heatsink materials with differing emissivity values, as I will show. [The *emissivity value* is the ratio of the level of IR radiated by a material with respect to the level radiated by a blackbody, both being measured at the same temperature. — *Ed.*]

Different materials, at the same temperature, may radiate different levels of IR. Pieces of the same material, with different surface treatments, may also radiate IR at different levels. Emissivity can vary from 0 (mirror, no IR radiation) to 1.0 (blackbody, a perfect IR radiator).

Table 1 shows the emissivity of various surfaces.[1,2] Of particular interest here are the emissivity values of metals typically used for heatsinks. This table indicates that IR thermometer measurements can be grossly in error. There was no user information packed with the instrument I bought, so I would guess that this information is not well known to the casual user of the IR thermometer.

As an illustration of these possible errors, refer to **Figures 1** and **2**. They show measurements on a uniformly heated, aluminum-sheet surface. This piece of aluminum is the same area as the stove burner heating it. It has been allowed to stabilize at the readings seen here. The right half of the surface is plain, as fabricated aluminum sheet. On the left half of the surface, an area of flat gray paint has been applied. A bimetal thermocouple has been attached to this gray area to measure the true surface temperature. It is connected to the digital multimeter that is displaying the temperature on the right-hand side of each figure.

Table 1
Surface Emissivity Ratios

Surface	Emissivity Ratio
Aluminum plate, polished	0.040
Aluminum, anodized	0.90
Brass plate, polished	0.096
Brass plate, oxidized	0.59
Brick	0.90
Concrete, rough	0.91
Copper, polished	0.04
Copper, oxidized	0.87
Glass	0.95
Paint	0.90
Paper, black or white	0.88 – 0.86
Water, pure	0.96

Summarizing these figures, the true temperature is 102.5°C. The bare metal gives an IR temperature reading of 30°C, and the painted area 90°C. This degree of error is *not acceptable* when trying to determine a sufficiently sized heatsink for power transistors or similar devices.

With the data taken using the setup of Figure 2, the emissivity of the painted surface has been calculated to be in the range of 0.8 – 0.84. A patch of black electrician's tape gave a slightly better result, its emissivity probably approaching 0.88 – 0.90. From this, one concludes that test patches should be used regardless of heatsink surface type, unless you are working with anodized or factory-painted heatsinks. Even then, for IR thermometer measurements, a correction factor must be added. Even with an improved emissivity surface as suggested here, one should add about 10 – 12% to the

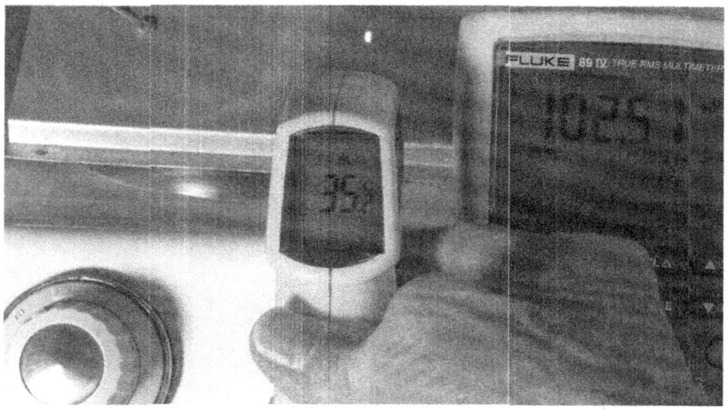

Figure 1 — The handheld thermometer shows the IR temperature measurement of the bare metal area of the aluminum sheet. The thermocouple measurement of the actual surface temperature is shown on the meter. [Ralph Crumrine, NØKC, photo]

Figure 2 — This shows the IR temperature measurement of the painted area of the aluminum sheet, demonstrating a significant difference in emissivity between the surfaces. [Ralph Crumrine, NØKC, photo]

IR thermometer readings to get close to a true temperature value. [Heatsink temperature should be well above ambient for accurate readings. — *Ed.*]

Trying to make sense of readings on the natural metal surfaces is out of the question. It is probably beyond the knowledge of the technician to know exactly what the emissivity is for an arbitrary surface appearance and what correction factor should be applied to the IR thermometer reading.

These tests and resulting data were taken holding the IR thermometer as close to the test surface as was practical, so the surface being measured completely filled the detector's aperture. For more information on the use of IR thermometers and some discussion of other possible sources of errors, go to **www.raytek.com/Raytek/en-r0/IREducation**. — *73, Ralph Crumrine, NØKC,* **n0kc@arrl.net**

Notes

[1]C. C. Dilio and E. P. Nye, "Thermal Engineering," International Textbook Co., 1959, p. 219.

[2]M. Q. Brewster, "Thermal Radiative Transfer & Properties," John Wiley, 1992, p. 56.

Weatherproofing a Coaxial Switch

Though my quad vertical array (QVA) was serving me well, I wanted to try an antenna that had some directionality, so I erected a 140-foot dipole to operate on 80 – 6 meters. With a single feed line coming from the shack, I needed a way to switch between the dipole and the QVA. The remote-controlled coax switches, though weatherproof, were expensive. I searched for a manual weatherproof coax switch, but came up empty, so I decided to build a weatherproof enclosure for a manual coaxial switch.

I purchased a three-way coaxial switch, anticipating a possible third antenna, and established a few requirements for the enclosure:

- Withstand driving rain, ice, snow, and freezing temperatures
- Provide easy access to the switch
- Provide easy resealing, without leaks, under all weather conditions

The Enclosure

After researching a variety of enclosures, I selected a 9-cup, 2.1-liter Rubbermaid food storage container, which measures 8½ inches square and 3¼ inches deep. Its removable rubber lid is tight-fitting, waterproof, and provides easy access to the switch. I planned to mount the enclosed switch on a panel (a kitchen cutting board) along with a balun, a remote antenna tuner, and lightning arrester. A preliminary layout is shown in **Figure 3**.

Figure 3 — This figure shows (clockwise from the upper left) the positions of a future balun/unun, balun, remote antenna tuner, a cardboard template representing the lightning arrester mounting bracket, and the switch/enclosure all sitting on a kitchen cutting board. The preliminary layout was used to establish the locations of the switch port holes in the enclosure and mounting holes for the other components. [Stan Ekiert, K3KKH, photo]

To access the switch ports, four large holes were cut in the enclosure, along with three smaller mounting holes. The switch port holes are ⅞ inches in diameter. In order to weatherproof these ports, I used the tops of plastic liter bottles. I selected bottles with large ⅞-inch inside diameter tops that will allow RG-213 with a PL-259 connector to be easily inserted through the top.

The tops of the bottles were cut off at the bottom of the threads just below the flange. The flange diameter is 1¼ inches with an inside diameter of ⅞ inches. Each hole was lightly sanded, the tops were positioned on the outside of the enclosure over each hole, and the tops were glued in place with DAP Weldwood Contact Cement (Gorilla Glue did not bond to the enclosure). The glued tops were clamped for 15 hours to allow the glue to cure. **Figure 4** shows the enclosure with the attached bottle tops.

To seal the tops against rain, holes were drilled in the bottle caps and inserted over each of the coaxial lines before attaching the PL-259 connector. During final assembly, the tops and bottle caps were filled with silicone caulk, and the bottle caps tightened against the top to form a watertight seal. **Figure 5** shows a close-up of the top-coax-bottle cap assembly during a fit check.

Mounting the Enclosure

The switch enclosure was attached to the mounting panel with three 6-32 × 2-inch screws. Holes for the screws were drilled through the enclosure and panel. To seal these holes, three ½-inch diameter plastic kitchen cabinet bumpers, with ⅛-inch holes, were placed on the inside of the enclosure over each hole; the switch sits on top of two of the bumpers. On the underside of the enclosure, three ¾-inch square plastic furniture feet, with ⅛-inch holes, were placed over each of the mounting holes (see **Figure 6**).

During assembly, the 6-32 screws were inserted through the switch, bumper, enclo-

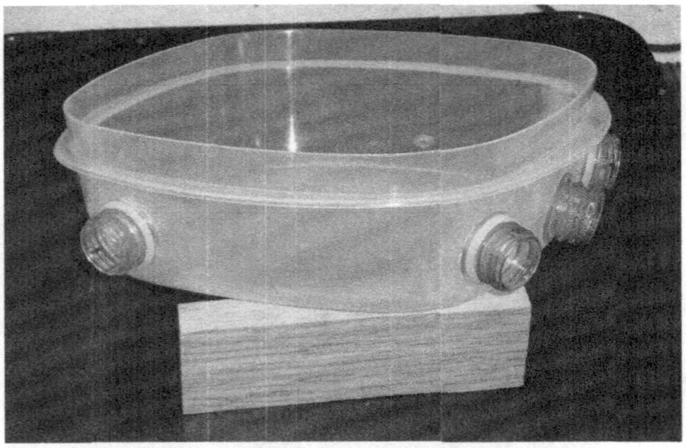

Figure 4 — This figure shows the enclosure with the four bottle tops attached. In retrospect, it would have been prudent to place a bead of hot glue around the flange-enclosure interface of each top. This would increase the mechanical strength of the attachments. [Stan Ekiert, K3KKH, photo]

Figure 5 — Close-up of the top-coax-bottle cap assembly during a fit check. To ensure a snug fit between the bottle cap and coax, a ¼-inch pilot hole was first drilled in each bottle cap, then a conical hand reamer was used to enlarge the hole to its final diameter. [Stan Ekiert, K3KKH, photo]

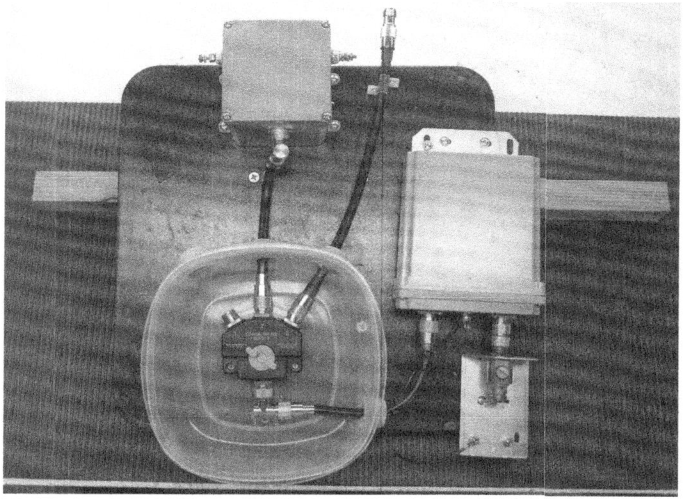

Figure 6 — Final assembly shows the mounted switch within its weatherproof enclosure with the cover removed. [Stan Ekiert, K3KKH, photo]

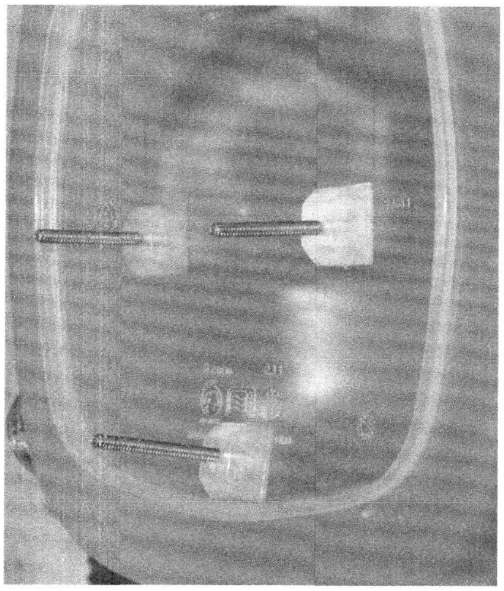

Figure 7 — Three ¾-inch furniture feet were placed over the mounting screws on the underside of the enclosure. [Stan Ekiert, K3KKH, photo]

sure, furniture foot, and panel. On the back side of the panel, another plastic furniture foot was inserted over each mounting screw and secured with a washer, lock washer, and nut. The third mounting screw was similarly attached, except that it did not pass through the coaxial switch. During the assembly, silicone caulk was applied on both sides of the furniture feet to seal the enclosure-panel assembly. The final assembly is shown in **Figure 7**.

The Garden Hose Test

No project is complete without a test, so I subjected the final assembly to 5 minutes of water spray from my garden hose fitted with a shower head. At the end of 5 minutes, there was no trace of water within the enclosure.

Conclusion

The enclosure has been operational since September 2015 and has protected the coax switch against numerous rain storms and the Jonas blizzard of 2016, which dropped 18 inches of snow on my New Jersey location. In addition, I have verified that the container easily reseals with a tight fit down to –2°F. — *73, Stan Ekiert, K3KKH,* **roadspyder@msn.com**

[1]C. C. Dilio and E. P. Nye, "Thermal Engineering," International Textbook Co., 1959, p. 219.
[2]M. Q. Brewster, "Thermal Radiative Transfer & Properties," John Wiley, 1992, p. 56.
[3]T. Brock-Fisher, K1KP, "Can Home Solar Power and Ham Radio Coexist?," *QST*, Apr. 2016, pp. 33 – 37.
[4]T. Fowle, WA6IVG, "Protoboard Battery Safety," *QST*, Jul. 2016, p. 59.

Tuning Extension Arm

As a contester, I've found that leaning over my operating desk for the hours needed to compete makes my back ache. I searched for software and hardware solutions to alleviate my discomfort, but had no success.

When I made my radio desk, I made the desktop in front of the radios deep so I could put more things there. One day, I was sitting at the desk, leaning back, and thinking of a way to "bring" the tuning knob to me. Being a DIY person, it occurred to me that I needed a long extension arm with a flexible joint.

I removed the tuning knob from my Icom and found that it mated to a ¼-inch coupler. I made a flexible joint out of a universal joint for a ¼-inch socket set by grinding the male end of the universal joint round enough to fit in the coupler. Next, I took a ⅜-inch oak dowel and ground one end square so it would slide into the female end of the universal joint (see **Figure 8**).

To finish the project, all I needed was a knob and a rest for the dowel. I glued three pieces of oak scrap together, used a wood lathe to make it round, drilled a ⅜-inch hole in the middle, and pressed it on the end of the dowel. The dowel rest is made from wood, with a groove at the top for the dowel to sit in (see **Figure 9**).

My extension arm gimmick works great (see **Figure 10**). I moved the computer mice and the Logitech G13 gameboard that I use with my contest logging program to the edge of the desk, and I can now lean back and work a contest in comfort. — *73, David Fuseler, NJ4F,* **nj4f@arrl.net**

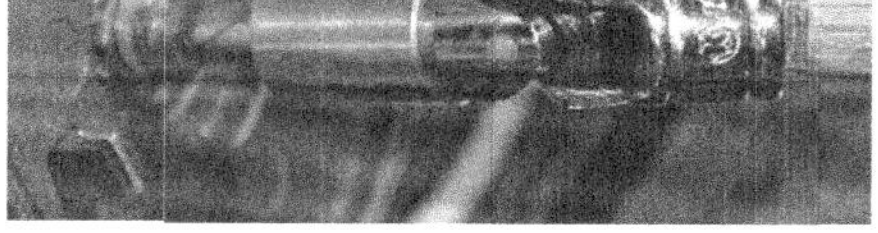

Figure 8 — A brass coupling and a flexible joint from a socket set allow a tuning knob and extension arm to couple to the Icom tuning shaft while allowing the business end to be positioned for maximum comfort. [David Fuseler, NJ4F, photo]

Figure 9 — The knob was made from scrap wood, shaped on a lathe, and drilled out to accept the extension arm, which is prepared from a wooden dowel. [David Fuseler, NJ4F, photo]

Figure 10 — The extension arm permits the operator to maintain a more comfortable and relaxed position, which greatly improves performance during long hours of contest operation. [David Fuseler, NJ4F, photo]

Rope Reel

People toss old hose reels in the trash when the **O**-ring seals start leaking. But they're a great (and free) way to keep long lengths of rope, wire, or coaxial cable under control (see **Figure 11**). Just pick it up and take it with you. Need some cable or rope? Just pull. Then simply turn the crank to reel it back up. — *73, Bob Bastone, WC3O,* **wc3o@arrl.net**

Figure 11 — These old hose reels are an easy way to keep long lengths of rope, wire, or coaxial cable organized. [Bob Bastone, WC3O, photo]

Call Sign License Plate

By displaying a license plate with your call sign on it, you can show pride and promote our hobby. You could, of course, get a personalized license plate, but that can be expensive at license renewal time. Here's how I make my own ham call sign "license plates" to display.

Go to a sign-making shop to buy a blank license plate used for signs. They're not expensive and can be easily obtained. Note that they are available in either white plastic or painted metal. I have tried both.

Print your call sign using Microsoft *Word*, written in Times New Roman font. Go to the Layout menu and select "Page Setup." Use "Landscape" format so the maximum size limits can be used. Select "Page Width" from the zoom pulldown or View menu. The font size for my call sign (KK5R) is 225, and the "Ham Radio" message is font size 100. Don't worry if the Text Size pulldown does not offer this option; click on the Text Size, type in the number, and hit enter. You could also select the text after typing it and use the Text Size option to change it; this allows you to adjust the width of the call sign or the "Ham Radio" message to your paper size. For the separation between the two text messages, use a blank line with a smaller font size to space the call sign and "Ham Radio" text. To do this, you have to select the line first and then use the Text Size option. You could also cut the two messages out and place them manually instead, as explained below. When printing, use bold font to get the maximum black-to-white ratio. Do not use the Draft option or the lettering may not be as dark as it should be.

After printing this master copy, take it to a business center, such as FedEx or UPS, for printing on a laser printer. If you have a laser printer, this can be done at home. It's important to use a laser copy because a bubble-jet copy does not weather as well, and isn't as resistant to sunlight.

The next step is to get the master laminated once you see that the message fits the plate. Cut the top and bottom lines apart, and arrange them so they don't interfere with the license plate mounting screw holes by placing the lettering on the blank license plate to locate it. Then laminate it. Once laminated, trim the lamination to fit the plate and punch a small hole in the lamination to pass the mounting screws, which should be #12/14 sheet metal screws about $^3/_8$ inches long. Place carefully, because it may be hard to rearrange later. Leave about a $^1/_2$ inch of clear space between the license plate and the "Ham Radio" plate.

Figure 12 — The homemade call sign "license plate." Note that this example is the lamination alone, without the plastic or metal plate used. [Bob Lunsford, KK5R, photo]

Making a plate took me less than an hour. The only real expenses are the plates (if used) and laser copies/lamination. The finished product hasn't shown effects of weathering after being used for a few months (see **Figure 12**), but replacement will be easy if necessary. This technique can also be adapted for other uses, such as signs for a sales table at a hamfest, or for a wall plaque. — *73, Bob Lunsford, KK5R,* **nocrud222@yahoo.com**

A Makeshift Faraday Cage

As Amateur Radio operators, we're supposed to be able to communicate in emergencies. But if there's ever a substantial electromagnetic pulse (EMP), whether from nature or from some enemy, our radios will be useful perhaps as paperweights, but nothing else. We need to protect our equipment — and I found a way to do it simply and on the cheap.

I keep my handheld transceiver and an extra charged battery in a plastic (therefore non-conductive) bucket with the metal handle removed. I put the entire package in a small, all-metal pail (see **Figure 13**). After I ensured that nothing extended past the plastic bucket to the pail's metal, I put the pail's cover on, and the inside was sealed. I believe this may serve as a Faraday cage, which should protect any electronic equipment from the effects of an EMP. Hopefully, if such a thing were to occur, I'd at least be able to transmit. — *73, Steve Sachs, K6SMS,* **dr.sachs@sachsscientific.com**

Figure 13 — A makeshift Faraday cage constructed from a plastic bucket inside a metal pail. [Steve Sachs, K6SMS, photo]

Mount for Radios with Bottom-Fire Speakers

You've just found the next radio for your collection at the local swap meet, and you can't wait to get home and fire it up. However, there's one small problem: it has a bottom-fire speaker, and the mounting bracket for the radio was not included. If the speaker mount is flat (i.e., the speaker fires straight down), the sound coming out of it will be muffled by the shelf the radio is sitting on. If the speaker is angled up a bit (see **Figure 14**), you might be able to hear what's coming out of the speaker, but it won't lie flat like the rest of the radios on your desk.

Both situations can be addressed with the same solution: a radio stand. For radios with the standard screws on the side that hold the radio on the bracket, a simple **L** bracket from the hardware store will work. But what if the radio has a different mounting style? A new approach is needed.

The only critical dimension involves the two bottom pieces sitting on the shelf. They must be as long as the radio without the radio falling through the cross braces. Because the remainder of the dimensions are not critical, component selection becomes simple. The wood I used came from an antenna I had purchased; it had been the box stiffener so that the antenna would not get bent during shipping. The **L** brackets I used were also not critical. The only requirement was that they held the cross braces up high enough to allow for speaker clearance and allow sound to escape.

Figure 15 — The finished bracket supporting a matching set of Midland radios. [Klaus Spies, WB9YBM, photo]

To finish things off, I installed an **L** channel on top of the cross braces with double-sided sticky tape. I used tape from an automotive supply store, because it's a bit more rugged than what's available at the office supply store. This will keep the radios from sliding off (especially when being pulled by the microphone cable).

The finished bracket is seen in **Figure 15**, showing two matching radios (one operating on 144 MHz, and its twin on 220 MHz). As long as you have enough wood, you can make the bracket for as many radios as you'd like. The more industrious builder may want to add a finish to the wood, but it's not mandatory, because the mount will be used indoors.

I noticed an additional benefit to this setup when I saw there was enough clearance below the radios to allow for some paperwork (like a logbook) to be stored there without impeding the function of the shelf. This approach also allows a bit of flexibility in height and width, to suit radios of any size. The entirety of this setup was placed on a bookshelf custom-made for my mobile radios. The back was left open to allow for air circulation around the heatsinks, and the sides and bottom provided a bit of a baffle to enhance the sound quality from the speakers. — *73, Klaus Spies, WB9YBM,* **wb9ybm1@yahoo.com**

Circuit Board Badges

Our club, the Riverside Radio Amateurs (**www.wa8rra.org**), recently adopted a logo that was designed by one of our members, Scott Beeman, NE8E. I converted the logo graphic into a printed circuit board design, and ordered a batch of circuit boards that I made into club badges.

To make the badges, I used Inkscape (which is available for free at **www.inkscape.org**) to convert Scott's original SVG file of the logo into images of the copper and solder mask layers. I imported those images into a circuit board design using Eagle design software, which is available at **www.autodesk.com/products/eagle**, and is also free for non-commercial use with boards up to 100 × 80 millimeters. I put a rectangular opening in the solder mask on the back side so I could solder on a pin clasp, and I added through-holes so I could make the little antenna in our logo using actual copper wires. I then generated Gerber files of the board design and sent them off to a PC board manufacturer to have the boards made.

After soldering the pin backs on and adding the antenna wires, I sprayed the front with clear enamel to secure the wires and to resist smudges. Finally, I personalized each badge with a gold-on-clear label showing the member's name and call sign (see **Figure 16**). Even with gold-plated traces, the total cost was under $5 per badge, including the circuit board and all materials. You can watch the whole process in my YouTube video, at **www.youtube.com/watch?v=1g1jHmrTVyc**. — *73, Jay McClellan, K8DC,* **jay@brainright.com**

Figure 16 — These WA8RRA club badges were made from printed circuit boards, converted from the club's logo design. [Jay McClellan, K8DC, photo]

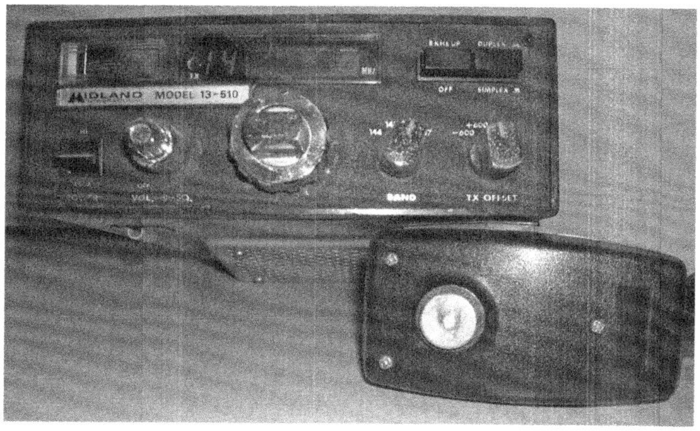

Figure 14 — If the bottom-fire speaker is mounted on an angle, you could probably hear it, but it won't lie flat like the other radios at your station. [Klaus Spies, WB9YBM, photo]

Headset Interface

Cell phone headsets are very common and inexpensive. They usually consist of an earphone, an electret microphone element, and a two-circuit 2½-millimeter (3⁄32-inch) plug. This interface connects the cell phone headset to an Icom RJ45 jack and provides a push-to-talk (PTT) pushbutton switch. Provision is made for an external PTT connection for a foot pedal or similar.

As shown in **Figure 17**, J1 is a 2½-millimeter (3⁄32-inch) two-circuit jack for the headset. "Tip" is for the MIC, and "ring" is the earphone. The ring terminal is connected to the ring and tip of P1, a 3½-millimeter (⅛-inch) two-circuit plug for the radio headphone jack. The tip terminal of J1 goes to the 2.2 kΩ resistor shunted by a 0.1 μF capacitor, then to Pin 6 of the plug P2.

The RJ45 plug is used for microphone connection on many Icom radios. The pins shown are believed to be universal, but check your manual before applying power (see **Figure 18**). The easiest source is to cut a CAT 5 cable as a direct connection to an RJ45 plug, which requires a special tool. Pin 7 is ground, and Pin 4 goes to ground through the SPST PTT switch on the box. This pin can also be taken to ground through J2 for a foot pedal or other type of remote PTT.

I have used this interface with my Icom IC-703 HF rig and IC-2720 VHF rigs and the results have been very satisfactory. — *73, Scott McCann, W3MEO,* **achess@juno.com**

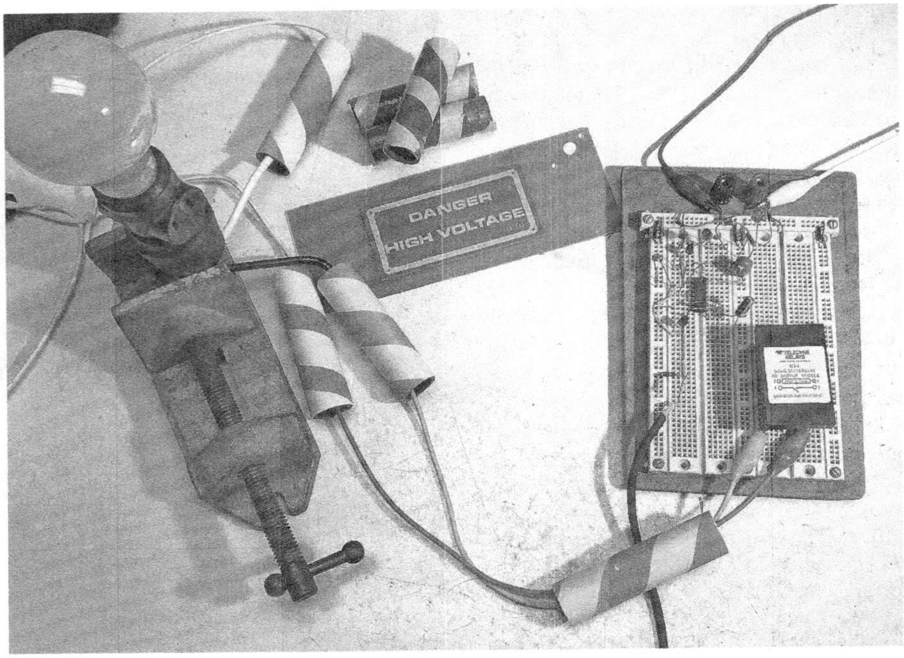

Figure 19 — These covers mark the high-voltage connectors when developing a circuit. [Terry Simonds, WB4FXD, photo]

Figure 17 — The headset interface. [Scott McCann, W3MEO, photo]

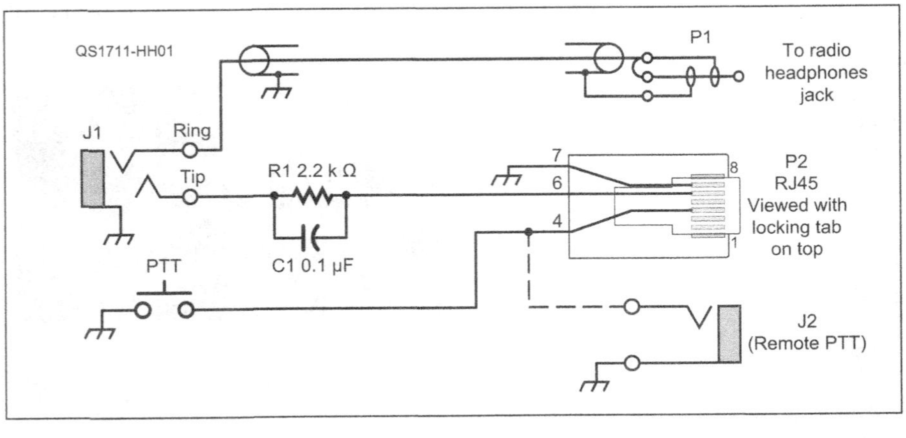

Figure 18 — A schematic of the headset interface, meant to connect a cell phone headset to an Icom RJ45 jack, while providing a push-to-talk (PTT) pushbutton switch.

C1 — 0.1 μF capacitor
J1 — two-circuit 2½-millimeter (3⁄32-inch) jack
J2 — optional jack for remote PTT
P1 — two-circuit 3½-millimeter (⅛-inch) plug tip and ring shorted together
P2 — RJ45 eight-conductor microphone plug for Icom (CAT 5 network cable suggested)
R1 — 2.2 kΩ resistor

High-Voltage Connection Covers

To prevent being hit with 120 V ac while developing a circuit, I've slipped covers over bare high-voltage connections leading to the breadboard. This way, I won't start a fire on my bench from an accidentally dropped tool, or put myself out of commission.

To make these, cut as many safety covers as you'll need from some ½-inch PVC pipe, about 3 – 4 inches long. Then wrap each with masking tape, so that you leave a spiral open area on the pipe. Spray each cover with paint in contrasting colors and let them dry overnight. The next day, peel off the tape, and you'll have something that will shout at you from the bench, reminding you to be cautious.

Say you want to connect the hot side of a component on the breadboard to the 120 V ac line. In my case, I used a jumper with alligator clips from the ac-hot (black) to a component on the board and safety covers over the neutral (white). Slip two safety covers over the jumper, connect each end with the clips, then slide the covers over the exposed wiring. In **Figure 19**, I've used one safety cover for the hot and load lines near a solid-state relay on the board. — *73, Terry Simonds, WB4FXD,* **wb4fxd@arrl.net**

Under-the-Desk Organization

I recently acquired another PC. I already had several other PCs under my radio desk — one for my primary PC work, one for APRS, RTTY, SSTV, JT65, and so on, and a third to experiment with. In **Figure 20**, you can see my current under-the-desk setup, made up of a 35 A power supply and the four Dell PCs (arranged from newest *Windows 10*

Figure 20 — The view under the author's radio desk, showing his PCs, 35 A power supply, and UPS on a plywood plank sitting on a furniture dolly. [Dennis Merritt, W6UHQ, photo]

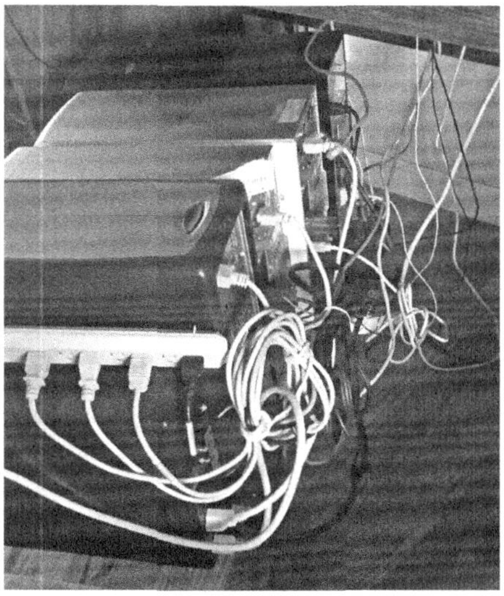

Figure 21 — A closeup of the pedestal swung out to the left, giving access to the backplane of all the equipment. [Dennis Merritt, W6UHQ, photo]

to oldest *XP*), plus a beefy UPS to keep things running until they can be shut down properly.

The issue was figuring out a way to squeeze yet another PC under the desk with all my electronics, and the inevitable spaghetti bowl of wires that follows every PC installation — ac power, speaker out, mic in, Cat 5 ethernet, USB devices of all kinds, and even an old parallel-port printer. It's not easy to try and swap out anything behind that wall of electronics. I knew I needed something to help me access these areas quickly and with little fuss.

Now, all the equipment is sitting on top of a ¾-inch piece of plywood. That plywood pedestal is resting on a furniture dolly. This way, I can easily roll out the entire equipment complement for easy maintenance. The key to this working well is to allow enough slack in each of the various lines so that all the lines pivot from one side or the other in a smooth arc when swinging out the platform. Also, it helps to have a pretty good KVM (keyboard-, video-, mouse-sharing) device. I opted to have three of the PCs share peripherals, while the fourth is my ham radio PC and has its own set of peripherals (see **Figure 21**).

While equipment maintenance always comes with some level of stress, with this pedestal on a furniture dolly, access to the backplanes has become simple. And the price was right. The plywood was scrap, about 2 by 4 feet on the Harbor Freight dolly, which cost about $20. The KVM switch cost around $40, and was well worth it. — *73, Dennis Merritt, W6UHQ,* **demerit@comcast.net**

A Handheld Holder

I recently bought a Tytera MD-380 handheld transceiver to allow me to get into the world of digital mobile radio (DMR). My job requires that I drive to various sites in several cities and states, so listening to DMR serves as great entertainment for the ride.

However, to get good coverage, you need an external antenna for the radio. I picked up a cable to match the SMA female on the MD-380 to a PL-259 connector, which worked great. I didn't like the way that the coax kept getting tied up in things on the truck seat when I used it, and I like to use an external mic to minimize the strain on the antenna connector, so I had to find a holder. I was unable to find anything that worked like I wanted it to, so I made one myself (see **Figure 22**).

I picked up a 2 × 2 inch piece of PVC and a 2-inch coupling at the home improvement store. I also picked up a can of spray paint, bringing the cost to about $6. PVC pipe is a thermoplastic material that, when heated to between 250 – 300 °F, turns into a thick, rubber-like material that can be bent and held until cool and it will retain the new shape.

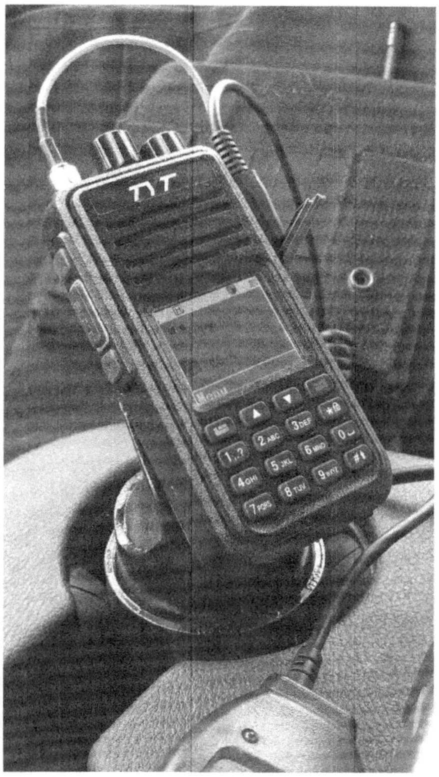

Figure 22 — The Tytera MD-380 handheld transceiver in its homebrew holder. [John Rusmiselle, KG4VHV, photo]

Remember to wear gloves to prevent burns, and be sure to heat it in the open so the fumes can dissipate.

I put the coupling on the pipe and placed the assembly in the cup holder in my truck, where it fit nicely in the rubber grips of the holder. I put a red pencil mark on the PVC pipe ½ inch above the cup holder to raise the radio up off the armrest a bit. I continued this mark around the pipe to give me a bottom cut line.

Next, I placed the bottom of the MD-380 on the ½-inch line and marked the width of the radio on the PVC pipe. To figure out where to cut, I made a mark at the top of the belt clip on the pipe. That way, I could round off the corner where the vertical radio width mark met the ½-inch mark around the bottom, to give it strength and a nicer look. A saw made short work of roughing out the PVC, and a file and sandpaper cleaned it to make a smooth holder.

Keep in mind that the PVC pipe will get soft when gently heated with a heat gun. The idea is to form the shape and angle of the actual part that holds the radio when the plastic is soft, and then hold it until it cools. I used a round dowel rod and a vise. Wearing gloves, I gently warmed the PVC at the point where I wanted the angle until it started to bend. The PVC was then put in the vise to flatten out the part of the pipe that the radio

Figure 23 — The holder bent at the necessary angle after heating the PVC. [John Rusmiselle, KG4VHV, photo]

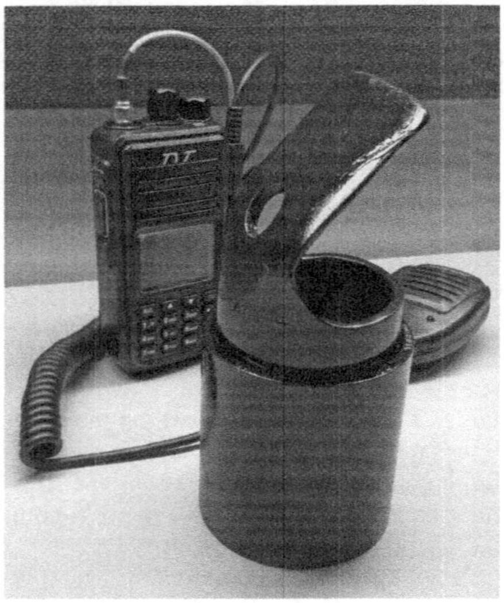

Figure 24 — The finished holder after being painted to match the vehicle's interior. [John Rusmiselle, KG4VHV, photo]

would rest against, and gave me something to hold while I set down the heat gun and bent the base to the angle I wanted to reach (see **Figure 23**). I then used the dowel to help make a nice corner and keep the PVC from kinking.

When this assembly cooled, I was ready to drill the hole for the belt clip for the radio. This belt clip has a lip on the bottom so the radio will not just slip up and off the belt. This feature also helps to hold the radio on the holder when it's done. I slipped the radio on the holder, checked the angle, and marked the place that a hole would allow the bottom lip of the belt clip to go through the PVC that I just bent. I then drilled the hole and cleaned the whole thing up with a file and sandpaper until I was happy with it. If you do not like the angle, a quick reheat and bend will work — just don't let the PVC burn. For the final touch, I painted the holder to match my vehicle (see **Figure 24**). — *73, John Rusmiselle, KG4VHV,* **mrjohnac@gmail.com**

Battery Tote Bag

My grocery store sells reusable fabric tote bags with dividers inside for carrying bottles of wine. I've found that they make great transport bags for 7.5 Ah batteries. For ARRL Field Day, I carry multiple batteries, and I used to have to cover a terminal on each battery with electrician's tape. With these tote bags, however, the bag's dividers do the work for me, and the bag has nice handles. If you aren't carrying lots of batteries, the compartments can hold meters or other small items for your setup. — *73, Randy Tompson, NC4RT,* **rtompson@triad.rr.com**

Turnbuckle Mobile Mount

I wanted to install my Icom IC-2720 transceiver (which has a separate body and control head) in my car without making any modifications to the car itself. The control head was easily mounted by using a cup holder type mount.

To mount the body of the transceiver in the trunk without drilling any holes in the car was more challenging. In coming up with ways to use several existing holes in the underside of the package shelf, it occurred to me that I could use turnbuckles — which have a hook on one end and an eye on the other — and create a three-point suspension.

The hooks go in three existing holes and a bolt through the eyes mounts to the bracket for the transceiver body. Two of the three legs can be seen in **Figure 25**. One leg required two turnbuckles to reach the existing hole. A stiff piece of rubber mat between the turnbuckles and the package shelf added stability.

Of course, every installation is different, but this is an idea of one way to mount a mobile radio without drilling any holes in the car. (Special thanks to Alan Giles, who was the man in the trunk.) — *73, Wayne Waller, W4TZB,* **w4tzb@arrl.net**

Anderson Powerpoles on Keys

I dislike putting ¼- or ⅛-inch adapters on my CW keys. As soon as I take the time to solder a ¼-inch plug on a key, I want to plug it into a ⅛-inch rig, or vice versa (and the adapters have a way of hiding from me). My solution is to put Anderson Powerpoles on all my keys. That way, I make one plug for each rig, terminate it in a Powerpole, then swap the keys in and out without worrying about adapters. —*73, Randy Tompson, NC4RT,* **rtompson@triad.rr.com**

Figure 26 — A CW key placed on non-slip tape designed for rugs stops sliding during sending. [Lionel Booth, N5LB, photo]

Stop Keys from Sliding

The surface of my newly built radio desk has a smooth finish. Unfortunately, my Kent key kept sliding around, which meant that my sending was awful. My solution was to use tape designed to stop rugs from slipping. This tape is adhesive on one side and has gridded raised rubber on the other side. The tape comes in a roll and can be found in the flooring section of home improvement stores. I just fastened the tape firmly in place, and now there's no more sliding (see **Figure 26**). — *73, Lionel Booth, N5LB,* **n5lb@outlook.com**

Color-Coded Gear

I have eight handheld radios. Because each radio has a microphone, two or three antennas, batteries, a battery charger, a program-ming cable, and belt clips, this makes for a large collection (see **Figure 27**).

Keeping track of it all isn't easy. It is rare that any two radios use the same accessories,

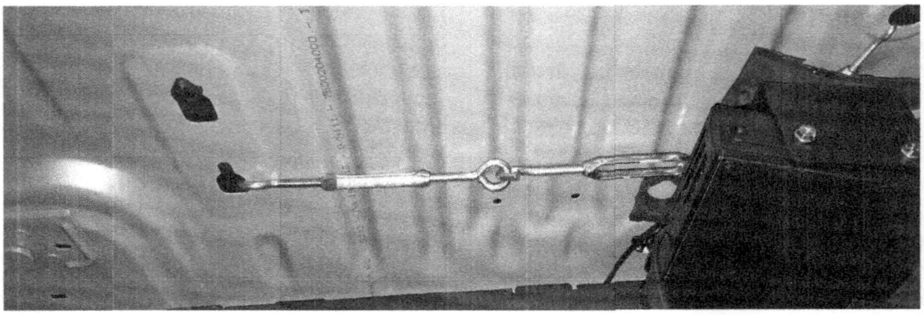

Figure 25 — A radio can be mounted in the trunk of a car using existing holes and some turnbuckles. [Wayne Waller, W4TZB, photo]

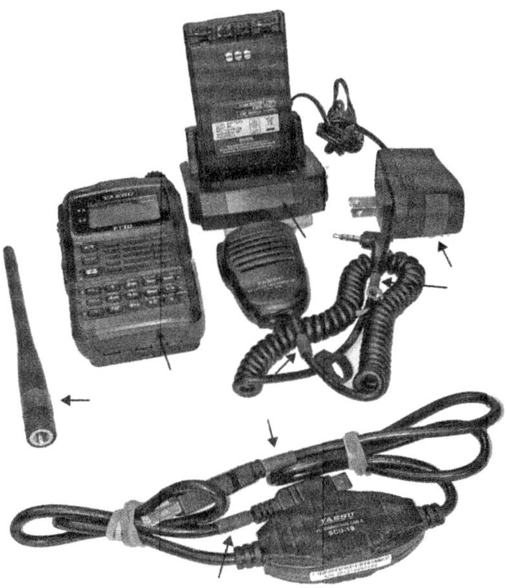

Figure 27 — One handheld radio, with all its accompanying gear. [Richard Koharik, W8RWK, photo]

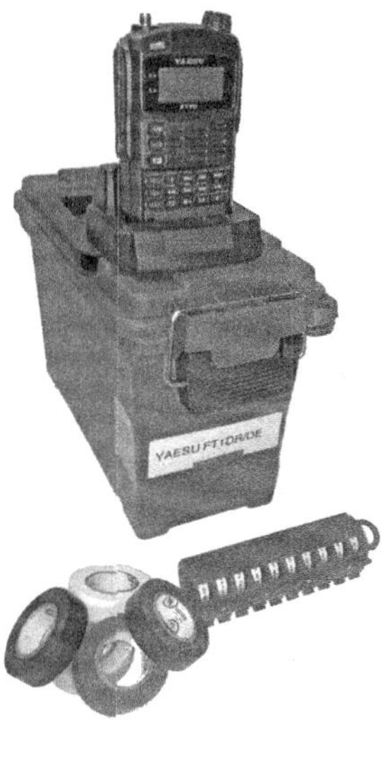

Figure 28 — Each radio and its accessories and wires can be organized by color using vinyl electrical tape. [Richard Koharik, W8RWK, photo]

and each radio tends to be unique in its programming and operation. After a lot of mismatched antennas and mics, I came up with a simple labelling and storage solution to keep every accessory sorted so they're ready to use in an emergency.

I started out with a color-coded system using vinyl electrical tape. (Some popular wire markers are made of paper and the color rubs off very quickly, so vinyl was the best option for me.) I used a different color for each radio manufacturer (see **Figure 28**). Packages of 10 rolls in various colors are available at many retail and online sources.

If you own more than 10 brands of radios, you can mark them with two colors. If you have more than one radio from a manufacturer, you can use wire marking numbers to denote radio "blue brand" 1 and 2. Mark programming and power cords the same way as the radios. You can identify paper documents like quick guides and operating manuals with color markers.

My plan also involves storing antennas, hand mics, batteries, and manuals in one spot. It's always good to have everything in its place. If you are active in ARES or SKYWARN, you know you may not have time to search when the lights go out.

A backpack is great for the soft stuff and snack bars. A water-resistant hard case is best for radio gear. Label your box and mark it with the same color tape as the radio gear inside. Now that your radios are super organized, you can go ahead and buy that new handheld with a lot less guilt. — *73, Richard Koharik, W8RWK,* **w8rwk@arrl.net**

Copper Wall Plate

I needed an inside wall bulkhead plate for coax that could accommodate several connectors, as well as connect the station grounding conductor to the external ground system. While perusing the local hardware store, I came across a 10 × 4 × 0.025 inch copper plate that seemed like a possibility, especially because it meant there'd be no cutting or bending involved.

For an investment of about $8 and half an hour in my shop, I built a copper wall plate (see **Figure 29**). I carefully used an 11/16th bi-metal drill bit by clamping the plate between 1/2-inch plywood, then drilling through the wood, the plate, and into the bottom piece of wood.

Work is ongoing for proper grounding and cable routing, but this is the basic design. A larger 3/8-inch copper plate serves as the outside barrier for cable entrances. I measured the resistance from the inside plate to the coax shield at a few milliohms. — *73, Lionel Booth, N5LB,* **n5lb@outlook.com**

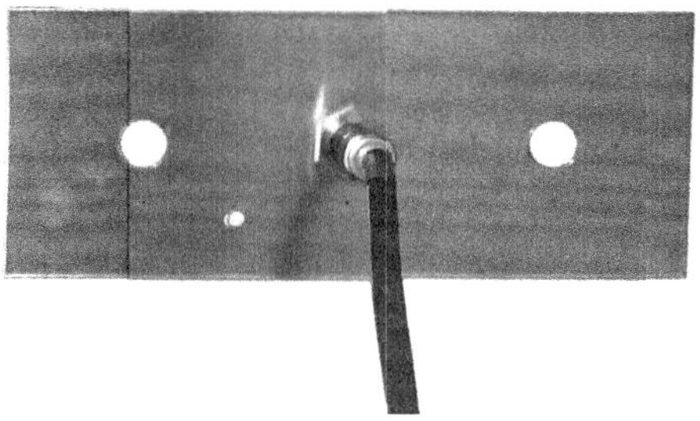

Figure 29 — The copper wall plate. [Lionel Booth, N5LB, photo]

An Inexpensive Mobile Scanner Mount

I needed a cheap and reliable bracket for the Uniden BCD536HP scanner that I keep in my truck. I looked at commercially made ones, but none that suited my needs were able to be installed in the close area between the center seat and my dual-band without obstructions. I also needed it out of the way while driving and needed to be able to see what was being received while parked.

I decided that plumbing materials were the answer. I went to my local plumbing supply store and purchased two 3/4-inch galvanized floor flanges; one 3/4 × 6 inch galvanized threaded nipple; four #12 1-inch self-tapping screws, and two 3/4-inch locknuts.

I installed the flanges on either side of the pipe, then tightened and mounted it to the transmission hump with the self-tappers. I took the mobile bracket that came with my scanner and drilled a 3/4-inch hole in the center of the bracket, installed one of the locknuts onto the pipe, and bolted it to the top of the pipe with the other locknut. Finally, I mounted the scanner to the bracket with the hardware.

After a few adjustments, I had a mobile mount with no obstructions and great visibility for only $4 (see **Figure 30**). This mount can be used with a variety of radios, using the radio's mobile bracket. — *73, Al Wasielewski, WA2VJL,* **wa2vjl@yahoo.com**

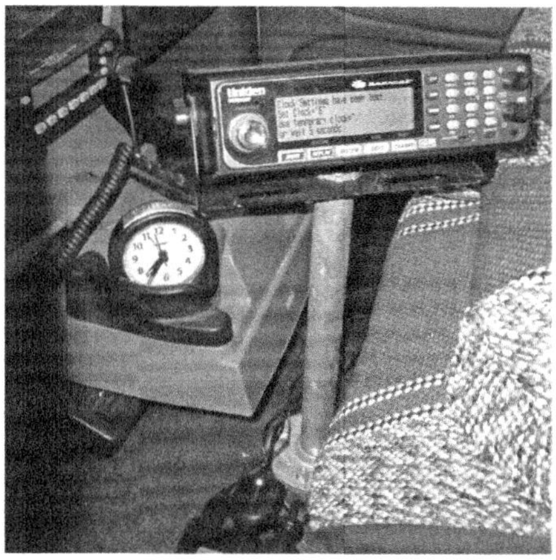

Figure 30 — The finished scanner mobile mount, made from inexpensive plumbing supplies. [Al Wasielewski, WA2VJL, photo]

Portable Racks for Radio Gear

For me, constructing a go-box and other radio projects calls for a lightweight, rigid rack to hold equipment. As I built my latest portable equipment rack, my goals included the ability to operate HF, VHF, and UHF, in addition to voice and HF digital; being able to immediately connect to antennas; 120 V power and battery power (with easy access to plug these in); the ability to remove the

Figure 31 — On the left are the cage clips that attached the shelves to the walls. On the right are the soft wire ties used to organize any loose antenna or radio wires in the go-box. [Jim Halfpenny, K9YNP, photo]

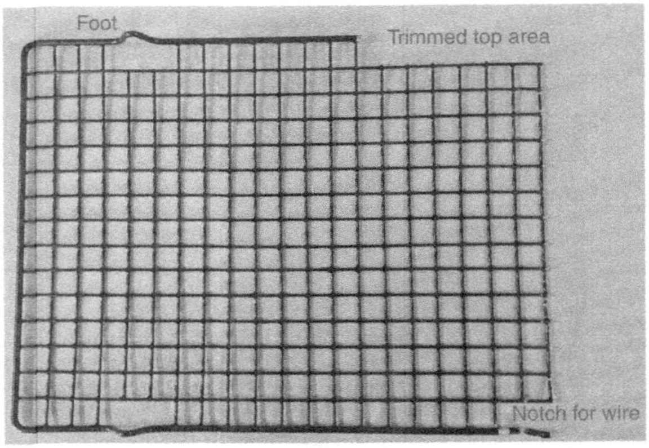

Figure 32 — The modified cooling rack, cut to make room for wires and to fit into the Spud 7 utility box. [Jim Halfpenny, K9YNP, photo]

rack for operation, and a lightweight but rugged structure.

My solution came in the form of several interesting products, including:

- The Spud 7 Utility Dry Box from MTM Molded Products Company (**www.mtmcase-gard.com**).
- Six cooling racks for cooking, purchased from a department store. They were lightweight, strong, and could be trimmed to size and bent into broad U shapes for shelves.
- Wire cage clips, designed for a rabbit hutch (model ACC1, **www.miller-mfg.com**), allowed for joining wires in parallel to build the rack. Needle-nose pliers work, but you could also use Pet Lodge pliers, designed specifically for these clips.
- Soft wire ties — flexible wire covered in soft rubber — were perfect for organizing radio wires (see **Figure 31**). They're sold by Miracle-Gro or your local gardening store.
- Industrial-strength Scotch Bundling Wrap (a type of hook-and-loop fastener tape) mounted all components on the rack.

My go-box included a Yaesu FT-857 radio; my Samsung Galaxy Note 3 cell phone; DroidPSK software (Wolphi LLC); a WolphiLink Smartphone-to-Ham Radio Interface; wires to connect instruments; an antenna tuner; antenna extension wires (two 9-foot SO-239 male-to-male extension wires — one for HF and one for VHF/UHF); a Bioenno Power 12 V, 12 Ah LFP battery; a Bioenno 14.6 V, 2 A ac-to-dc charger, and a Powerwerx Switching Power Supply.

Construction

To build the rack, I started by measuring the Spud 7 box. I had to shorten the walls of the rack by cutting material from the back. The Spud 7 is larger at the top and smaller at the bottom, so one edge of each wall must be clipped to create a sloping polygon. The clipped edge would be the top. Also, the two walls are mirror images that are reversed. I placed the wall "feet" on the bottom of the cooling rack away from the walls of the Spud 7, as this would facilitate connecting the walls to the shelves later. On the bottom edge, I cut a corner notch to accommodate antenna wires (see **Figure 32**). Sharp edges were later filed smooth and covered with electrical tape. This way, the HF antenna wire could be directed out under the corner on one side, while the VHF/UHF wire runs out the other side.

The cooling racks were longer than the width of the Spud 7 and had to be bent at both ends for the shelves to fit. The bent ends provided additional support and facilitated binding the shelves to the walls. While each wire could be bent with pliers, there were many wires to bend. If you build this rack, I suggest using a bending break to make all bends at once.

I attached the shelves to the walls using the wire cage clips. For the bottom shelf, I made sure the cooling rack feet pointed up. For the other shelves, the feet pointed down. The top shelf sloped down to the back to match the slope of the Spud 7. The bottom edge of the walls sat flat on the bottom of the Spud 7, while the top shelf edge sloped down.

Figure 33 — The open rectangles in the cooling rack are large, so I cut pieces of aluminum to make square washers large enough to cover the holes. Washers could also be constructed out of hard balsam wood as a lightweight option. [Jim Halfpenny, K9YNP, photo]

For the FT-857 shelf spacing, I wanted the bottom space to have 3 inches between shelves, with the middle space being 3½ inches. The remaining space was allocated to the top space. More space was needed in the middle to allow for attaching the FT-857 to the shelf. The rack for the FT-857 was mounted with bolts from the bottom. Because the open rectangles in the cooling rack are large, you may not have washers large enough for bolts. I made square washers by cutting pieces of aluminum large enough to cover the holes (see **Figure 33**). Also, for saving weight, the "washer" can be a cut piece of hard balsam wood.

I used the bundling tape to mount the components into the completed rack. From outside the rack, I stuck the bundling tape through the wire shelf and wrapped it around

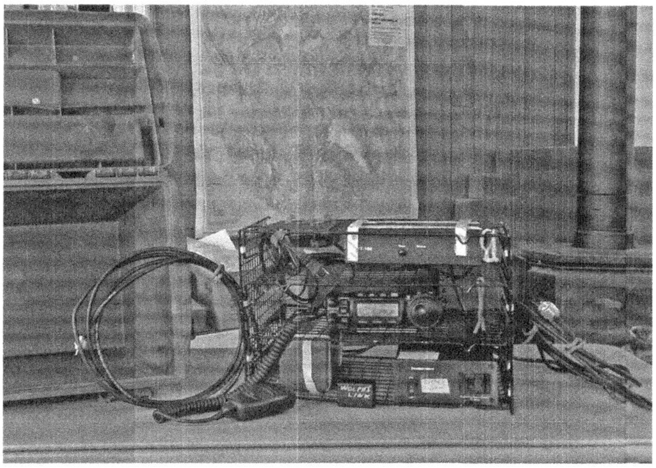

Figure 34 — All the components in place for operating outside the Spud 7 utility box in which the rack will be stored and transported. There is room to the right and left of the rack to store 9-foot antenna extension cords, and my manual and papers can be stored on top of the rack. [Jim Halfpenny, K9YNP, photo]

the component, then back to the outside. Then, holding the component against the shelf and holding one end of the tape, I pulled the tape tight across the shelf before sticking it.

Components on the bottom shelf (in the bottom space) had the tape cinched tight on the bottom of the rack, while those in the top space had it cinched on the top of the rack. This allowed plenty of space to work in while placing and securing the components.

Once components were placed in the rack, I used the soft wire ties to secure loose wires in place. I ran antenna extension wires out the back lower corners. Power wires from the battery charger and for Powerwerx could also run out under the lower back corners or the front, as I did for the battery charger (see **Figure 34**). — *73, Jim Halfpenny, K9YNP,* **trackdoctor@tracknature.com**

Install or Remove Mobile Radios Without Tools

Every so often, I need to remove my mobile radio to use it in the shack, or for security reasons. I didn't like needing a screwdriver nearby and keeping track of the small screws every time I did this. I wanted to find a way to remove the screws by hand.

For my Yaesu FT-1900, the screws are metric, but I was able to find the screws needed at a home improvement store. I used Phillips panhead machine screws, size M4-.7 × 10, and purchased the wing nuts with threads to match. I simply inserted the screws backwards through the wing nuts. I also put a little glue under the screw head to keep it from backing out during use (see **Figure 35**), but a lock washer would also work.

This has made the process of removing and reinstalling my mobile radio much more convenient for me. — *73, Joe Hopkins, K4BKI,* **jhopkins1938@gmail.com**

Figure 35 — The wing nut installed for tool-free removal and installation. [Joe Hopkins, K4BKI, photo]

An Easy, Portable External Speaker

The internal speakers in radios need to be small enough to fit inside the radio enclosure and are often on the top or bottom of the radio, sacrificing a bit of sound quality due to the location. External speakers, however, are often used to improve the audio and have the option of a front-firing speaker, which can enhance audio and make it easier to hear.

I decided to construct an external speaker that was easy to take into the field for public service events or Field Day outings (see **Figure 36**). I planned out what I was looking for. A handle would help with easy portability, and a rugged enclosure would help keep the speaker from getting beaten up in the field. The parts to build the unit would have to be readily available at a reasonable cost. And, of course, it had to sound good.

First, I needed to find a speaker. I wanted a speaker that was self-enclosed to make construction easier. I kept in mind how I'd be mounting it in the enclosure and the need for some sort of protective grill in front of the speaker cone to keep it from getting damaged. I found that Parts Express (**www.parts-express.com**) sold a 5-inch sealed back mid-range speaker (model number GRS 5SBM-8) that fit the bill. It also had a built-in mounting ring with holes to easily mount the speaker unit to the enclosure.

As for the enclosure, a small plastic ammo box seemed like a simple, inexpensive solution. The plastic would also make it easier to cut out the hole to fit the speaker into the box.

Finally, I needed to select a connector to connect the speaker to the radio equipment. I chose red and black binding posts from Parts Express (part number 090-475) for their versatility, and easily identifiable polarity (see **Figure 37**). They only required two ⅛-inch holes, spaced ¾ inch apart, to mount them to the box. If desired, you could use a pigtail type of connection cable to attach the required connector to mate with the radio's external speaker jack. The pigtail could be stored in the ammo box enclosure when not in use. Of course, any other user-preferred connection could be employed to connect the external speaker to the radio equipment.

In summary, the project worked out well. The external speaker met my design goals and looks appealing. It's been used with different radio equipment and sounds good. It may fit your need for a portable external speaker for Field Day or for everyday use. — *73, Don Varner, WB3CEH,* **wb3ceh@yahoo.com**

Figure 36 — The finished portable external speaker. [Don Varner, WB3CEH, photo]

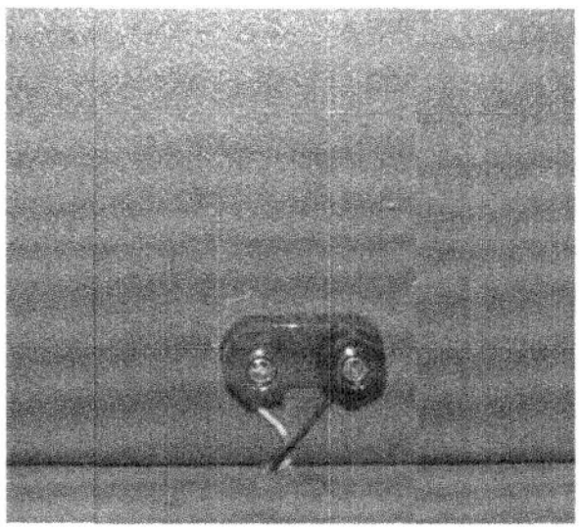

Figure 37 — The red and black binding posts on the rear of the speaker unit connect the speaker to the author's radio equipment. [Don Varner, WB3CEH, photo]

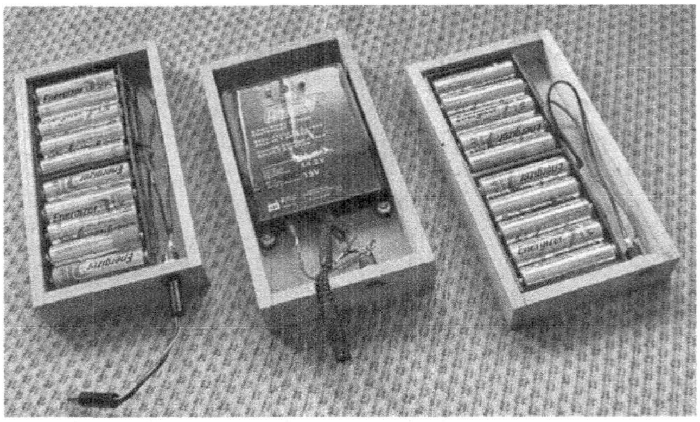

Figure 38 — The author's homemade wooden boxes help protect the battery holders (left and right) and solar panel controller (center) from being knocked around during handling. [Charles Hooker, VE3CQH, photo]

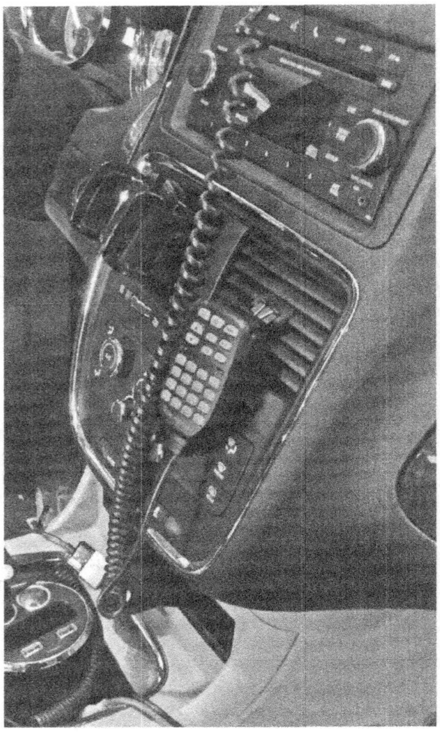

Figure 40 — The author's mic hooked to a car vent using a bent binder clip. [Bill Kerber, WA2ILY, photo]

Wooden Battery Pack Holders

Ever since becoming devoted to QRP, I found it difficult to make many CW contacts at my slow Morse speed. I decided to try and go for the Field Day bonus points awarded for alternative energy (i.e., my solar panel).

I bought lithium-ion batteries and found ten-cell holders at American Battery Company (**www.americanbatterycompany.com**). The holders needed some protection from my rough handling, so I constructed wooden boxes for them, about 6½ × 3½ × 1 inches (outside dimensions). My solar panel's controller coincidentally had the same spatial needs, though a bit thicker. Its box is 1¼ inches high.

The boxes include space for dc power plugs. A lid was unnecessary for my use. The patch cord for my tribanders is connected to one battery pack. These boxes have helped keep my battery holders and controller safe from bumping around too much (see **Figure 38**). — *73, Charles Hooker, VE3CQH,* **chuckcynthia@gmail.com**

A Safety Vest Antenna Sleeve

Amateur Radio operators need to wear safety vests when they volunteer in support of fun runs, marathons, and other public service events. The vest not only provides personal safety, but also serves as the carrier for handheld radios when moving around the event. However, with the radio simply clipped onto the vest, the antenna is at risk of snagging on branches or hitting the operator in the face or eye. I wanted to find a solution that provided better security for my radio, a stronger signal, and prevented any eye injuries or antenna snagging. I came up with an antenna sleeve that allows the operator to keep their handheld secure in a vest pocket and use longer antennas without safety risks.

Construction of the antenna sleeve was easy. I fashioned an open-ended cloth tube — 2.6 inches long and ½ inch in diameter — from material similar to that of the safety jacket (see **Figure 39**). The sleeve is several times the diameter of the antenna to allow vertical movement and easy removal of the antenna. I sewed the tube along a vertical line just lateral to the right breast pocket. With the radio in the lower pocket location, a longer antenna can be employed safely, providing a stronger signal. Also, a raincoat can be donned over the antenna and radio, protecting both while still allowing operation. This small sleeve holds the antenna in a vertical position, against the chest and away from the face.

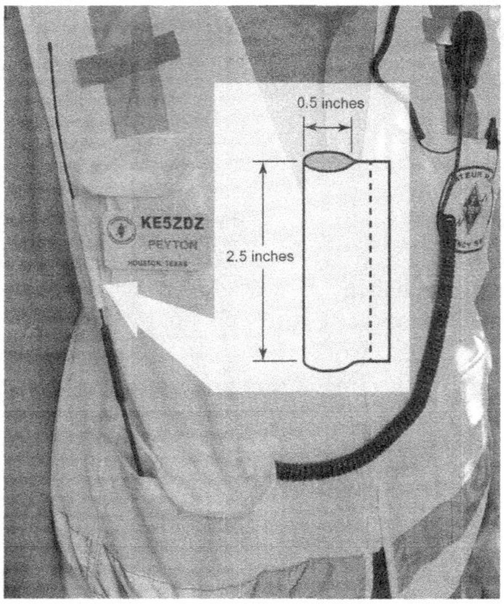

Figure 39 — A small antenna sleeve sewn alongside a vest pocket allows for the use of longer antennas without potential safety risks. [Peyton Barnes, KE5ZDZ, photo]

In the field, this antenna sleeve allows me freedom of movement and effective communication. I hope this is useful for any amateurs who like to go out into the field this Field Day and during public service events throughout the year. — *73, Peyton Barnes, KE5ZDZ,* **skaevola@hotmail.com**

No-Fuss Mic Hooks

I've found a way to make a quick and easy mic hook for a car or home station using binder clips. Simply bend one end up on the clip and attach it to the vent in the car or anywhere else you can hang your mic (see **Figure 40**). — *73, Bill Kerber, WA2ILY,* **wa2ily@charter.net**

Easy Keyer Tips

I've seen articles discussing how to make electronic keyers' output circuits key either a negative (grid block) circuit or the standard 5 V type. I've seen different configurations that accommodate either type with just a flip of a switch. Over the years, I've built a few keyers, using both types of transmitters. I wanted to make it as easy on myself as possible, so here are some of my tips.

My method is simply to put an emitter follower (EF) on the code bus before it exits to the keyer and feed a reed relay from the emitter of the EF. This only requires one transistor, one resistor, and the reed relay. Almost any NPN transistor will do the trick (2N2222 or the like).

I put a 1,300 Ω resistor in the emitter. Now it will key either polarity. The reed relays are fast — usually less than 1 ms, which will handle Morse code faster than I could dream of sending. Just be sure to put a diode across the coil to suppress the back EMF from the relay coil. Any small rectifier will work just fine. I use the Clare PRMA series relays, but there are lots of others available. Some even have the diode built into the relay case. They also come in dual contacts, so you might key two transmitters without any changing effort at all.

A few years ago, I bought an MFJ Enterprises keyer because it was considerably smaller than my homebrew keyers. I added the transistor and resistor (EF) and relay, suspended in midair inside the case using solid copper wire, because they were so small and lightweight. I also ran a high-intensity LED from the EF and poked it out of the front of the keyer. The LED just flashes with the code (and looks cool). I put a 150 Ω resistor in series with the LED. I like to keep the current around 15 mA through the LED.

I have also connected both positive and negative keyed transmitters simultaneously using steering diodes. This is the simplest and easiest solution; just don't put both transmitters in transmit mode at the same time.

One other thing I've added is another EF feeding a delay chip (9602), which also pulls in a reed relay that connects to the PTT of my transmitter. With the proper R/C added and a potentiometer on the front panel, I can now easily adjust the break-in time delay for CW.

There are also other devices, optically coupled: the VTL2C3, which is an LED to photo conductive cell that goes from 50k dark to 1k with light. In addition, there's the MOC3031, a phototriac device that will key positive or negative voltages, or even ac. I have used both.

If you are going to use any of the solid-state devices, hold **KEY DOWN**, short the key input to your transmitter with a clip lead, and see if there is any noticeable increase in power. This will tell you if the keying resistance is too high. — *73, George R. Keller, W2DM,* **georgenann@aol.com**

Tuners in Series

I have an MFJ auto tuner a few feet from the center connection of my dipole. I also have an MFJ Versa II manual tuner that can sometimes do what the auto tuner can't.

My solution is to put them both in series (not at the same time, however). The manual tuner has a switch that controls which antenna to use and bypasses the tuner circuits and feeds the transmitter directly to the antenna (or, in my case, to the dc through the coax control box of the auto tuner).

Because the auto tuner is direct feed-through when the dc is off, I can use the manual tuner in the normal fashion. When I use the auto tuner, I switch the manual tuner to the direct feed-through and the auto tuner functions normally.

With this setup, I can conveniently select which tuner is best without any changing of coax lines or using a separate coax switch. — *73, Doug McCray K2QWQ,* **k2qwq@comcast.net**

Shoe Rack Radio Organizer

When my station was getting cluttered with equipment, my wife suggested I consider buying a shoe rack, because they're basically just shelves, after all. At the store, I found a metal shoe rack with metal mesh shelves. With two shelves sturdy enough for my accessories and completely open for air flow, this was perfect for me.

The shelves are sturdy, although I wouldn't try placing an amplifier or base rig on them. I did raise the lower shelf up about 2 inches by drilling new holes. There was a small hole on the inside of the end supports for the threaded rods to pass through and a larger diameter hole on the outside of the frame for the thumb screws to counter sink. I measured these out and made sure all four holes lined up. I drilled the four smaller holes all the way to the end supports so I would have a starting hole for the larger outer holes, which were then lined up for the threaded rods.

This project only cost me about $20 and came out way better than anything I would have built on my own (see **Figure 41**). — *73, Scott Phillips, KK4RT,* **sfphilli@gmail.com**

Figure 41 — The author's radio gear organized on an expensive shoe rack. [Scott Phillips, KK4RT, photo]

Colorblind LED Hack

As a colorblind ham, one of my greatest frustrations has always been color coding. Most LEDs on chargers simply change color from red to green to indicate the state of charge. I got tired of asking my wife or grandchild to come to my radio room and tell me if my battery was charged, so I found a simple solution that didn't involve changing the LEDs.

I simply taped a piece of shiny, transparent red cellophane paper (typically used to wrap Christmas gifts) over the LED. While the light is red, it is bright and clear to the red/green colorblind person. When the light switches to green, it disappears. Everyone else can still see it, but colorblind people can't, so I know the charge is complete. I have six chargers and it works on each one. I no longer need to ask anyone to tell me if my batteries are charged. — *73, Ken Munford, N7KM,* **n7km123@gmail.com**

Labelling Coax with Bread Clips

The flat plastic clips that are used as closures on bread and other supermarket packages are very useful for labeling coaxial cable. They clip directly on RG-58 and other similarly sized coax.

There are a few ways to mark the clips to use as labels. First, you could simply use a label maker or felt-tip pen. Second, you can implement a standard color-code system — red for 2 meters, green for five elements, and so on. And finally, you could edge notch in Morse code. For example, on a dipole cable reading down, you could notch *dah-dit-dit* — "D" for dipole (see **Figure 42**). This method would be useful in reduced light or for the visually impaired. — *73, H. Scott McCann, W3MEO,* **achess@juno.com**

Figure 42 — Notching bread clips to identify what they're labelled with could help locate coax when it's difficult to see the label. [H. Scott McCann, W3MEO, photo]

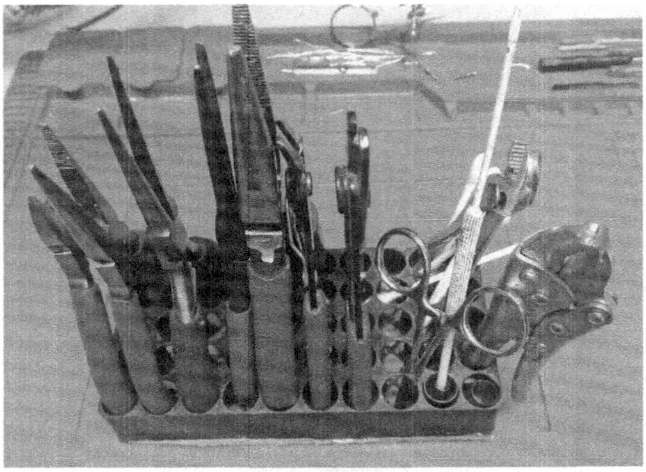

Figure 43 — An empty ammo cartridge box with cardboard glued to the underside makes a handy tool holder while working on projects. [Jim Garrett, K5BTV, photo]

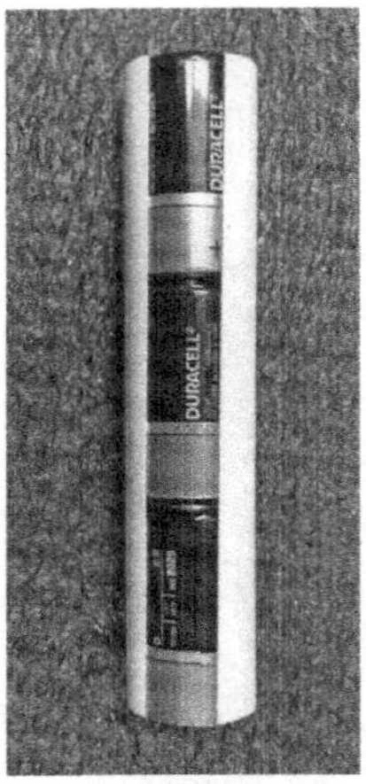

Figure 45 — An inexpensive plastic shower rod cover holds batteries in place without added bulk. [Clint Millett, VE3CMQ, photo]

Tool Holder

For years, I've organized my tools using spent ammo cartridge boxes (see **Figure 43**). They keep my tools close at hand for building projects. If you use this tip, remember to glue cardboard to the underside of the boxes so you can pick them up without any tools poking out the bottom. — *73, Jim Garrett, K5BTV,* **jgk5btv@gmail.com**

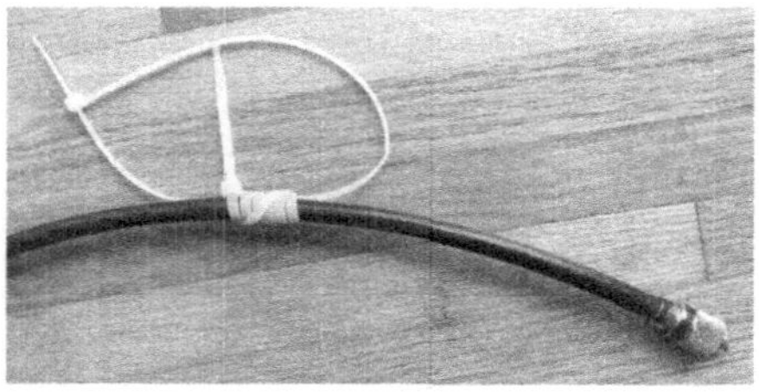

Figure 44 — Two zip ties help provide strain relief by anchoring the coax. [Terry White, VE5TLW, photo]

Anchoring Coax with Zip Ties

I've discovered an easy method of coax strain relief. I wrap one zip tie around the line of coax, then loop another through the first zip tie (see **Figure 44**). This is similar to the method described in Lyle Nelson's, ABØDZ, item, "Anchoring Coaxial Feed Line," which appeared in the August 2009 "Hints & Hacks" column. If more support is required, additional zip ties may be wrapped around the coax.

For illustration purposes, I used long white zip ties. Using wide, black, weather-resistant zip ties would be preferable because they help spread the force over a wider area. — *73, Terry White, VE5TLW,* **twhite@sasktel.net**

DIY Battery Holder

I finally got an old Spilsbury SBX-11 working, but when I looked at the battery case, I realized there was no way the batteries could be held in position. They needed three tubes to hold all nine of them together.

My solution was to buy a plastic cover for a shower rod from my local hardware store. I simply cut the tube to length with scissors and wrapped it around the batteries. It provided a tight fit without adding much to the diameter of the battery pack (see **Figure 45**).

The plastic cover only cost $2.49 for a 6-foot length, so I was able to make the three holders for D-cell batteries with around 4 feet left over. They would also work for C cells, 18650 Lithium-ion batteries (although some tape might be needed to hold the tube tight), and more. I am going to use this trick to replace the paper tubes used by some manufacturers, as well as the solid plastic tubes that were used in other equipment and eventually disintegrated. — *73, Clint Millett, VE3CMQ,* **ve3cmq@secrs.com**

Using Various Keys with Different Rigs

In the April 2018 "Hints & Hacks" column, I described an easy way to use short adapter cables to reverse the "dot" and "dash" keyer paddle orientation without rewiring the paddles. Since then, I have discovered other methods that are easier to implement and provide a more flexible way to use your paddles with a variety of rigs and electronic keyers.

The easiest method is to attach two or more cables to your paddle, with different connector plugs, selected to mate with various rigs (see **Figure 46**). Each can be wired with the left-right orientation to match the keying circuit. The disadvantage is having several cables connected, and the likelihood of tangles and an untidy appearance.

Another way to accommodate various paddles and keyers/rigs is to attach a stereo cable from each contact of the paddle (dot, dash, and ground), each terminated in an RCA-type phono plug. I used RCA-to-RCA two-conductor audio cables because I had some in my junk box. Then all that's left to do is to carefully strip the cable at the cut end and connect the ground and two wires to the terminals on the key.

Audio adapters are available inexpensively online in various configurations. Insert the appropriate adapter plug for each of your keyers and transmitters. Remember to select a stereo adapter on the rig end and mono on the input end (see **Figure 47**). Then, just plug the RCA plugs into the adapter in the rig for either "dot left" or "dot right" operation. Of course, instead of RCA plugs, other input plugs can be used with appropriate adapters, such as ¼- or ⅛-inch if you have those on hand. It's a quick, flexible way of utilizing various keys with various rigs. — *73, Ron Pollack, K2RP,* **k2rp@arrl.net**

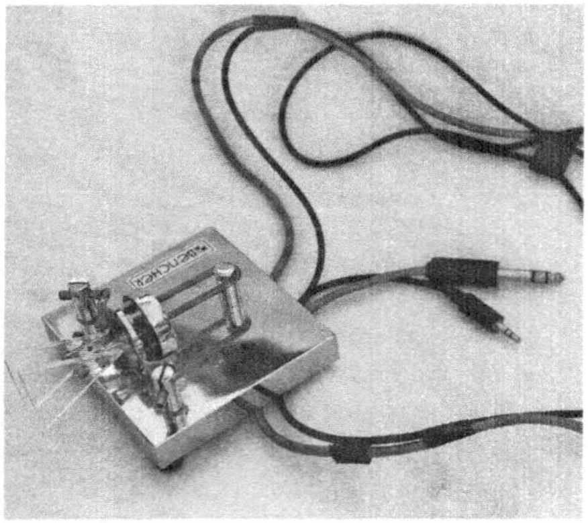

Figure 46 — The easiest way to use your key with a variety of rigs is to attach cables to your paddle selected to mate with various rigs, then wire them with the left-right orientation to match the keying circuit. [Ron Pollack, K2RP, photo]

Figure 47 — Using inexpensive audio adapters is another way to connect a key to a different rig. [Ron Pollack, K2RP, photo]

Anti-Fly Hamfest Fliers

Have you ever placed your club's hamfest fliers on a table at a prior hamfest to promote your event, only to later find that your fliers have fallen onto the floor? At the Skyview Amateur Radio Society, we have eliminated this issue by binding our fliers into handy tear-off tablets.

Organize a batch of fliers into a neat stack and place them at the edge of a table. Then place something heavy (such as a power supply) at the top edge of the stack. Apply Elmer's glue with a small brush (or even your finger), coating the edge of the stack, as shown in **Figure 48**. Come back in a few hours and — *presto!* — you now have a tablet that will hold all the fliers in place until interested visitors pull them off. — *Bob Bastone, WC3O,* **wc3o@arrl.net**

Figure 48 — Assemble a stack of fliers and align them at the edge of a table, using a heavy weight (a power supply in this example) to compress the paper and hold it in place. Using your finger, apply Elmer's glue along the entire edge of the stack. Allow to dry completely. [Bob Bastone, WC3O, photo]

Concealing Radios Deters Theft

As a retired chief of police, I always encourage people to do three sim-ple things to help prevent thefts from their vehicles:

1) Lock the car.

2) Remove items of value from the vehicle.

3) Park in a lighted area (if possible).

Still, I am amazed at the items people leave in plain sight within their unlocked cars — wallets, purses, laptops, and even full sets of golf clubs.

A transceiver in a car is like a beacon to thieves (see **Figure 49**), but taking it in and out every day is usually impractical. If the rig must remain in your vehicle, you need to reduce its visual profile. Some transceivers feature designs that allow their control heads to be mounted in the open while the rest of the radio is under a seat or in the trunk. This is useful because only the control head is visible, so it is the only thing that would need to be replaced in the event of theft.

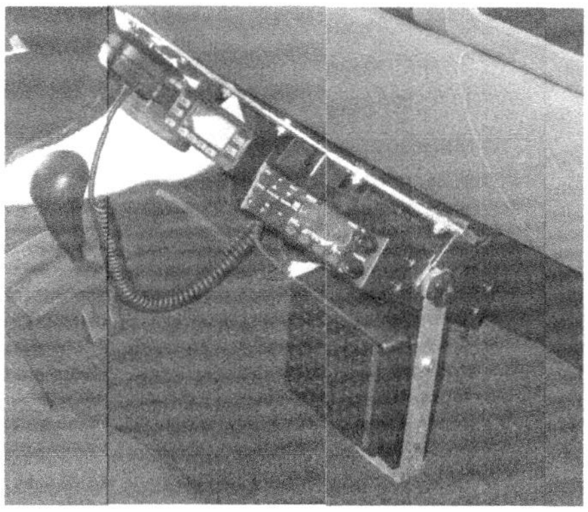

Figure 49 — A set of radios in plain sight is a burglary waiting to happen. [Mike Burg, N8QQN, photo]

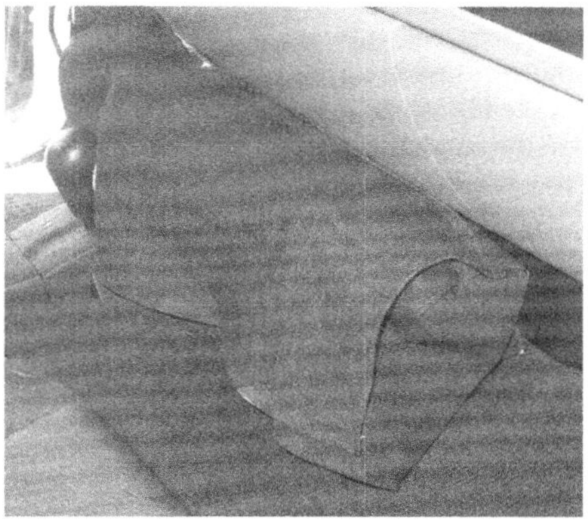

Figure 50 — Something as simple as a piece of black denim may be all it takes to make a potential thief less interested in the contents of your vehicle. [Mike Burg, N8QQN, photo]

If this transceiver arrangement isn't an option for you, consider at least concealing the entire radio. In my truck, the radios are close to the floor and the carpet is black. My wife made a cover for my transceiver out of black denim (see **Figure 50**). It's very easy to remove and put back on. In many instances, simply hiding the rig is sufficient to prevent it being stolen.

Another clever bit of camouflage is to take an old power cord and a piece of coax with a PL-259 connector on it and sew it to the bottom of the fabric cover to make it look as if the radio has been removed.

Be sure to write down the makes, models, and serial numbers of everything in your mobile installation. Take photos of your equipment as well. This information will be helpful to law enforcement in the identification and recovery of your equipment if it's taken, and your insurance company will appreciate it as well. And, of course, a car alarm is always a worthwhile investment. — *Mike Burg, N8QQN,* **mburg@neo.rr.com**

A Cordless Headset and Microphone

If you have a cordless phone and a phone patch in your station, you are closer than you realize to having a voice-operated cordless headset and microphone.

I had a Panasonic 900 MHz cord-less telephone with headset and Kenwood phone patch. All I needed to complete this installation was a two-line A/B switch and assorted telephone cables.

To use this telephone interface with a half-duplex radio requires the separation of transmit and receive audio into two separate paths. Combined with some switching and level adjustments, this is the major function of a phone patch. A variety of phone patch units can be purchased on eBay and from other used equipment sources in the $25 to $50 range.

Two wires (red and green for line 1) from the cordless telephone are connected to **LINE IN** of the phone patch and to the swinger of the A/B switch (see **Figure 51**). The A/B switch simply connects or disconnects the incoming phone line from the system. When in the A position, the wireless phone can be used as a normal telephone (with patch **OFF**), or as an actual phone patch (with patch **ON**). In position B, the local telephone instrument, patch, and HF radio are isolated from the incoming telephone line and (with patch switch **ON**) become your cordless headset and microphone.

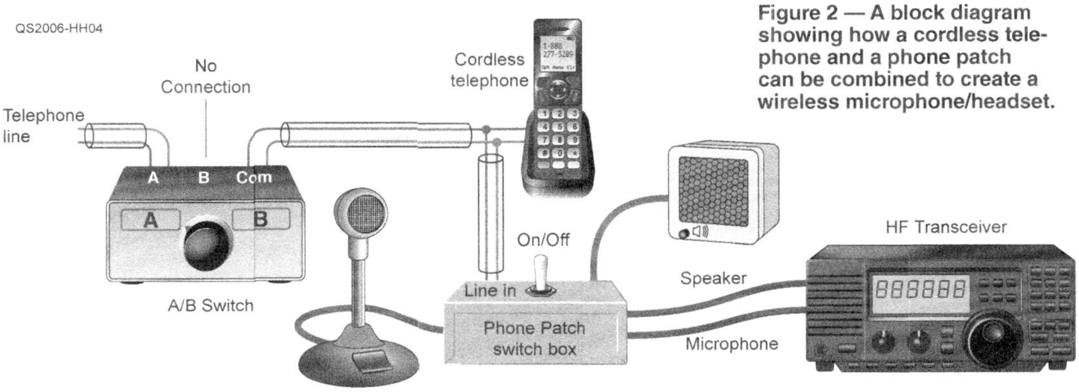

Figure 2 — A block diagram showing how a cordless telephone and a phone patch can be combined to create a wireless microphone/headset.

There is no provision for push-to-talk on the cordless handset, so the HF transceiver is set for VOX mode and the **MUTE** button on the handset serves as the VOX on and off functions. All phone patch level and balance controls on the phone patch still perform normally. — *73, David Byrd, KD7VA,* **kd7va@arrl.net**

Gripping Coax Connectors

It's usually best to finger-tighten the threaded shells of coaxial connectors, so they can be easily removed. That's good advice, until you need to unscrew an especially stubborn shell. Even a shell that's been finger-tightened can become difficult to remove over time.

My solution is to use one of those rubber pads that are intended to help open ultra-tightened food jars and bottles. They go by many names and are available at grocery stores. The grippy rubber pads are inexpensive and are ideal for loosening difficult connectors (see **Figure 52**). They also come in handy when it's time to attach a connector, allowing you to finger-tighten with the best of them. — *73, Cameron Bailey, KT3A,* **kt3a@arrl.net**

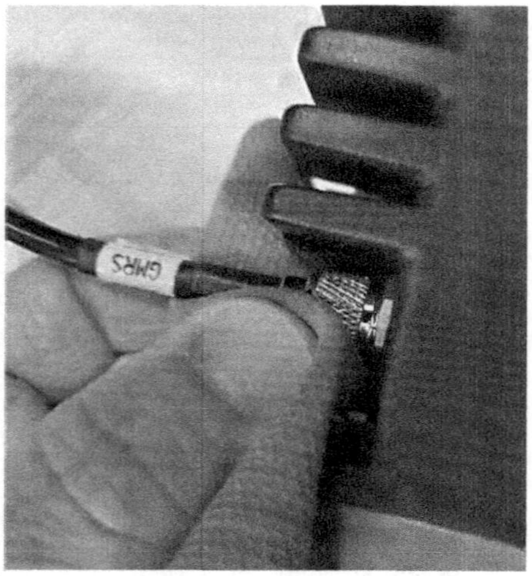

Figure 52 — Inexpensive rubber jar-opening pads are ideal for dealing with overtightened connectors. [Cameron Bailey, KT3A, photo]

Figure 53 — Cut away the top of the can with scissors or any other suitable tool. [Karl Schwab, KO8S, photo]

Figure 54 — With pliers, grip the remaining metal at the top of the can and roll the pliers around the circumference of the lid, removing all the metal. [Karl Schwab, KO8S, photo]

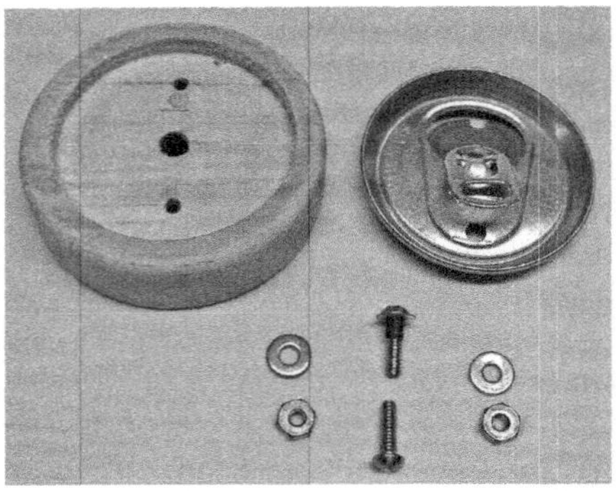

Figure 55 — The can top is secured to the base with two 6-32 × ½ inch long machine bolts. [Karl Schwab, KO8S, photo]

Figure 56 — The finished key. [Karl Schwab, KO8S, photo]

The Soda Can Key

To create this key, you need to cut the top of the beverage container free, just below the lip, exposing sharp edges. The best way I found to deal with this hazard was to carefully cut all around the can and about ½ inch below the rim with a sharp utility knife (see **Figure 53**). Metal cutting shears or scissors may work as well. Once the top has separated, make a single cut through the metal to the rim at the top. Then, using long-nose pliers, grip the metal at this cut and roll the pliers around the circumference of the lid, removing all the metal (see **Figure 54**). The goal is to have a can top with none of the can sides remaining, so it can be handled safely and fit properly into the wooden base.

To fashion the wooden base, you'll need a piece of pine at least ¾ inch thick, 2¾ inches wide, and a foot or so long (the length is just to give you an area to apply pressure with your hand while drilling). Draw a circle on the wood using a compass with its diameter just less than the width of the board (in this case, about 2¾ inches) and mark its center. At the center point, drill a hole all the way through using a 3/32-inch-diameter bit. This hole will accurately spot the center of both circles onto the opposite side. With a 2-inch-diameter Forstner drill bit (which you can find at any home improvement store), drill the side that will be the top of the base to a depth of 3/16 inches. Test fit the can top to this drilled hole to make sure it seats properly. Turn the board over and, using the same 2-inch-diameter bit centered in the 3/32-inch hole you drilled previously, drill to a depth of about 7/16 inches. What should be left is about 1/8-inch thickness of wood. Choose the point where you want the wire cable to be brought into the side of the lower part of the base and drill a hole for it to pass though.

The can top is retained in the base with two 6-32 × ½ inch long machine bolts. In addition to securing the top, they also function as electrical contacts (see **Figure 55**). One bolt must be insulated from the metal and located under the movable arm so that when the arm is depressed it touches the head of the bolt. The other bolt is not insulated.

Place the top in the base and, using an ice pick or equivalent, punch through the aluminum under the arm just enough to mark a drill location in the wood. Do the same 180 degrees on the opposite side.

The hole under the arm (in the top only) needs to be 3/16 inches in diameter to electrically clear the bolt. The opposite bolt is at the junction of the can opening and the rest of the top. This hole is made by filing a 1/8-inch-diameter notch, or just large enough for the 6-32 bolt to pass through.

Drill two 1/8-inch-diameter holes through the base, at the two previously marked positions for the two 6-32 bolts. Now you can cut off the surplus wood and leave it in a square shape or round off the edges into a circular shape, as I did. Once the sanding was complete, I stained the wood with varnish (see **Figure 56**).

As you install the bolts in the can top to attach it to the base, be sure to insulate the bolt underneath the arm. You don't want it making electrical contact with the top or bottom. I used clear plastic washers and placed a tiny length of heat-shrink tubing on the threads of the bolt itself.

With the cable attached to the bolts on the underside of the base, use a digital voltmeter to make sure there is no continuity between the two bolts until the arm is pressed down (touching the bolt head). — *73, Karl Schwab, KO8S,* **ko8s@arrl.net**

Portfolio Protection

After recently remodeling my station, I wanted to find a way of cataloging and protecting my DXCC certificates without putting them in frames. The certificates are 11 × 14 inches, so it took some searching to find a solution. My answer turned out to be the Itoya 11 × 14 Art Presentation Portfolio (see **Figure 57**). This product, and some similar ones, are available on Amazon for less than $20. It was an ideal fit for the awards (see **Figure 58**). — *73, Mike Flowers, K6MKF,* **mike.flowers@gmail.com**

Figure 57 — The Itoya Art Presentation Portfolio. [Mike Flowers, K6MKF, photo]

Figure 58 — (right) DXCC certificates are protected and treasured. [Mike Flowers, K6MKF, photo]

A D-104 Mic Adapter

When it came time to connect my venerable Astatic D-104 microphone to my newly acquired Knight T-150 transceiver, the thought of changing the mic's four-pin connector to a standard ¼-inch plug required by the T-150 gave me pause. If I wanted to switch the D-104 back to my other rig, I would have to pull out the soldering iron and replace the four-pin connector. This prompted me to come up with a quick modification.

A quick check with the ohmmeter revealed which of the four pins was audio and which was ground. Looking at the D-104 plug from the female pin end, I noticed a notch. The pins are numbered clockwise from that notch, 1 through 4.

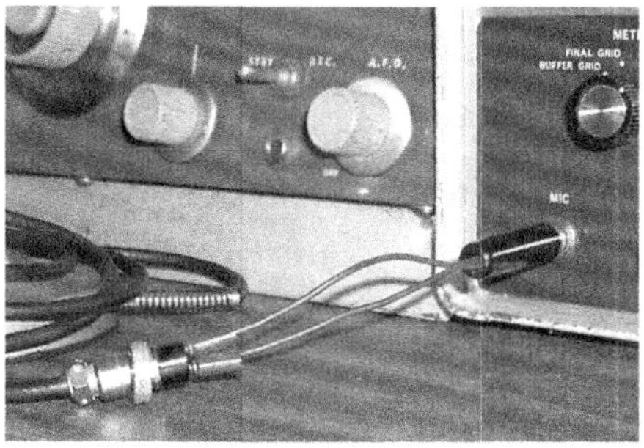

Figure 59 — This simple adapter can be used to connect a multi-pin female audio connector to a ¼-inch phone plug. [Barry Fox, W1HFN, photo]

I cut an old mini banana plug cable in half. I inserted one banana plug into Pin 1 (mic audio) and soldered it to the "tip" contact on the ¼-inch mic plug. Then I inserted a second banana plug in Pin 4 (audio ground) and soldered it to the "sleeve" contact. Pins 2 and 3 are not used (see **Figure 59**). Now I can easily get the excellent D-104 audio on either of my rigs with just a quick swap of the adapter. — *73, Barry Fox, W1HFN,* **w1hfn@arrl.net**

Remote Audio

I can't always bring my ham gear when I'm working on other projects, like yard work or car maintenance. However, with a bit of creativity, I can at least listen to my favorite frequencies. The key player in this scheme is a low-power FM stereo transmitter (see **Figure 60**), of the type that were popular before CD players were installed in car stereos. These battery-powered devices, still offered today by retailers like Amazon, are plugged into portable CD players and can easily transmit from anywhere in the car to the vehicle's FM broadcast receiver, and probably a few yards further (especially when not shielded by the metal body of a car).

Instead of plugging a CD player into the transmitter, however, I used a cable to connect the transmitter input to the external speaker jack of a ham transceiver. Because these FM units transmit in stereo, I can get creative by separating the right and left audio input lines and routing one cable to, for example, a 2-meter rig, and another to an HF radio. Now I can simultaneously listen to both devices with the FM receiver of my choosing.

Many of these inexpensive transmitters operate on battery power. I chose to power my setup with a wall wart. Because these are Part 15 low-power transmitters, it helps to maximize their coverage. The antenna adage "height is might" applies here. If your ham station is on the upper floor of your home, you're covered. Otherwise, discreet audio wires may need to be routed upstairs to where this stereo transmitter is located. Following the path of telephone lines from an operating position on the lower level of the house to a room upstairs is safer than other possible options.

Once you've located the stereo transmitter on an upper floor, or as high as you can get it, tape the transmitter's wire antenna onto a wall, but make sure there isn't metal con-duit behind it. Taping the antenna to a window will likely provide the best coverage, but pick a window facing the direction where you'll do most of your listening (like the back yard or driveway). Some windows have metal screens, so tape the antenna on the glass above if you can't move the screen out of the way.

If you own wireless Bluetooth headphones, ear buds, or portable speakers, consider using a Bluetooth transmitter instead. They're available from several retailers and also have the capability to transmit in stereo. They are the modern-day functional equivalents of low-powered FM stereo transmitters and will serve the same purpose. — *73, Klaus Spies, WB9YBM,* **wb9ybm1@yahoo.com**

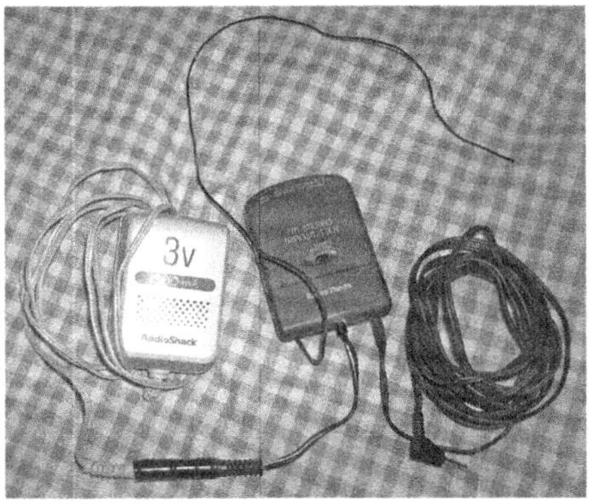

Figure 60 — You can repurpose a low-power FM stereo transmitter to rebroadcast audio from your station. You can even re-transmit two audio sources at once. [Klaus Spies, WB9YBM, photo]

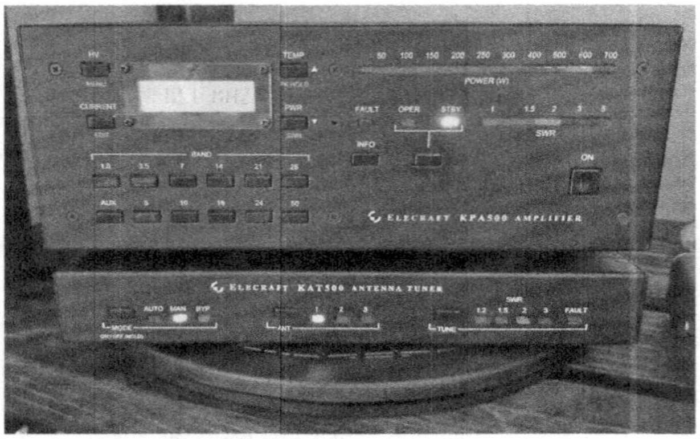

Figure 61 — A rotating tray is an affordable, easy way to access the rear panels of equipment when necessary. [Roberto Sadkowski, AJ6CY, photo]

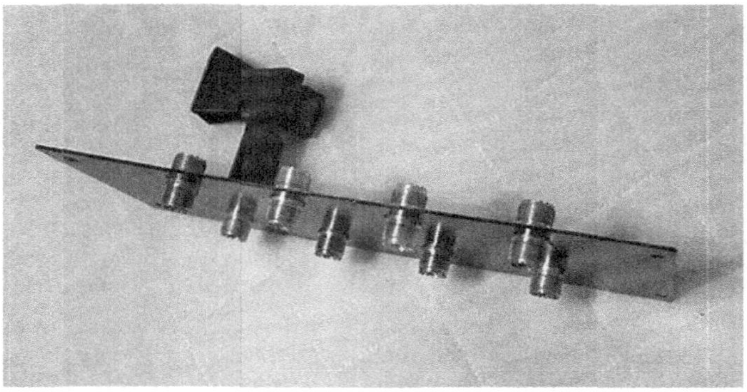

Figure 64 — The front and back view of the assembled jumper panel (the black object was used to support the panel for the photo). [Jerry Sobel, KØMBB, photo]

Rotating Heavy Equipment

I recently acquired a 500 W amplifier and antenna tuner. Together they weigh around 40 pounds. The equipment is located on a shelf at about 4 feet high. Every time I needed to add or modify a connection to the back panel, it was a strain to lift the equipment, rotate it, and access the connectors.

I figured a heavy-duty rotating tray (also called a "Lazy Susan") would allow me to rotate the equipment with ease. I found one in an online store that was rated for 150 pounds. The weight is supposed to go in the center of the plate, but it was robust enough to withstand the load even with the weight distributed to its legs (see **Figure 61**). Now I can easily rotate the equipment and reach the back panels when needed. — *73, Roberto Sadkowski, K6KM,* **rsadkowski@gmail.com**

Figure 62 — Two bungee cords temporarily hold the HF antenna mast in place. [Art Horovitch, VE3AIH, photo]

Temporary RV Attachments

When roving in my recreational vehicle (RV), I prefer to use antennas that I can deploy quickly and temporarily, rather than relying on permanent installations that could lower the resale value of the vehicle. I use masts to support my antennas and I've found that these can be set up and taken down in a hurry.

For HF operating, I have a wire vertical antenna supported by a 33-foot collapsible mast. The mast is attached to the front pin box of my fifth-wheel trailer with two bungee cords (see **Figure 62**). The cords should be at least 38 to 40 inches long and can be found at local hardware stores. The arrangement is quite flexible, as the mast tapers at the top and has withstood winds of up to 40 MPH.

Figure 63 — Metal clamps hold the pole that supports the dual-band VHF/UHF antenna. [Art Horovitch, VE3AIH, photo]

For VHF/UHF operating, I use a collapsible fiberglass/metal pole that was part of a pool-cleaning net. It supports a dual-band **J**-pole antenna. The pole is attached to the slideout of the trailer using metal clamps (see **Figure 63**), available from hardware stores in a package of various sizes for about $10. They have plastic pads to avoid damage to the fiberglass finish and the slideout's rubber seal. — *73, Art Horovitch, VE3AIH,* **suzanandart@gmail.com**

Patching (RF) Leaks

The article "'Leaky' Antenna Switches" by **Ellwood (Woody) Brem, K3YV**, published in the March 2020 issue of *QST*, prompted me to send along my solution for the same problem. Rather than dealing with potential switch faults, I built a coax jumper panel instead. [ARRL's Hiram Percy Maxim Memorial Station, W1AW, uses a similar solution. — *Ed.*]

I happened to have a quantity of SO-239 barrel connectors in my junk box. I also had a nice piece of 3⁄16-inch aluminum sheet that I primed and spray-painted after drilling. Of course, I also had to assemble some short coax jumpers. My connectors are 2½ inches apart vertically and each jumper is 6 inches long.

The eight holes were sized to be a press fit for the center of the barrel connectors and were further secured with superglue. You could also use longer connectors and secure them with nuts and lock washers. **Figure 64** shows the front and back of the panel.

The finished panel is shown in **Figure 65**, mounted next to my vintage Heathkit SB-200. I attached the selected input and output cables to the rear of the patch panel and placed labels on the front. I used inputs and outputs for the transceiver, amplifier, and antenna tuner, as well as HF, VHF, and UHF antennas. Of course, you may add as many connectors as you need and configure the connections however you prefer.

Figure 65 — The jumper panel installed and ready for use. [Jerry Sobel, KØMBB, photo]

While a jumper panel may not be as convenient as a switch, it has its advantages, such as total shielding of coaxial feed lines; reduced or no impedance bumps in the coaxial feed lines; easy total disconnection of antennas during storms; ability to easily test new equipment and antennas or isolate problems, and inexpensive assembly and disassembly. — *73, Jerry Sobel, KØMBB,* **arsk0mbb@aol.com**

Lock Your Rig

The Kensington Security Slot is an anti-theft lock commonly used to keep laptop computers from being stolen. It can also be used to secure your ham radio equipment.

The lock slot can be added to panels up to 3/16 inch thick. I added a slot to the heatsink on my Icom IC-7000 transceiver (see **Figures 66** and **67**). I drilled two 1/8-inch-diameter holes, with the centers 1/4 inch apart. Then I drilled additional holes between. I slowly worked the bit side to side to mill out the remaining material to create an appropriately sized slot. Finally, I inserted the lock and rotated 90°. All that's left to do after that is remember the combination. — *73, Bob Klaus, NØYWB,* **n0ywb@arrl.net**

Figure 66 — A small slot cut in the heatsink or metal frame of your radio is all that is needed to accommodate a Kensington lock. [Bob Klaus, NØYWB, photo]

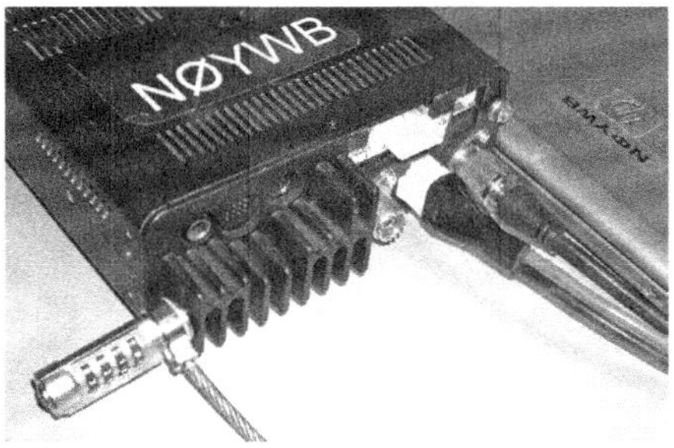

Figure 67 — A Kensington lock securing your rig will add an extra measure of security to any portable operation. [Bob Klaus, NØYWB, photo]

Slow Your Dits

Many medium-speed CW operators like to use bugs but have a hard time getting the dits under control. I have found that rather than adding more weight to the pendulum arm, if I extend the arm so that the weight is outside of the damper arm, I can get the dits exactly right. Best of all, the feel of the bug is better than it is with a lot of weight on the pendulum.

If you'd like to try this method, cut 2 inches of 5/32-inch brass rod and 2¼ inches of 3/16 × 0.014 brass tubing. Slide the piece of brass tubing on the pendulum arm of the bug and then slide the brass rod into the tubing. — *73, Lynn Kuluva, KØIMI,* **k0imi@arrl.net**

Cable Organization

Like many radio amateurs, I have a lot of wire and cable laying around, just waiting to be used in a project. However, this stuff can take up a lot of space, and a coil of coax can become a tangled mess if you're not careful. To keep things organized, I made a simple rack for spools of wire using ½-inch electrical conduit and metal curtain rod brackets from the hardware store. Each pair of brackets was under $3, and a 10-foot piece of conduit was about $10. For the coax and other large cables, I discovered that my local dollar store sold metal garden hose hangers for about $3 each, including mounting hardware. I mounted four of these on the wall and was able to get all my cable off the floor (see **Figure 68**). I spent about $30, which cleaned up the shack considerably and made accessing these materials for projects much less painful. — *73, Chris Parker, VA7PK,* **parkercac@gmail.com**

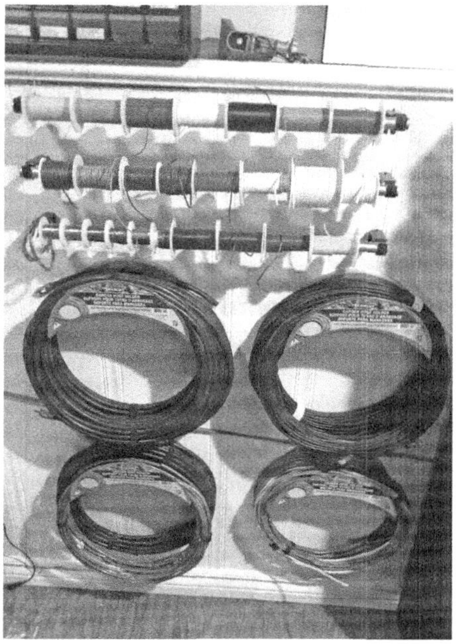

Figure 68 — A small investment in hardware was all it took to create a clean space for cable storage. [Chris Parker, VA7PK, photo]

Figure 69 — An attractive handheld transceiver stand crafted from a 4-inch-long piece of 1 × 3 inch scrap wood. [Stephen Peterson, KI7L, photo]

An Easy Handheld Stand

Handheld transceivers tend to be unstable when standing upright. There are several commercial stands available, but they generally cost about $20. I built a suitable stand out of a 4-inch-long piece of 1 × 3 inch scrap wood (see **Figure 69**).

The vertical section is a piece of 1/8-inch-thick hardboard (Masonite), cut to the width of the base block and about 5 inches long. Using a table saw, I made a kerf cut halfway through the base block at an angle of 15 to 20 degrees, about a third of the way from one end of the block. The kerf provides a snug fit for the Masonite, but a dab of glue may be necessary. A notch can be cut in the top of the Masonite to accept the radio's belt clip. Finishing is optional. — *73, Stephen Peterson, KI7L,* **scpki7l@gmail.com**

Figure 70 — A low-cost digital micrometer. [Theo Turk, WB8ADA, photo]

A Different Approach to QSL Counting

I've found an easy way to count stacks of QSL cards, which involves a precision measurement tool known as a micrometer. Micrometers have been around for ages and you can find modern versions with digital displays for less than $40 from sources like Amazon (see **Figure 70**). They're great to have around when you need to precisely measure the thickness of an object.

In this instance, use the micrometer to measure the thickness of a single QSL card in the stack. Write down the result. Then, using a millimeter ruler, stack the cards together side by side and measure the thickness of the entire stack. Divide this measurement by the thickness of the single card and you will discover the approximate number of cards in the stack. — *73, Theo Turk, WB8ADA,* **HammyTee@hotmail.com**

QSL Binders

Exchanging paper or cardboard QSL cards isn't as popular as it once was, but many amateurs still maintain the tradition. The problem, however, has always been finding a convenient place to store all the cards.

Of course, there are the plastic wall hangers that ham dealers still offer today, as well as the simple use of shoeboxes. I've come to rely on albums with clear plastic inserts (see **Figure 71**). These are popular among people who stored printed photos because they not only protect the photos, but they make it easy to browse the collection any time you want to.

Many of these albums consist of three-ring binders with an initial stock of plastic inserts. For postcard-sized inserts, each page can usually hold two cards per side. That means a single page can hold up to four cards, and quite a few of these pages can fit within a single binder. Depending on the size of the binder, it is not uncommon to have a single binder capable of storing more than 200 cards.

Just make sure the insert pages can accommodate the cards. This usually works if the album's pages are specified to hold 5 × 7 photographs. You can even find specially treated inserts intended for long-term archival storage. The chemical composition of their plastic is designed to avoid degrading photographic prints — or QSL cards.

Of course, there is still the option to use conventional card boxes such as the ones you'll find at Hobbymaster, at **www.hobbymaster.com**.— *73, Klaus Spies, WB9YBM,* **wb9ybm1@yahoo.com**

A Customized Station Clock

I've designed a unique station clock that simultaneously reads in both UTC and local time. The magic of this clock lies in its uncommon arrangement of numerals.

In **Figure 72**, you can see that there is an 8 where a 12 would normally be. This is because the "out-of-phase" numerals are UTC, not local time; you don't need numerals to determine local time. Therefore, the numbers on this clock display UTC only.

Try reading the clock yourself. Are you able to read the local time in Figure 1? Clearly, it is 8 minutes after 9. How about UTC? Again, it is easy to see that it is 8 minutes after 5.

Of course, the UTC numbers must be different for each local time zone. For my home in California, the UTC offset is local time plus 8 hours. That's why there's an 8 where a 12 would normally be. You will find number templates to download for the six main US time zones on the *QST* in Depth page at **www.arrl.org/qst-in-depth**.

The little numbers on the outer rim are also UTC, but only for when your time zone is observing Daylight Saving Time.

More Personalization

I like the idea of custom hands with a bit of ham radio flair. I own a 3D printer, so it was easy for me to make hands with a theme. (If you don't own a 3D printer, there may be a ham in your area who does.) I chose a multi-element Yagi antenna for the hour hand and a length of coax with a PL-239 connector for the minute hand. You may, if you wish, use stock hands. The 3D printer STL files for my custom hands are available at **www.arrl.org/qst-in-depth**.

As a final touch, and one which is very handy for operating, I added my Maidenhead grid square number and my latitude and longitude to the clock face.

This project uses an inexpensive clock you can find on Amazon for about $11. Just go to **www.amazon.com** and search for "Bernhard Products Black Wall Clock." Look for the 10-inch model.

However, this project does not use the quartz movement that comes with the clock. I wanted the additional accuracy of a WWV-synchronized clock for my station. An inexpensive 60 kHz WWV synchronized "atomic" replacement movement is available from **www.klockit.com**. It is a 1:1 snap-in fit for most quartz clocks (see **Figure 73**).

The atomic movement is also better than the stock unit in that it accepts much thicker modified or 3D-printed plastic hands. I used glue-on card stock cutouts to customize the hands on my initial prototype. You may do the same with this movement using any of the stock hands.

If you have experience with *Photoshop*

Figure 71 — Photo albums with plastic insert sheets can make excellent QSL card archives. [Klaus Spies, WB9YBM, photo]

Figure 72 — A customized station clock. It is based on an inexpensive Bernhard Products clock that has been fitted with a few personalized features. [John Portune, W6NBC, photo]

Figure 73 — The WWV-synchronized 60 kHz "atomic" replacement movement from Klockit. [John Portune, W6NBC, photo]

or similar graphics software, you may wish to add a background image. I used an azimuthal map of the world, centered on my location. Tom Epperly, NS6T, provides maps at his website at **ns6t.net/azimuth/azimuth.html**. On that site, you can customize and download an azimuthal map centered on your location, at a radius of your choice.

Align the numerals template for your time zone, the background image, and the text of your Maidenhead grid square and latitude and longitude. This can easily be done using the "Layers" functions of your graphics program. Print the result as the new face on heavy-weight card stock. If your $8\frac{1}{2} \times 11$ inch printer cuts off the image at the edge, resize the face to compensate, or join together two sheets of card stock for a face that is wider than your printer allows.

Construction

Disassemble the case. Pull off the hands and remove the existing face. Unscrew the motor nut and remove the movement from the mounting fixture on the back.

Glue the new face in place and cut out the center hole with a hobby knife. Push the new atomic movement into the fixture from the back and secure it in the case with the nut and washer supplied (see **Figure 74**). It is larger than the stock movement and must be mounted upside down. During initial setting of the atomic movement, as detailed in the supplied instructions, you will need to pull off and then reattach the hands in inverted positions because of the upside-down movement.

Using my 3D-printed hands, the Yagi-beam hour hand slides onto the largest shaft diameter of the Klockit motor using light pressure (see **Figure 75**). I recommend first manually running a $\frac{15}{64}$-inch drill bit carefully through the mounting hole in the hand.

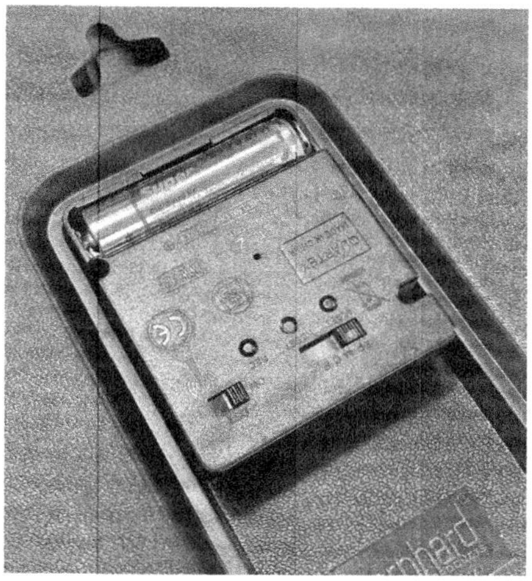

Figure 74 — The Klockit movement mounted in its inverted position in the support fixture. [John Portune, W6NBC, photo]

Figure 75 — Custom 3D-printed hands secured on the motor shaft. [John Portune, W6NBC, photo]

Go easy; the hand is strong but breakable. Do not use a drill motor or drill press. The minute hand also pushes on and is secured with the small, knurled nut supplied with the motor. Pay attention to the alignment flats in the mounting hole in the hand. Synchronizing instructions for the atomic movement come with the motor. — *73, John Portune, W6NBC,* **jportune@aol.com** *All photos by the author.*

Building a Bus

I needed a copper bus bar for my station RF ground, so I fabricated my own, as shown in **Figure 76**. Looking around my shop, I found a section of 2-inch-diameter heavy-wall copper pipe that I salvaged from the remodeling of our swimming pool. I cut a 14-inch section of pipe in half and drilled a line of evenly spaced $\frac{17}{64}$-inch holes. The

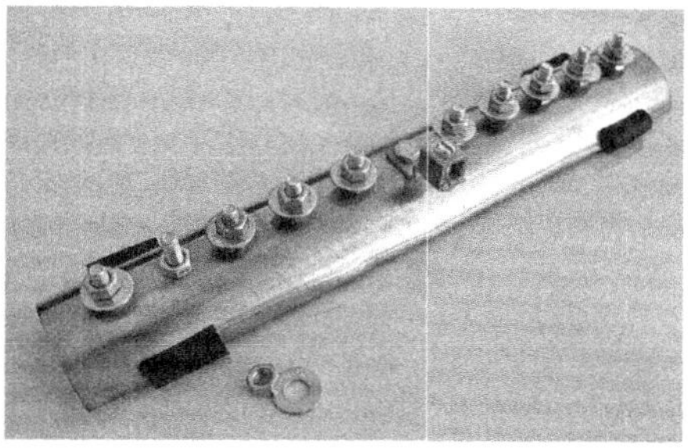

Figure 76 — A custom copper bus bar made from a 2-inch-diameter pipe. [John Quinzio, W7JMQ, photo]

fastening hardware comprises $\frac{1}{4}$ inch by 1-inch bolts, nuts, and washers. The bolts are fed through the bottom and secured with a nut. A second nut and a washer are used to secure the ground strap. Using two nuts makes it easy to get ground straps on and off. The rubber feet are made from a $1\frac{1}{2}$-inch section of $\frac{1}{4}$-inch fuel line hose, and I used a small drop of automotive trim glue to hold them in place. The ground wire lug can be placed anywhere along the bus bar. — *73, John Quinzio, W7JMQ,* **radio.w7jmq@gmail.com**

A Flexible Work Light for Mobile Equipment

When you need some illumination, particularly when operating mobile, a solution may be as close as the nearest drinking straw (see **Figure 77**). In my case, I use a plastic flexible drinking straw with an LED on the end. The design allows the straw to slide in and out of an outer tubing to extend and retract the light. The light can be rotated within the outer tubing, and with the combination of flexible portions of the straw assembly, the direction of the light can be adjusted to focus where it needs to be and remain in place. The light is controlled with a linear taper potentiometer equipped with a switch, which provides the ability to turn the light on and off as needed and adjust the intensity of the light.

Coming from the dc power source, a twisted pair of wires scavanged from a Cat 5 network cable seemed to be the best solution for a flexible, yet taut, medium to use

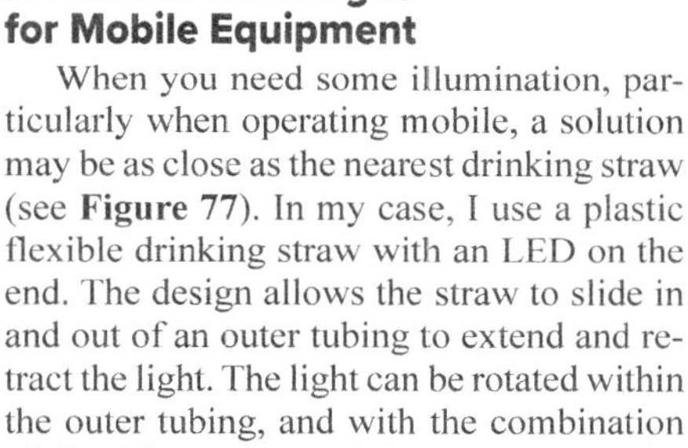

Figure 77 — Ordinary drinking straws can be repurposed to become a flexible mobile lamp. [Mark Kupferschmid, AC9PR, photo]

for the occasional in-and-out movement of the lamp assembly. The cable is cut to length so there is a minimum of slack when the lamp assembly is not in use, but just enough length when being used.

The circuit is shown in **Figure 78**. The brightness control is a 10 kΩ linear taper potentiometer with a switch. The positive side of the 13.8 V dc from the vehicle is protected by a fuse, and it connects to one side of the switch. A 680 Ω ¼-W current-limiting resistor connects between the other side of the switch and the wiper (center terminal) on the pot. Facing the rear of the pot, the right side of the wiper (right terminal) will connect to the orange wire (B+) of the twisted pair. The other end of the orange wire will connect to the anode (long lead) of the ultra-bright 5 V white LED. The white/orange wire of the twisted pair will connect from the ground bus (–) to the cathode (short lead) of the LED. I built the control components into a power control box of my own creation (**Figure 79**), but obviously this circuit can fit within a much smaller enclosure.

The outer tubing is a six-inch length of ¼-inch ID plastic tubing, the type typically used for a water faucet feed tube. The end of the tubing where the flexible straw will extend and retract from within the outer tubing was reamed out to a slightly larger diameter down to about a one-inch depth.

I also removed a flexible section from a second straw and secured it to the end of the first straw. I used two ⅜-inch cable clamps to secure the outer tubing to the equipment bracket. See **Figure 80**.

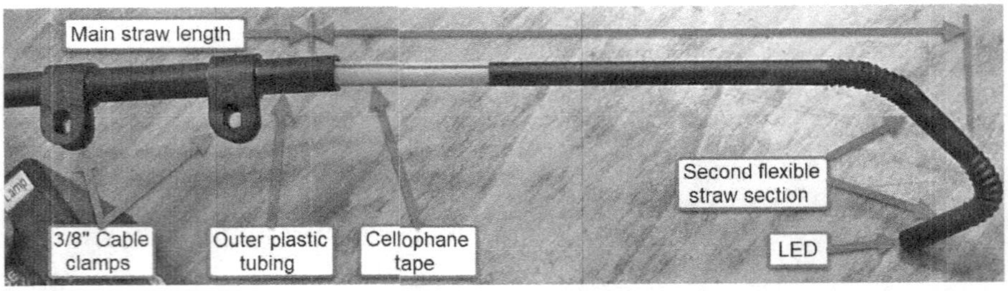

Figure 80 — The finished lamp, with components labelled. [Mark Kupferschmid, AC9PR, photo]

I developed a way to keep vehicle motion, vibration, and wire tension from pulling the lamp assembly back into the outer tubing. I used cellophane tape as a shim, so it would apply just enough tightness (or compression) between the plastic straw and the inside diameter of the outer tubing to hold any adjustment that I made to the lamp assembly in place until I adjusted it again. To get just the right amount of compression, I pulled the straw all the way out of the outer tubing, then carefully wrapped cellophane tape on the inner straw near its end. I then inserted the straw back inside the outer tubing and tightened the two cable clamps.

Next, I completed the wiring for the LED. The outer tubing must be secured in place, and the straw needs to be extended to its fullest point, without pulling out of the outer tubing. Make sure the second flexible straw portion is temporarily inserted onto the main straw. Feed the twisted wire pair through the outer tubing and straw and pull just enough of it through the straws to connect the LED, plus about an inch, and then cut the cable. Strip the wire ends just enough for soldering.

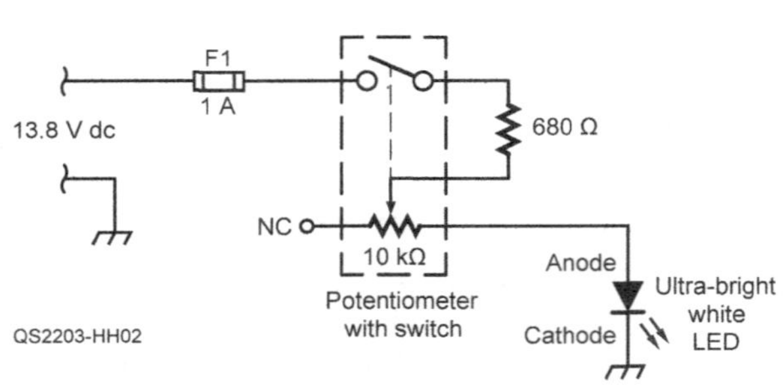

Figure 78 — Schematic diagram of the lamp circuit.

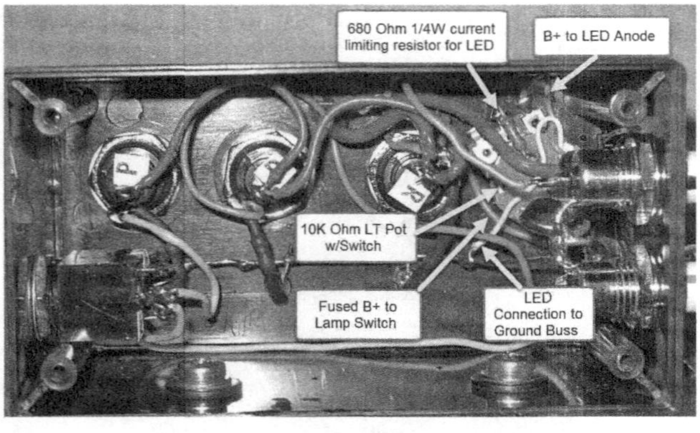

Figure 79 — I built the LED control components into a mobile power-control box of my own design, but any small enclosure will do. All you need is a box that you can mount near the radio so that you can easily adjust the potentiometer. [Mark Kupferschmid, AC9PR, photo]

Identify the LED anode (the long lead) and cut it short enough that it will not hamper wire movement through the flexible portion when assembly is completed. Slip a short piece of heat-shrink tubing over the orange wire and solder the wire to the LED anode. Repeat this step with the LED cathode and the white/orange wire. Slide the pieces of heat shrink tubing over the solder joints and exposed leads and shrink them in place.

Install the LED ring/lens onto the very end of the straw. I would suggest securing it with a small amount of clear sealant to the lens assembly to keep it in place at the end of the straw. After the sealant had set, I then applied a small amount of E6000 black adhesive to the two mating ends of the two flexible straw sections and let it cure overnight. The result worked perfectly (see **Figure 81**)! — *73, Mark Kupferschmid, AC9PR,* **mkkupferschmid@gmail.com**

A Better Desk Microphone Cable

We can probably all agree that there have not been many positive things to come out of the COVID-19 pandemic. However, one of the few good things for me was to get reacquainted with ham radio after being off the air for a few years. Unfortunately, I sold all my HF gear a few years ago, so I needed to find a new radio.

After some research I purchased the Icom IC-7300 transceiver. After using the IC-7300 with the Icom SM-30 desktop microphone for a few months, I noticed that I kept hitting the microphone connector because it pro-

Figure 81 — The lamp at its highest illumination setting. [Mark Kupferschmid, AC9PR, photo]

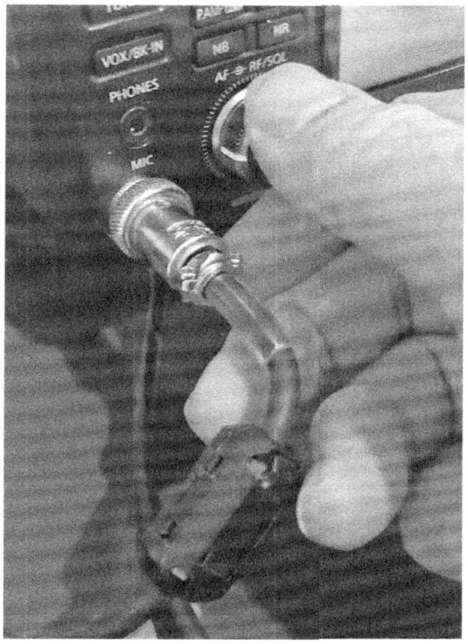

Figure 82 — The SM-30 desktop microphone connector protruded a few inches from the transceiver's front panel, making it difficult to access the volume/squelch control. [Scott Eckert, KK6DZ, photo]

truded several inches from the front panel. I was afraid I would eventually damage the connector, and it also interfered slightly with my access to the volume/squelch control. Also, because I have limited desk space in front of the radio, I just didn't like the cable sticking out so far from the front panel (see **Figure 82**).

I searched all the usual places for an adapter cable that would simply connect to my existing SM-30 microphone cable and terminate in a female 90-degree eight-pin connector that I could plug into the front panel connector. Surprisingly, I couldn't find the cable adapter I was looking for.

I decided I would simply replace the existing straight connector on the SM-30 cable with a 90-degree connec-tor, and after some online searching, I located the necessary connectors. However, after receiving the connectors, I again decided I didn't want to modify the existing SM-30 cable because it was under warranty. Changing the connector on the SM-30 is obviously the cheapest solution, if you don't mind modifying your microphone cable.

My ultimate solution was to buy an eight-pin microphone extension cable online that had male and female straight connectors for simply extending the existing cable. I would retain the straight male connector, but would replace the female connector with the 90-degree connector.

The extension cable I found was a 24-inch coiled cable, but because I wanted the adapter cable to be as short as possible, I simply cut the cable about 6 inches from the male connector end, right before the coils began.

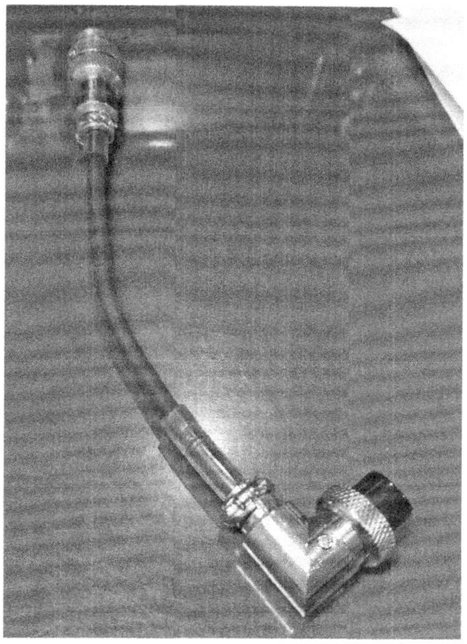

Figure 83 — The finished adapter cable. [Scott Eckert, KK6DZ, photo]

Figure 84 — Thanks to the adapter cable, my SM-30 desktop microphone can connect to my IC-7300 transceiver without the plug getting in the way. [Scott Eckert, KK6DZ, photo]

Replacing the connector was straightforward (see **Figure 83**). I only connected the five wires that are used for the SM-30 desktop microphone, not all eight that are used with the handheld microphone that comes with the IC-7300. To do this, I identified the three unused wires and cut them back when preparing to install the new 90-degree connector. The arrangement worked perfectly (see **Figure 84**).

Although this example highlights the IC-7300 transceiver, you can add a similar adapter to almost any radio. — *73, Scott Eckert, KK6DZ,* **sbeckert@verizon.net**

Build a Speaker-Stacking Bracket

Most transceivers tend to have internal speakers that face up. For improved clarity, I usually use a small external speaker with my rigs so I can direct the sound where I want it. One of my favorites is a 3¼ × 2½-inch speaker, sometimes sold as the Astatic VS4, Workman 717, Big Mouth, etc.

When I need to use two of these speakers with two separate radios, I prefer to save as much horizontal space as possible. My solution is to build a bracket that allows me to stack two speakers vertically. All that's needed is a solid base and a bracket that supports the two stacked speakers. I made a 4 × 2¼-inch base from a piece of ⅜-inch aluminum. The bracket was crafted from a 1⁄16 × ¾-inch aluminum strip, 6061 or similar. The finished result is shown in **Figure 85,** and a drawing of the bracket appears in **Figure 86**.

If you use other speaker sizes, just adjust the base and bracket dimensions accordingly. — *73, Don Dorward, VA3DDN,* **va3ddn@arrl.net**

Figure 85 — Two Big Mouth speakers stacked within the homebrewed bracket. [Don Dorward, VA3DDN, photo]

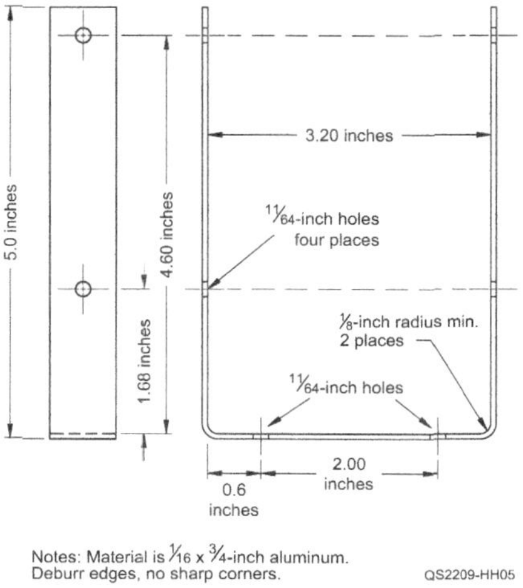

Figure 86 — A diagram of the speaker-stacking bracket.

Scanner Laws and Mobile Radios

We've all heard it before — a friend or family member borrows our vehicle and gets ticketed for speeding or some other infraction. One infraction you probably haven't thought about involves scanner laws.

You may be thinking: what does this have to do with me? I'm a licensed ham, and the FCC preempted all restrictive state and local laws. FCC PR 91-36 rules that it is legal for licensed amateur radio operators (who have a copy of their license in their possession) to operate a transceiver capable of receiving police and public service frequencies in their vehicle. However, it does not cover that family member or friend who borrowed your car and is not licensed, even if the radio is off when they're pulled over. Here in Minnesota, state law covers any radio capable of receiving emergency frequencies.

I asked a state trooper about the law regarding amateur radio equipment. He referenced Statute 299C.37, stating it is illegal for non-hams to operate a vehicle with such equipment installed, even if the radio is turned off. In some states, the equipment can be confiscated and destroyed, not to mention the additional fine. But what if you have a plug-in connector, making it impossible to be operated unless it's plugged in? If the officer can plug it in from inside the car to turn on the radio, it's still illegal. What if the connector is outside or under the hood, therefore not connected when you loan out your car? That is allowed because the radio can't be powered from inside the vehicle — hence, it would not be "installed" according to the law.

I devised a simple plug-in adapter using an old three-pronged extension cord (see **Figure 87**). A two-pronged extension cord would work, but I don't recommend it. I cut one end off for the radio, and the other end for the battery in my truck. Check the polarity of the prongs and holes for each wire, and then solder them to your radio and battery. I used the female end on the radio. The ground prong acts as the ground for the radio, and the other two prongs act as the positive; with a three-pronged plug, you're less likely to reverse polarity and short out your radio.

In my work truck, I have the plug-in just outside the passenger door, tucked under the cab and within my reach. At the end of my shift, I pull the plug. This way, should I have a day off and someone else uses my truck, they can't use my radio. If they get pulled over, they can show the officer that it cannot be used. Those who don't drive a truck can tuck in the plug under the hood.

If opening the hood every time is inconvenient, you can add a keyed switch inside the car that can disengage the radio's power source. I have a lawn mower ignition switch with a key and positions for on and off (see **Figure 88**). It is soldered to the negative wire going from the battery to the vehicle. When I am not using it, I simply turn the key, pull it out, and keep it with me. This way, I am still the only one who can turn on the radio. It is also visibly marked for officers to see that it cannot be tampered with. But if in doubt, check your local laws and loan out your vehicle at your own risk. — *73, Matthew Koscielski, KDØJZI,* **kd0jzi@live.com**

Figure 88 — An example of a two-way keyed switch. [Matthew Koscielski, KDØJZI, photo]

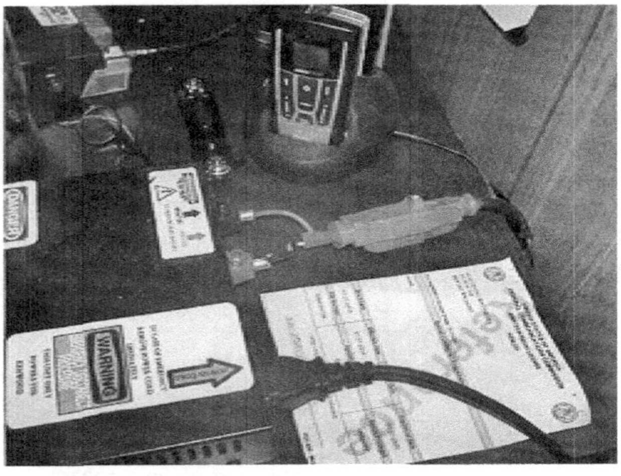

Figure 87 — A simple plug-in solution with a three-pronged extension cord. [Matthew Koscielski, KDØJZI, photo]

Antenna Switch for Three Remote Operating Locations

In the "Hints & Kinks" column in the October 2015 issue of *QST*, I described a wireless remote antenna switch that allowed me to share two different antennas between two operating locations in my home.

I am still using the same two Dow-Key Microwave 62-110 single-pole, double-throw (SPDT) coaxial RF relay switches, but in a different configuration. I can now share a single antenna, such as a dual- or tribander, among three operating locations in my home. These coaxial relays are rated to handle 150 W CW at HF and have low SWR to well above 1 GHz.

In **Figure 89**, the coaxial relays, labeled Relay 1 and Relay 2, are cascaded. In the default power-off condition, Rig 1 is connected to the antenna. With Relay 1 energized, Rig 1 is disconnected and Rig 2 is then connected to the antenna. When Relay 1 and Relay 2 are both energized, Rig 3 connects to the antenna.

This time, instead of using wireless control switches, I decided to use a simple switching solution based on 4 conductor #24 AWG communications cable. At each operating location, a small remote-control box fitted with a pushbutton switch and LED is connected to the 4 conductor cable.

Each relay has one side of its energizing coil connected to +12 V dc (red wire), so either one can be turned on by connecting the other coil lead to the common ground (black wire). This is done with pushbutton switches S1 and/or S2 (see **Figure 90**). Diode D1 functions as an AND connection, causing both Relay 1 and Relay 2 to energize when S2 is pressed.

Relay 1 is located in the ham shack near Rig 1 and the antenna connection being shared. Relay 2 is located about 25 feet away, where I split the antenna feed between the two remote locations.

For the ham shack connections, I mounted Relay 1 on a small board with the start of the 4 conductor cable connected to terminal block 1 (TB1). LEDs 1 and 2, in addition to a connector for the external 12 V dc, are also mounted on the board. For Remote 1, I mounted TB2 on a small piece of perf board. S1 and S2 can be any type of switch that you

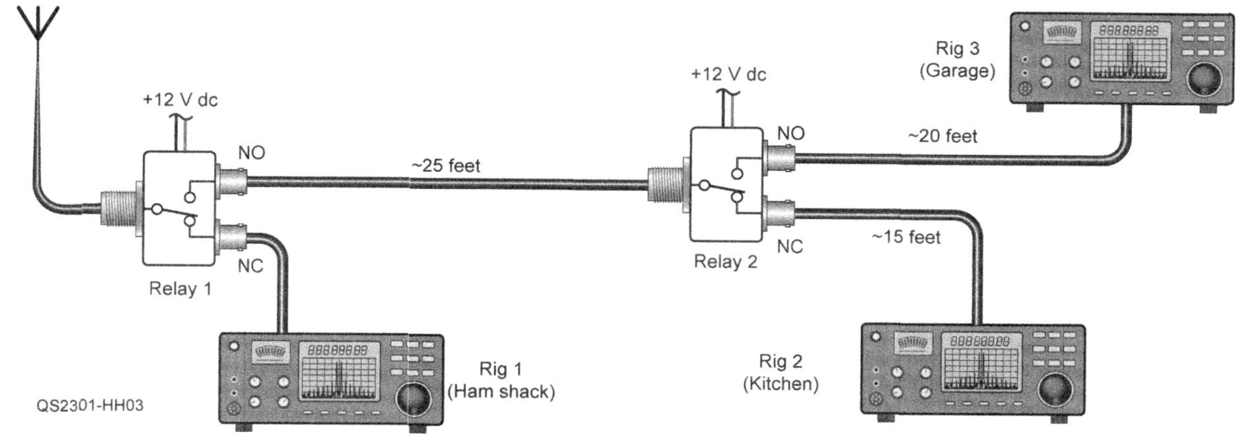

Figure 89 — The cable and switch layout.

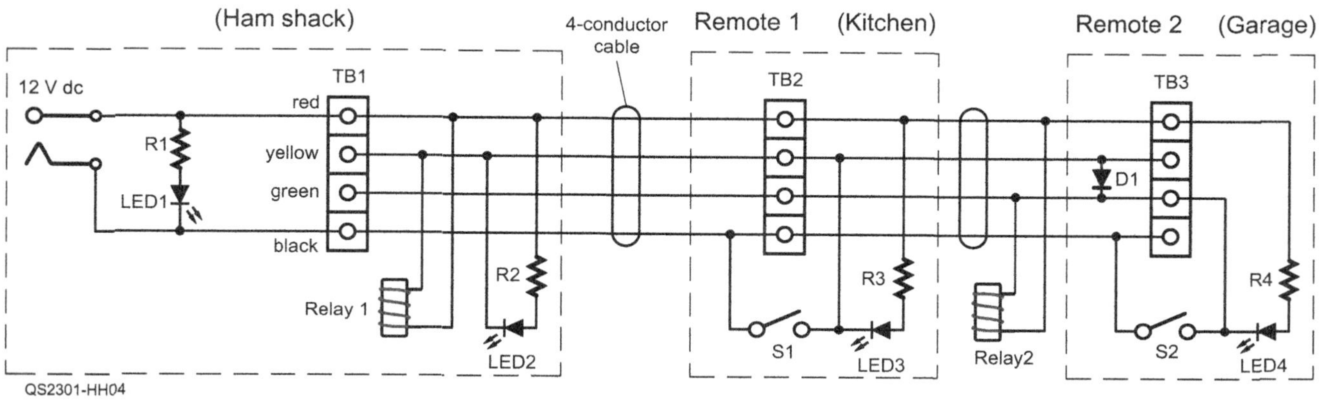

Figure 90 — Schematic diagram of the control wiring. Part references and descriptions are shown below:

R1, R2, R3, R4 — 2k2 for about 5 mA per LED
LED1 — 3 millimeters, green, indicating 12 V power on
LED2, LED3, LED4 — 3 or 5 millimeters, self flashing
Relay 1, Relay 2 — Dow-Key 62-110 or equivalent
TB1, TB2, TB3 — Terminal block, four position, screw terminals
S1, S2 — Switch, SPDT, push on/off, or toggle
D1 — 1N4002 or equivalent
Housing/case for remotes — Digi-Key 377-1175-ND or equivalent
12 V panel jack for barrel connector 5.5 × 2.1 millimeters
4 conductor cable, #24 AWG

like. I used the push-on/push-off locking style, but a small toggle switch is also okay. Remote 2 is similarly constructed, except for the switch wiring and the addition of diode D1. — *73, Don Dorward, VA3DDN*, **va3ddn@arrl.net**

Using Magnets to Slow Down a Vibroplex Key

After purchasing a 1945 Vibroplex bug at a recent hamfest, I experienced an asymmetric ratio between dits and dahs when using my particular device. At the slowest setting, the dits were in the 25 WPM range, and the dahs were as fast as my index finger could send — about 15 WPM. The favored 1:3 ratio was out of the question. I have seen some creative solutions to this type of problem, such as weights and arm-length extenders. I wanted a tidier, reversible fix for my bug, and it had to be made from materials that I already had in my shack. I had a stack of magnets on my shelf that would work for this project.

Magnets are not attracted to non-ferrous brass. A dab of clear RTV will hold a pair of magnets to each side of the brass weight. I selected the magnets by polarity so that they would attract each other while the RTV dried. The next day, I started to attach magnets to each side, thus increasing the mass on the horizontal pendulum (see **Figure 91**). I increased the number of magnets until dits could be sent at 15 WPM. Adding magnets slows the dit timing, while removing magnets speeds it up. The mechanical arm and my wrist are now in perfect harmony, and so is my ear while listening via sidetone speaker. — *73, Al Rabassa, NW2M*, **nw2m@qsl.net**

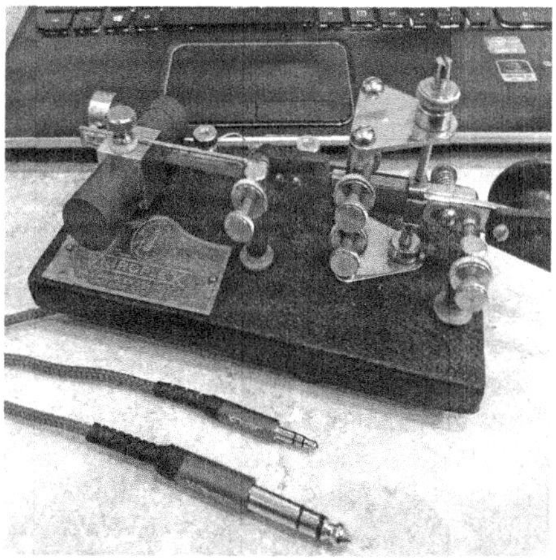

Figure 91 — The Vibroplex key with magnets attached. [Al Rabassa, NW2M, photo]

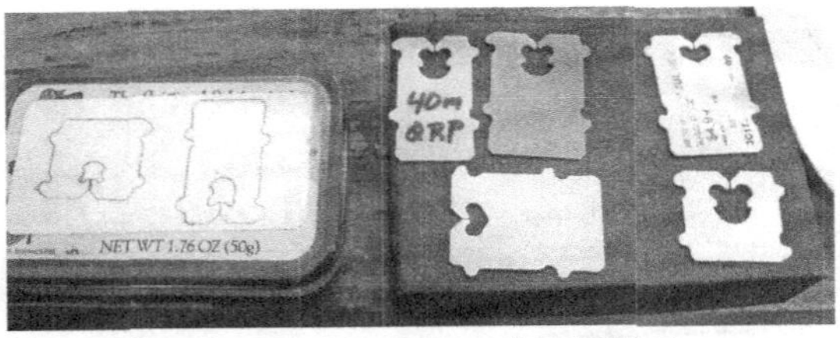

Figure 92 — Assorted food bag clips. [Phil Karras, KE3FL, photo]

Figure 93 — Food bag clips can be used to tag and identify the many coaxial cables in a typical ham shack. [Phil Karras, KE3FL, photo]

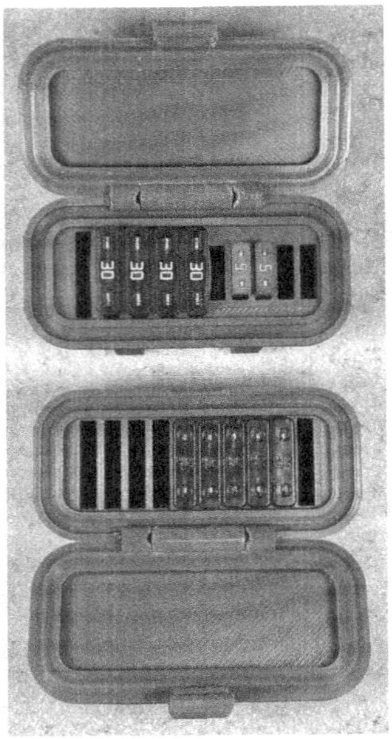

Figure 94 — Two versions of a small storage box — one with space for both mini and regular blade fuses, and the other for only standard-size blade fuses. [Warner Phelps, KN4POG, photo]

Figure 95 — This is a larger container for bulk storage of standard blade fuses. The 3D-printed p [Warner Phelps, KN4POG, photo] art is the insert for the Altoids® tin [Warner Phelps, KN4POG, photo].

Figure 96 — This storage container was originally designed for storing metal tins, but it works well for loose glass fuses. [Warner Phelps, KN4POG, photo]

Tagging Cables

I use plastic food bag clips to identify the power cables and wires for my wall warts, modem, router, and Ooma Telo box, to name a few (see **Figure 92**). If there are several wall warts plugged into an outlet or a power box, then I label each one. This way, if I need to remove a device, I know which cables and which power cord/wall wart I'm looking for. I also use these tags for coaxial cables connected to my radios (see **Figure 93**). The two tags in the foreground are for a 2-meter radio and a TEN-TEC Century 22 HF CW rig.

Referring to Figure 1, one of the tags is already marked for my MFJ-9040 40-meter radio's coax. Conversely, the white tag on the top right has its original printing on it. I have found several ways to remove this lettering. In some cases, rubbing it with your fingers will suffice, but you can also use a pencil eraser. If neither of these methods work, you can opt for fine-grit sandpaper or a sanding sponge.

Once the product printing has been removed, I use a permanent marker to mark the tag for its intended use. Sometimes, I place tags on both ends of the cable if I can't see where it is going or coming from. — *73, Phil Karras, KE3FL,* **ke3fl@yahoo.com**

Spare Fuse Storage

How many unique fuses are in your shack? Do you have spares for all of them? By asking myself these questions, I realized that my shack lacked spares of critically important fuses.

To make sure your station is fully prepared, take inventory of all of the fuses used in your equipment. Fuses may be found in radios, battery chargers, power supplies, go-kits, and battery boxes.

Radios may come with spare fuses, but other devices usually don't. My inventory consisted of mini and standard low-voltage blade fuses, as well as older glass fuses.

Once you've identified the fuses you need, I recommend storing them in labeled boxes for quick retrieval. If you have access to a 3D printer, you can easily make your own containers. There are several free designs to download. To find the latest 3D designs, visit **www.thingiverse.com** and search for fuse storage.

Figure 94 shows two versions of a small storage box. One version has space for both mini and standard blade fuses, and the other has space for only standard blade fuses. You can attach these fuse storage boxes to the inside of a go-kit or battery box with hook-and-loop fasteners. **Figure 95** shows a larger container for bulk storage of standard blade fuses. The 3D-printed part is the insert for the Altoids® tin. At present, there aren't any storage designs on Thingiverse meant exclusively for mini blade fuses.

If you have older-style glass fuses, **Figure 96** shows a storage container that will keep them organized and safe. It was designed for storing metal tins, but it works well for loose glass fuses. You can label the box for convenience; if labeling the top of the box, you may want to label the bottom, too, so that they are correctly oriented.

Currently, there are no downloadable storage boxes for a mixture of mini blade, regular blade, and glass fuses. Perhaps that's something a reader can design for the future.

— *73, Warner Phelps, KN4POG,* **kn4pog@arrl.net**

Adding Wheels to an Amplifier

My ham station is located on a desk in a cramped closet. Space is extremely limited, so to be able to access the back of my amplifier and the rest of my gear, I mounted my amplifier on wheels (see **Figure 97**). This simple addition made it much easier to service my equipment. — *73, Bill Kerber, WA2ILY,* **wa2ily@arrl.net**

Figure 97 — Adding a set of small wheels (circled in yellow) to my amplifier cabinet made it easy to maneuver in tight spaces. [Bill Kerber, WA2ILY, photo]

Microphone Hanger

I was having a tough time finding places to hang the transceiver microphone in my truck, so I produced an easy solution. I took a surplus car adapter (see **Figure 98**) and removed the interior wiring to prevent the screws from causing a short circuit. Then I attached a microphone connector to the top of the adapter (see **Figure 99**). After this, I was able to plug the adapter into a power outlet and hang my mic (see **Figure 100**). — *73, Dave Lane, KBØRFY,* **kb0rfy@q.com**

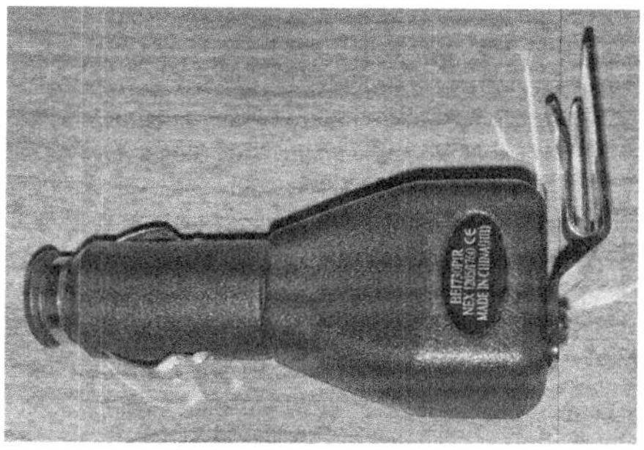

Figure 98 — Start with a surplus car adapter. It is likely that you have one in your junk box. [Dave Lane, KBØRFY, photo]

Figure 99 — Using small screws, attach the microphone connector to the top of the adapter. [Dave Lane, KBØRFY, photo]

Figure 100 — With the adapter inserted into a convenient outlet, my microphone will always have a place to reside. [Dave Lane, KBØRFY, photo]

Mic Cleaning

I offer a straightforward way to keep microphones clean. Unscrew the back case and remove the internal PC board (if present) while making sure not to lose any parts. If the microphone element isn't part of the PC board, remove that as well.

With the case free of components, use a soft-bristle toothbrush along with a bit of liquid dishwashing detergent and water to remove any accumulated dirt, oil, etc. Clean the front, back, and interior, and then use a microfiber cloth to absorb residual moisture. Put everything aside and allow all of the parts to dry before reassembling — *73, Rich Ballieu, WBØTML,* **ballieurichard1@gmail.com**

Remote Antenna Selection Made Easier

I have a remote station 70 miles west of my primary station. Thanks to software that lets me control the remote computer and a PowerUSB device (a plug strip controllable via a USB port), I can turn the transceiver power and rotator controller on and off. The remote transceiver is connected to a computer by a control cable that allows me to change frequencies as needed.

I have two antennas at the site: a 20/15/10-meter beam and an 80/40-meter dipole. I'd like to be able to select the antennas by using my connection to the remote site computer. There are high-end switches that would do the job, but they are expensive and more than I need.

My alternative is a separate PowerUSB Basic model switch controlled by the remote computer. This is a plug strip with four 120 V ac outlets, three of which can be turned on and off via the computer at the site. The strip is connected to the computer with a USB cable.

I purchased an MFJ-4712, which is a two-position remote antenna switch. This product provides a single-pole, double-throw antenna switch controlled by a control box in the shack that applies or removes 12 V dc from the coax. There is a manual slide switch on the control box that, in one position (antenna 1), does *not* send power down the coax. In the other position (antenna 2), the box applies voltage to the cable. The voltage then energizes the relays in the remote enclosure to select antenna 2.

But, I wanted to select antenna 1 or 2 with the computer at the remote site, not by hand. My solution was to plug the power supply for the control box into one of the USB-controllable outlets on the PowerUSB unit, and then leave the control box switch in the antenna 2 position. This allows me to select an antenna by simply sending a command to turn the MFJ-4712 power supply on or off. When the power supply is off, the remote switch defaults to antenna 1. When the power supply is on, the control box applies the relay voltage to select antenna 2. — *73, Bill Pellerin, KE5XV,* **ke5xv@arrl.net**

Overlooked Lightning Protection

I have studied many articles about grounding and lightning protection, but I can't recall seeing much about internet/cable TV grounding. The ground at the cable entrance can become faulty due to mechanical separation, corrosion, or expansion and contraction from seasonal temperature variations. It can also be the result of poor workmanship during the initial installation. Regardless, a missing ground might allow surges from nearby lightning strikes to enter your house and destroy your internet modem, the front ends of your TVs, and a host of other connected devices.

Figure 101 shows my reconditioned cable connection. Close inspection had revealed that the electrical connection to the ground wire was virtually nonexistent because of oxidation. To my astonishment, I also discovered that the other end of the ground wire was no longer attached to the house ground!

I disassembled the ground attachment at the cable junction, removed oxidation from the wire, and then secured the wire beneath the grounding screw. I finished by cleaning and reattaching the other end of the wire to the house ground. Every year, remember to check your outdoor connections. — *73, Sam Patterson, N4DRV,* **n4drv@arrl.net**

Figure 101 — The ground connection at the cable/internet entrance after refurbishment and replacement. Not only was the ground wire connection badly oxidized and corroded, but the other end of the wire had become disconnected from the house ground. [Sam Patterson, N4DRV, photo]

CHAPTER 3

Batteries and Other Power Sources

Pico-fuse Project Protection

I would like to extend my thanks to Tom Fowle, WA6IVG, for his "Technical Correspondence" contribution, "Protoboard Battery Safety."[1] His note would seem to state the obvious, underscoring the importance of adding a fuse to homebrew projects. For years, I have made it a practice to incorporate a pico-fuse into my projects. These fuses are inexpensive and no bigger than a resistor, yet will save your project every time.

I power all my projects off a 35 A bench-top power supply, which has more than enough current capacity to do some significant damage. As an example, I built a rotator controller based on the Arduino, which included some support circuits for the Arduino to power the motors. Everything worked fine until I mounted my project into a chassis. Then a +13.8 V supply rail found its way under a mounting stud. There was no direct short across the dc power supply, but the positive rail was essentially grounded through my computer's USB port.

The pico-fuse opened in an instant, giving me a second chance to chase down the problem. I can only imagine that, without the circuit protection, I would have burned ground traces in my computer. Safety concerns for our homebrew projects cannot be overstated. — *73, Steven Appleton, KC2LNC,* **kc2lnc@arrl.net**

Notes

[1] T. Fowle, WA6IVG, "Protoboard Battery Safety," *QST*, Jul. 2016, p. 59.

Power Line Neutral Fault

I've encountered an unusual safety problem, which seems to happen mostly in older houses. This problem is indicated by appliances not working properly and/or lights flickering. This isn't a complete loss of power, but seemingly random over and under voltages around the house.

This type of problem could be caused by an open ac neutral line in the power feed to your house. Normally, power enters the house as a two-wire 220 V circuit, a neutral line, and a ground. The neutral divides the 220 V circuit into two 110 V circuits. Without the neutral, working devices are randomly connected in series across the main 220 V circuit.

This random connecting of loads alters the current flow through the devices, which can result in the applied voltage varying considerably from the 110 V standard. This can cause damage to all electrical equipment in your home, including ham gear.

If the home has cable TV, try putting a clamp-on ammeter around the cable feed line, before it enters the house and the grounding block. If the neutral is open, the ac will return to the neutral at the pole through the shield of the cable, and you will see current on the meter — not a good thing. This is because, electrically, the cable shield is effectively parallel to the neutral going to the house. So if the neutral is open, the whole neutral current is now being carried by the cable TV shield and its connectors.

This is not a universal test for detecting open neutrals, but it may show up a problem neutral before it fails. Note that if you have a fiber optic cable connection, this test will not help, as most fiber feeds contain no shield or ground connections to the ac mains. If a neutral line problem is discovered, notify your cable provider to check for burned connectors on each end of the cable.

If you are experiencing unusual power problems and can't make the above check, it is best to get your electrician to inspect the system. [*Warning: The interior of a breaker panel contains* lethal *voltages. The inspection and repair of panel wiring should be left to an electrician. — Ed.*] — *73, George Keller, W2DM,* **georgenann@aol.com**

Generator Cable Pass-Through

I recently had my 40-year-old breaker box replaced because two of the breakers were starting to fail. When the electrician replaced it, I also had a power transfer switch installed to connect the generator to the house circuits.

My generator is a 5 kW gasoline-powered unit with four 110 V outlets and a four-prong 220 V, 30 A connection. I keep it in a garden shed for protection from the weather and to attenuate the noise when it is running. I needed a way to safely connect the generator outside to the transfer switch inside.

I did not want to permanently wire the generator to the transfer switch because I frequently take it to remote camps. The power transfer switch will handle up to eight circuits and also has a four-prong 220 V connector. After doing a little research at the home improvement store and on the internet, I purchased some cables with the correct current carrying capacity and the right connectors. [See the "Component Data and References" chapter of the *The ARRL® Handbook* for a table listing copper wire data. — *Ed.*]

Once I had the cable, I needed a way to get it to the transfer switch inside the garage. The access method would have to meet a few basic criteria:

- The generator needed to remain in the garden shed, protected from the weather and away from the house for fire and carbon monoxide safety.
- It had to provide a permanent weather-tight pass-through into the garage.
- It had to keep animals out.

The electrician suggested I hire a carpenter to cut a small, hinged flapper in the outside wall of the garage, but I was not sure that would seal well or keep animals out. I hit on the idea of using a PVC threaded connector with a cap. The plugs on the cable are 2¼ inches in diameter. This is too large for 2-inch pipe, so I had to use a 3-inch clean-out fitting.

My garage is finished with perforated board on the inside and sheathed with plywood and vinyl siding on the outside. I carefully picked a spot on the inside garage wall near the power transfer switch. It would be

Figure 1 — The clean-out mounted in the garage wall and closed off with the plug. [Ralph Phillips, KE5HDF, photo]

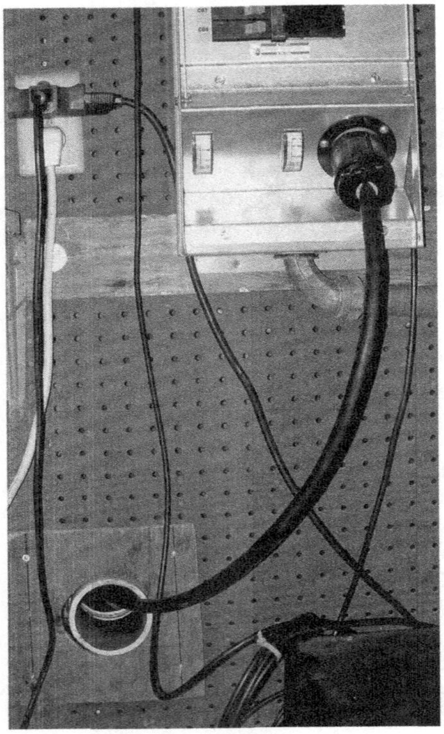

Figure 2 — Here, the cable has been pulled through the clean-out and plugged into the transfer switch. Note the plywood brace supporting the pipe. [Ralph Phillips, KE5HDF, photo]

away from any wires, but the edge of the finished hole would be only 2 – 3 inches from a stud. Using a small hole saw, I cut a hole all the way through the wall, centered about 5 inches from a wall stud. I positioned the clean-out fitting over the cutout and used a felt-tip pen to trace around the clean-out fitting on the outside wall. (I put the fitting on the wall so it just touched the top of my cutout). I used a sabre saw to cut the traced hole, leaving a hole all the way through the sheathing. (Alternatively, you could use a 4¼-inch hole saw to make a penetration of the correct size.)

Once I had a large enough hole for the fitting, I put a bead of adhesive caulk around the opening, pushed in the fitting, and added another bead of caulk around the perimeter of the cutout. This provided a weather-tight seal. Once the caulk dried, I added the threaded plug (see **Figure 1**).

I added a short piece of 3-inch PVC piping to bridge between the clean-out fitting and the interior wall face. [Be sure to sand the edge of the PVC pipe smooth to prevent it from cutting into the cable over time. — *Ed.*] (I elected not to use PVC glue. The joint is held in place by friction.) The perforated board is old and somewhat brittle, so I made a brace for it by cutting a round hole in a small piece of plywood. The brace is positioned a little high to maintain a slight downward slope of the PVC toward the outside to keep rain water out when the pass-through is open. The plywood overlaps a wall stud and is secured with 1-inch wood screws (see **Figure 2**). This keeps the pass-through secure and stable. I plan to use a large sponge to close the pass-through if hard rain or winds blow water into the garage.

Now I have a permanent way to pass the generator cable to the transfer switch that can be closed tight when the generator is not being used. — *73, Ralph Phillips, KE5HDF,* **ke5hdf@arrl.net**

Battery Isolators for Safe Charging

In the November 2016 edition of "Hints & Kinks," Scott, N5MJQ, wrote about the real danger of the crowbar circuit in a power supply shorting out a backup battery placed across the output of the supply. I believe it is poor practice to place a battery directly across a power supply that is not designed for floating a battery. Many years ago, when I first tried a backup battery across a power supply, I noticed that there was current flow from the battery into the power supply when the power supply was off. Not even considering the crowbar danger, it was not the result I would want if my house lost power and the power supply slowly drained the battery.

The easy solution is to use a boat/RV/auto battery isolator, which is essentially two high-current diodes in a heat-sunk case, designed for isolating batteries from an alternator (see **Figure 3**). The power supply should then be adjusted to compensate for the diode drop of the isolator, and the voltage at the output of the isolator set for the appropriated battery float voltage. The backup battery is now totally isolated from the supply. The supply could fail in any number of ways and still the battery would not cause problems. Here's how I went about making a battery isolator.

A battery isolator is just as it sounds — a way to isolate two batteries from each other. In a motorhome, one battery is the start battery and the other is the house battery. Without the isolator, you may run both batteries down when dry camping. With the isolator, the deep-cycle house battery may be depleted, but the start battery (being isolated) remains in a charged state.[1, 2]

A more sophisticated system has the alternator charge-sense wire connected to one of the batteries, driving the alternator voltage higher, accounting for the diode drop in the isolator. In my situation, all the isolator is doing is act-

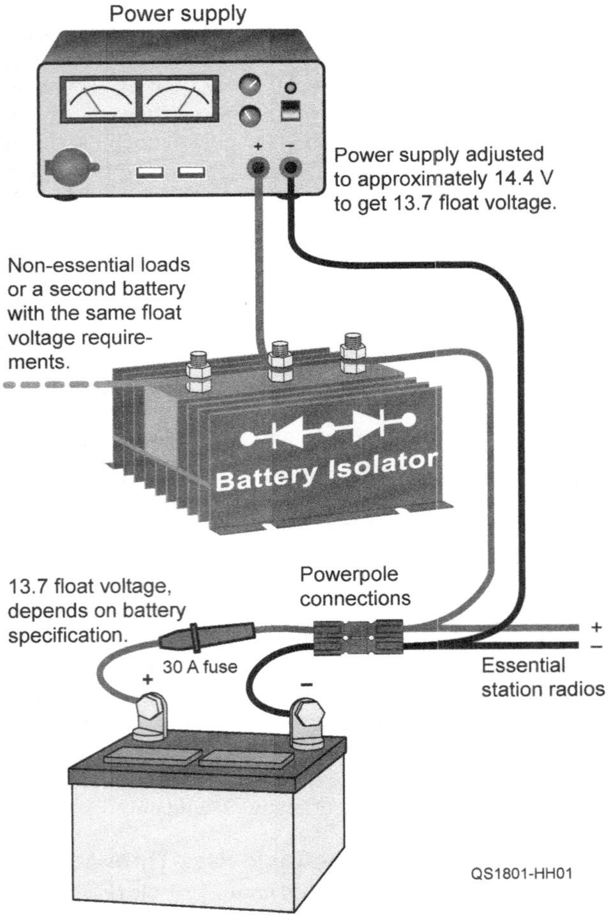

Figure 3 — A diagram of how this simple battery backup works.

ing as a diode in a heatsink package and preventing the battery draining back through the power supply when the ac fails. From the other terminal (AUX), non-essential loads can be attached. That way, when the power is lost, the non-essential loads are not powered by the battery.

I use an Alinco DM-330FXT switching power supply that delivers 25 A continuous or 30 A surge, which can be set for a voltage output of 9 to 15 V dc. Previously, I used an Astron, for which the output voltage is also adjustable (however, the adjustment is inside the case).

Setup

1) Adjust the power supply without a battery, using a light load like an auto headlight (3 A) or a base station VHF radio on receive (about 400 mA). Once the voltage across the load is adjusted to 13.7 to 13.8 V (depending on the battery float voltage rating), you can shut everything off and connect the battery.

2) After the battery is connected without additional loads, continuously monitor the battery voltage. When the current on the power supply drops to zero, the battery should be at its float voltage.

3) Once at float, connect your radios up to the battery and turn the power supply back on.

With this circuit, the radios will work normally with or without a battery connected. — *73, Allen Wolff, KC7O,* **kc7o@arrl.net**

Notes

1 See **www.globalserve.net/~jrivers/jr/pictures/aux-batt.htm** for more details on wiring an auxiliary battery relay that will charge while driving.

2 Visit **www.bluewatermarinesvc.com/html/bat_isolator.html** to learn more about battery isolators.

Pencil Box Power Pack

Handheld VHF/UHF FM radios are great for emergency communications. The snap-on battery packs have enough milliampere hours (mAh) to see you through activities for a day. But what if an emergency continues through multiple days and nights? I've come up with a robust, easy-to-build, external power pack that will keep you powered up for 3 days and nights. A lot of the build material comes from the supermarket and the office supply store.

Alkaline cells are a great choice for power packs that have to perform when you need them. D cells pack quite a punch. At very low drain levels, these cells can deliver 17 Ah. The capacity drops to about 10 Ah when several amperes of current are required. I estimate that the useful energy is in the order of 14 Ah for handheld operation with intermittent transmit. A set of nine cells will give you a voltage of just under 14 V. The voltage at pack depletion will be about 9 V. This range is perfect for handhelds, which typically have a useful supply voltage range of 6 – 16 V.

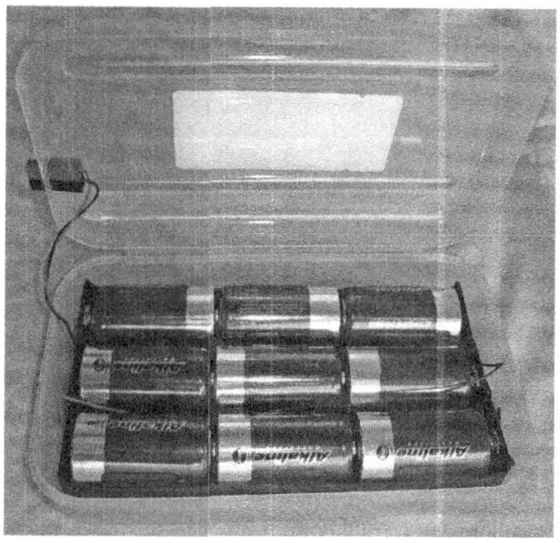

Figure 4 — The completed pencil box battery pack. [Scott Reaser, K6TAR, photo]

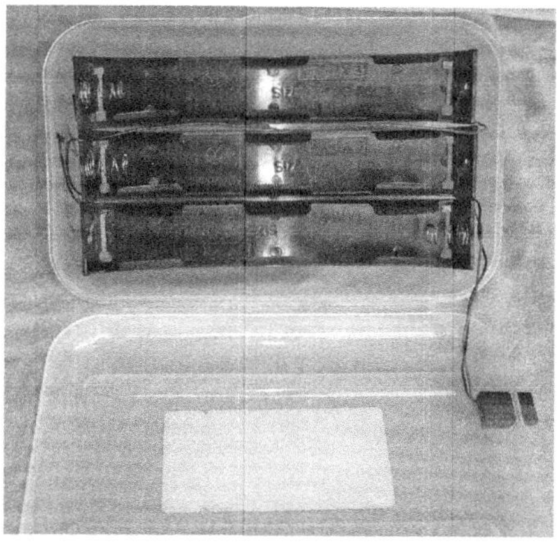

Figure 5 — The battery holders are secured to the pencil case using cable ties at each end. [Scott Reaser, K6TAR, photo]

I have packed the group of nine cells in a pencil box (Staples #22859, $8\frac{1}{4} \times 5\frac{5}{8} \times 2\frac{1}{4}$ inches). The fit is perfect (see **Figure 4**). The three triple battery holders are DigiKey #BH13DW-ND. These are secured with small nylon tie-wraps, as shown in **Figure 5**. The total construction cost, including Anderson Powerpole® connectors, is under $10. A set of batteries is about $12.

You just locate the battery holders and mark the corner mount holes from the inside. The case is translucent, and the marks show from the outside. You can use a ⅛-inch drill to make tie-wrap pass-through holes. I arranged the holders so that the red and black wires come together at one corner. These wires go to the pair of Powerpoles mounted in the lid.

I made the Powerpole cutout so it is just above the box flange when it is closed. The connector pair is placed to sit between the back and middle battery racks. I made the cutout $\frac{9}{16} \times \frac{9}{32}$ inches. The best way to make the cutout is to mark the outline of a mated pair of Powerpoles with a fine-line ink marker. Then cut inside of this marking with a hobby knife with a new blade. Take your time and exercise patience. I found that mildly, and repeatedly, heating the blade with a heat gun will help you cut through the polyethylene. You want to trim the mounting hole by trial fitting the connectors until the pair will snap into position. The short edges of the cutout engage the Powerpole notches. You should slightly radius the corners of the cutout before final connector installation, or else small cracks will develop.

My original plan was to connect a power pack to a handheld via the external power jack. The interface was a Powerpole to car power socket adapter. This then mated the external power connector to the car accessory plug adaptor offered with most handhelds. I found problems in doing this setup. You can see the issues involved in **Figure 6**. External power is routed in handhelds by at least three different methods. The first method (A) uses a relay switchover. I found that the problem with that is the relay will probably drop out significantly before the power pack energy is expended. In case

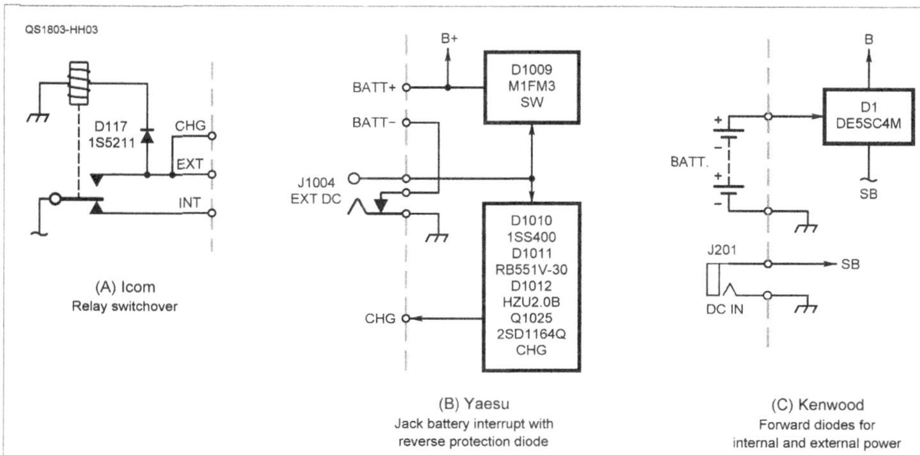

Figure 6 — Schematics detailing the three main methods used by manufactures for the DC IN connection on handheld transceivers.

(B), plugging in the external power connection disconnects the internal battery (good), but you are left with the voltage drop (and power robbing) reverse-polarity protection diode. And in case (C), both internal and external voltage sources feed through a diode pair. The external voltage source diode step down is unavoidable without modifying the radio innards.

I found that the workable solution for (A) and (B) is to connect the pencil power pack with a dummy handheld battery case. You can make these from an expired battery pack or an alkaline battery case. These devices are also found on sites like eBay, advertised as battery eliminators. The homemade versions save you the cost of a car accessory plug that you do not need.

Alkaline battery packs are rated for about 2½ years of shelf life for aircraft emergency locator beacon use. I note that the supermarket batteries show a "best before" date that works out to 9 years shelf life. My replacement estimate is 3 years. At that point, the retired batteries can be used for flashlights, electric lanterns, and the like. The shelf life is comparable with that of rechargeable batteries. — *73, Scott Reaser, K6TAR,* **sreaser@verizon.net**

Simple +/– Power Supplies

Many semiconductor circuits require both positive and negative voltages with respect to ground. For example, op-amps require positive and negative voltages of equal magnitudes and PMOS integrated circuits, often require +5 V and –12 V.

Most often, power supplies to provide these voltages use transformers with center-tapped secondary windings; one side provides the positive voltage, and the other side the negative. **Figure 7** shows a typical example, with positive and negative three-terminal regulators for the output voltages.

A simpler alternative uses two half-wave rectifiers and their associated filters in parallel across a transformer's secondary. [Due to half-wave rectification, these circuits cannot supply as much current (for a given value of filter capacitor) without falling out of regulation due to the larger ripple voltage. — *Ed.*] One rectifier provides the positive voltage, the other the negative, relative to ground at the other end of the secondary. **Figure 8** presents such a circuit, with the filters again followed by three-terminal regulators for the desired output voltages. Note that the two rectifiers are installed in opposite directions.

The latter arrangement places simpler demands on either the transformer or the rectifiers and filters. For example, for outputs of ±15 V, a center-tapped secondary must provide at least 30 V plus the losses associated with rectification and regulation, or at least 17 V or so followed by a pair of voltage doublers. Our alternative requires only 15 V plus the losses, and simple rectifiers and filters.

Figure 9 shows two equivalent rectifiers and filters. The left unit requires a center-tapped secondary and implements two half-wave voltage doublers. The right unit, with half as many components, implements the scheme described here and doesn't not require a center tap on the transformer's secondary. — *73, Bryant Julstrom, KCØZNG,* **kc0zng@arrl.net**

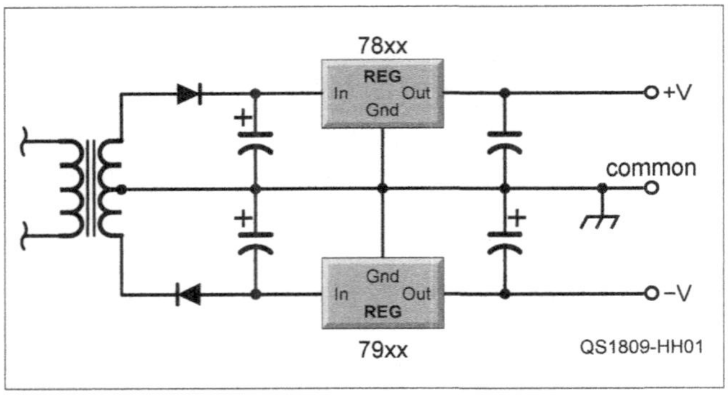

Figure 7 — Outline of a typical positive- and negative-voltage supply. [Bryant Julstrom, KCØZNG, photo]

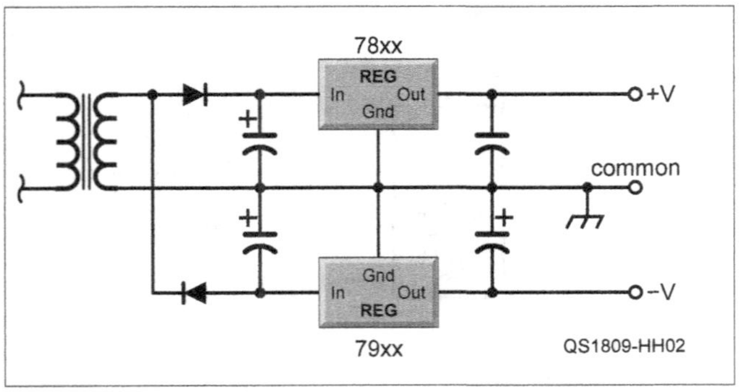

Figure 8 — Outline of an alternative circuit. [Bryant Julstrom, KCØZNG, photo]

"Free" 5 V Power Supplies

I have a large assortment of USB wall chargers. Most of these "wall warts" are from items that were recycled long ago. Most of them will charge anything. Some of them, although small, will provide up to 2.5 A at 5 V dc — perfect for just about any project. The bonus is they are very small and easy to implement. Also, these adapters almost always work from 80 to 240 V ac, which is perfect if you travel.

I've been disassembling some of these power supplies and removing the small printed circuit board inside. These units are often glue together, so disassembly can require a vise or hammer. With the circuit boards removed from the enclosure, I can include them in whatever homebrew project I'm working on at the time.

Because the circuit boards are so small, they rarely have holes in them for mounting in my project enclosure. My solution is to cut a small piece of rubber (slightly larger than the circuit board) and secure the rubber to the chassis with room-temperature vulcanizing (RTV) sealant. Then I use the RTV to hold the board to the rubber.

This method hasn't failed me yet. These boards are small enough to fit in some small, low-power transceivers and are capable of up to 2 W or so of continuous power. If you need 12 V, ± 5 V, ± 12 V, or even higher, you can use small board-mounted dc/dc converters. Quality dc/dc converters are available for under $10 from many distributors.— *73, Clint Millett, VE3CMQ,* **ve3cmq@secrs.com** *All photos by Rex Lester, G8UBJ.*

Figure 9 — Rectifiers and filters for ± voltage supplies. The scheme suggested here is implemented on the right. [Bryant Julstrom, KCØZNG, photo]

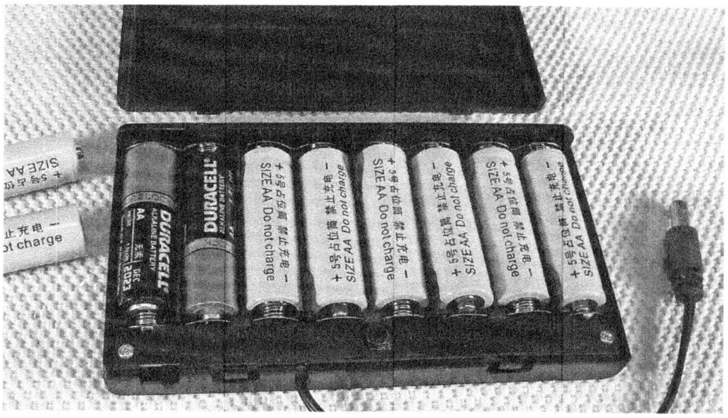

Figure 10 — A 3 V power pack comprised of two AA alkaline batteries and six AA dummy batteries. [Karl Schwab, KO8S, photo]

Figure 11— The new battery was an exact replacement for the old one, except for the wire jumper added at the upper right on the positive terminal. [Bob Lee, KN6PXS, photo]

AA "Dummy" Batteries

So-called "dummy" batteries are intended to be used as "shorting bars" when a regular battery is not required, yet a path to complete the circuit is still necessary. The dummy battery provides this path.

I often use dummy batteries in eight-cell AA battery holders — the types that have on/off switches along with a male coaxial connector on a short lead. I find these to be very handy for projects that use 12 V dc or less.

Let's say that 3 V is required. In this case, two used batteries are inserted, and the rest of the battery spaces are then filled with dummy batteries to complete the circuit in this battery holder (see **Figure 10**). When individual cells drop below their designated volt-age, a dummy cell may be removed, and an additional AA cell of like voltage can be added to maintain the desired output voltage. This method allows me to get more service life from my AA cells.

These dummy batteries can be found on the internet by searching, "No Power AA Dummy Battery." They are marked to indicate that they are electrical shorting bars and not true batteries. — *73, Karl Schwab, KO8S,* **ko8s@arrl.net**

Rejuvenating Portable Power

Portable jump starters are popular 12 V dc sources for short-term field operations, especially if they have an accessory socket for plugging in a radio. Many models offer other useful features, such as an air compressor, an LED work light, or USB charging. But after about 4 to 6 years of service, you may find that your jump starter never fully recharges, its maximum voltage after charging is low (less than 12 V), or it depletes faster than it used to. These are symptoms of the internal rechargeable battery having reached the end of its life.

That isn't cause to throw the unit away and buy a new one. The designers of these devices use off-the-shelf sealed lead acid (SLA) or absorbent glass mat batteries in industry-standard sizes and configurations, which makes them easy to replace. The charging circuitry seldom fails, so a new battery almost always rejuvenates the device for more years as a power source, and for much less money than a new one.

The first step of replacing a battery is to open the jump starter and identify the battery type. The battery is prob-ably not the only off-the-shelf component engineered into your jump starter. In my unit, the prongs on the side, where the ac power cord plugs in for recharging, were part of a 16 V dc wall wart that drives the charging circuitry. The battery inside was a common 12 V, 18 Ah SLA type with screw terminals and secured with a screwed-down metal restraining strap. I found compatible models online from many different manufacturers.

The new battery was the exact same shape and size as the original (see **Figure 11**). One thing kept this project from being a drop-in replacement: the new battery's terminals were located slightly inward from where they were on the original, so I couldn't reuse the short steel piece connecting the positive terminal to the **STANDBY/READY** safety switch. I fashioned a short jumper with #12 AWG copper wire and crimped-on ring terminals to replace it. With the battery connected, I reinstalled the restraining strap and reassembled the two halves of the jump starter.

New SLA batteries are usually sold with at least a partial charge. After you reassemble the jump starter and plug it in to recharge, it should reach a full charge in just a few hours. Instead of discarding the old battery in the trash, process it at a suitable recycling or reclamation facility. — *73, Bob Lee, KN6PXS,* **kn6pxs@arrl.net**

CHAPTER 4

Construction/Maintenance

Tiny Screws

There are occasions when I have to deal with tiny screws, or screws that have to be placed in a hard-to-access spot in a piece of equipment. I found that beeswax is a great help in dealing with these aggravations. Place a tiny dab of the wax on the tip of the screwdriver (it usually doesn't take much to hold a screw). With the screw on the screwdriver tip, it becomes easy to place the problem screw in its proper spot. This is also handy for removing and retrieving screws. They don't fall off and disappear!

Beeswax is usually available at fabric/sewing stores. Store it in an old Altoids tin or something similar. If you need to remove any residue, just wipe it off with a mild solvent, such as paint thinner. — *73, Steve Mollman, KD9HL,* **kd9hl@yahoo.com**

Microphone Cord Repair

I have quite a few microphones in my collection, and something that always bothered me was how the coiled cords looked after several years. I finally decided to see what I could do about it.

I used a piece of threaded rod (wood dowel is actually better) that was about ⅜ inches in diameter. Coil the microphone cable around the dowel and tie the starting end to the dowel with a wire tie. Then, start tightly wrapping the coiled cord around the dowel, keeping the kinks out. When finished, tie the other end to the dowel (see **Figure 1**).

Using a heat gun, warm up the cable along the entire length. I found that heating the cord to the point where it was almost too hot to touch worked well. Let the cable cool for 10 minutes or so.

The offset for the mic connector and the microphone itself isn't there. If you want that, turn the dowel so it is vertical and the mic/connector is hanging down. Heat this as well and let it cool for 15 minutes or so.

The end result is pretty good. It is not as good as it was originally, but it looks much better than it did. — *73, Clint Millett, VE6CMM,* **ve6cmm@secrs.com**

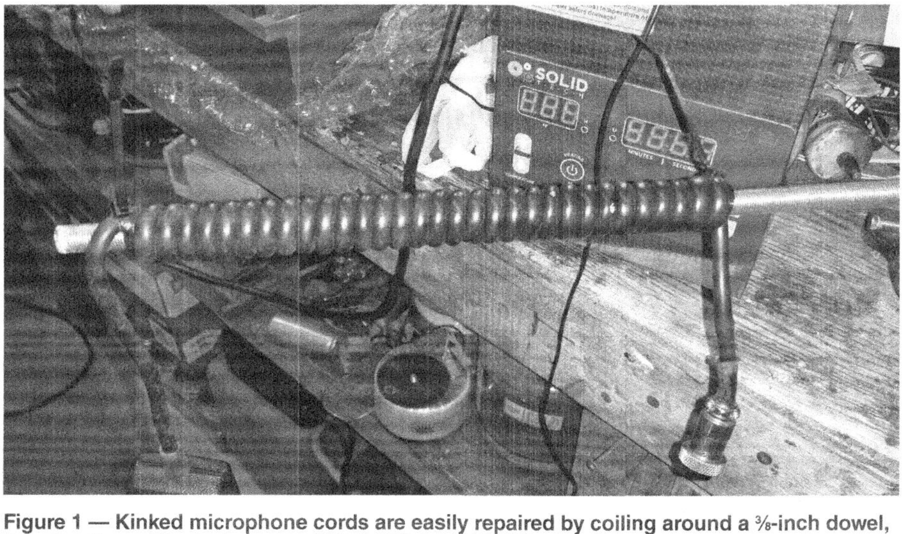

Figure 1 — Kinked microphone cords are easily repaired by coiling around a ⅜-inch dowel, then heating with a hot-air gun. [Clint Millett, VE6CMM, photo]

Figure 2 — The combination jig and heatsink holds the connections in place for soldering, without allowing excessive heat to soften the plastic pin housing and loosen the pins. [Lynn Burlingame, N7CFO, photo]

Heatsink and Jig for Microphone and DIN Connectors

Soldering cables to microphone and DIN connectors is difficult because they are hard to hold in place, and excessive heat will soften the plastic pin housing and loosen the pins.

This combination jig and heatsink will solve both problems. Solder a short piece of ⅛-inch brazing rod to the solder pins on the back side of a chassis mount microphone or DIN connector, then insert the connector to be soldered into the mount (see **Figure 2**). [*Copper is a better heat conductor than brazing rod, which might be hard to obtain for some. Using several pieces of bare #12 copper house wire as the heatsinks, perhaps one piece soldered to each pin, could also work — Ed.*] The brazing rod serves as a heatsink and the jig can be held in the optimum soldering position with a tilting head vise. — *73, Lynn Burlingame, N7CFO,* **n7cfo@n7cfo.com**

Plastic Case Shielding

Plastic boxes are easy to work with, but lack one thing that aluminum boxes have — RF shielding. I came up with a fast, simple, and inexpensive way to fix this.

I utilized a roll of aluminum foil tape used to seal HVAC ducts against air leakage. First, cut the tape to fit inside the plastic project box. Peel off the paper backing, and carefully apply the tape to the plastic box. Smooth

Figure 3 — Nylon socks stretch over headphone earpieces for a cheap, easy cushion repair. [Ralph Phillips, KE5HDF, photo]

it out and you're done. You now have the shielding a standard plastic box lacks. — *73, Rich Russo, KB3VZL,* **kb3vzl@verizon.net**

Headphone Cushion Repair

When the earpiece cover of my headphones began to degrade, I looked for a supply of matching replacement covers. Finding none, I looked around the house and saw an old pair of my wife's nylon socks. They are black, soft, transparent to sound, and elastic. They stretch over the earpieces and form nicely to them (see **Figure 3**). They are the most comfortable (and inexpensive!) earphone covers I have found. — *73, Ralph Phillips, KE5HDF,* **ke5hdf@sbcglobal.net**

A Tool for Cutting PVC

Many hams experiment with antennas. A good source of material for making insulators and mounts are the many PVC items (tubing and fittings) at home improvement stores and at most good hardware stores. However, it can be difficult to hold the item and make precise cuts to shape PVC into what we want. Hacksaws are known for drifting, so it's sometimes necessary to make multiple passes with a saw, which results in wasted material and time. There is a solution to this that results in more precise and easier cutting and is less time-consuming. The solution is using a tool called an oscillating blade cutter.

Oscillating blade cutters are relatively safe compared to other cutting devices. In fact, the blade can be held against the finger or palm of the hand without cutting because soft tissue is too elastic for the blade's teeth to take hold. Of course, I do not advocate trying this because there can always be exceptions to any rule.

Other power cutting techniques result in the PVC material heating up and forming an ugly mess in the teeth of the cutter. In fact, some power cutting results in so much heating of the material that it actually "glues" itself back into a solid that must be cut again. This goo can also stick to the cutting blade of other devices and render them useless. However, this does not occur with the oscillating cutter.

The trick is to first mark the location of the desired cut or cuts on the PVC piece with a fine-tip permanent marker or felt-tip pen. Then you need to enter the piece at a 45-degree angle and keep the cutting action approximately in the center of the cutting part of the blade. It only takes a few seconds to grasp this technique.

I have made antenna parts this way that have fewer material losses due to a misguided cut or due to the cutting blade "roaming" into unwanted areas. It's easier to make a tentative cut to try the piece and then make the final cut for more precision. The result is no wasted material.

Knowing that it is a work in progress until you have what you want and need is more comforting than finding that you have to start over with a new PVC item because the one you cut and tried removed excessive material. Also, it is much faster and easier to make several cuts with the oscillating blade cutter than trying it with a hacksaw.

The "oscillating multitool" is relatively inexpensive. The one I use is from Harbor Freight and it works well, but I'm sure the tools can be obtained from other sources. The model I use is available at **https://www.harborfreight.com/oscillating-multi-tool-68861.html**. Because it comes without a "standard blade," the blade I use can be found here: **https://www.harborfreight.com/1-38-in-high-carbon-steel-multi-tool-plunge-blade-61816.html**.

My cutter is used on PVC and not wood, so the blade lasts longer. My blade is still cutting like new after a couple of years of use. The cutter is also a good tool to have around the house. I trimmed off the edge of a dragging door without taking it off the hinges, so it can also be a time-saver in other areas.

Whenever you need to cut or shape PVC or any other similar plastic material, this tool can provide better results. In fact, I've reduced the number of my quick-and-ready power tools in the radio room to just the oscillating multitool and a hand drill. — *73, Bob Lunsford, KK5R,* **nocrud222@yahoo.com**

More Powerpole Conversions

I experienced a similar issue to the one addressed by Ken, K9CAX, in the November 2016 edition of "Hints & Kinks." I was also having intermittent power issues, so I modified both my Icom IC-756PROII and the backup Kenwood TS-440 to Powerpoles to solve the problem.

My method was somewhat different than Ken's. It looked to me as though the four-wire factory power connection was required by the current limitations of the Molex connector. I converted the power wiring to #14 AWG, so that only a single set of Powerpoles was required.

I retained the Molex housing, drilling $^3/_{16}$-inch clearance holes for the #14 AWG wire

Figure 4 — On the Icom IC-756PROII, the replacement #14 AWG wires pass through the modified Molex housing, which acts as a grommet. [James Joyce, NB6TE, photo]

Figure 5 — The new #14 AWG positive lead of the TS-440 is connected to the original dual leads of the transceiver. The ring connector on the new negative lead is mounted to the chassis screw, which is located under the wiring bundle. [James Joyce, NB6TE, photo]

(see **Figure 4**). That gave me room to allow the Powerpole connection to lie flat on a shelf rather than bear the weight of the lead harness. After feeding the wire through, I sealed the Molex housing with silicone caulk. As you can see, the conversion to a single wire inside the PROII was easy.

As for the TS-440S Powerpole modification, clearance is tighter in the 440 and the positive (red) lead is short. You can see that the two positive leads are only about 2 inches long in **Figure 5.** They are long enough to do the job, if you work carefully. Using a plastic probe to hold the wire bundle aside, you can see the negative terminal screw. The original negative (black) double lead is crimped into a ring connector, then screwed in place. Just remove it, crimp a ring connector to one end of a piece of black #14 AWG wire, and screw it back in place. Then mount a 30 A Powerpole connector on the other end.

I didn't have a grommet handy, so I double shrink-wrapped both the negative and positive leads, wrapped them together with silicone insulating tape, and then applied two pieces of heat shrink to combine them into one cable. — *73, James Joyce, NB6TE,* **nb6te@mcn.org**

Heliax Cutting Tool

Cutting Heliax cable needs to be neat and precise. A hacksaw doesn't do a great job, a wire cutter will smash the cable end, and commercial tools are always specialized and expensive. Here's a trick I learned for making clean cuts in hardline.

Go to a home improvement store and buy a PVC pipe cutter — the ratcheting type. I use a $1\frac{5}{8}$-inch model BrassCraft. To make the cut, place the cutter on the ridge (not the valley) of the Heliax outer shield. Don't cut like it's PVC pipe, but gently squeeze while turning the tool. You'll end up with a cleanly cut end, ready for the connector. — *73, Glen Pitts, K6KJQ,* **kjq45@att.net**

Crossover Crimper

When Powerpole connectors started to appear as the prime connector for ARES and RACES, I thought it was time to make the change over from Molex connectors. Having no success in finding a die set for my Paladin CrimpALL tool, I purchased an early version of a crimp tool and die from Powerwerx, thinking I could interchange the dies and use the CrimpALL. The die set would not line up in the CrimpALL as the mounting holes were different for the Powerwerx model, so I ended up buying the Powerwerx crimper.

A few years ago, I noticed that DX Engineering was not only selling a crimp tool but also individual die sets for Powerpoles, RG-8, RG-8X, and uninsulated and insulated wire connectors. After an exchange of e-mails, I ordered the Powerpole die set. DX

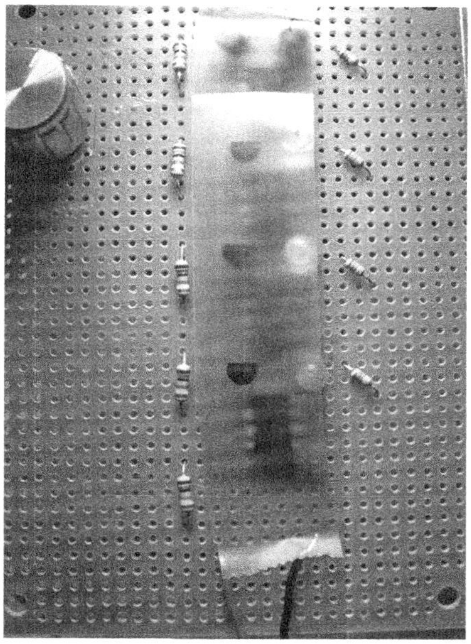

Figure 6 — A piece of clear tape will hold components on a circuit board, simplifying the soldering process. [Bruce MacAlister, W4BRU, photo]

Engineering was not sure it would fit my crimp tool, but said I could return it if it did not. The DX Engineering dies did fit my CrimpALL perfectly. Subsequently, I purchased the RG-8 and RG-8X die sets for UHF and BNC connectors. In summary, anyone who owns a Paladin/Greenlee 8000 series CrimpALL tool can safely order dies from DX Engineering. — *73, Joseph Walc, W4EEI,* **w4eei@arrl.net**

Holding Components on a Perfboard

In the past, I secured components to a perfboard or printed circuit board by bending the leads on the bottom to keep the components from falling off when I turn the board over to solder. All those leads can be a mess with lots of closely spaced items. What I now use is clear tape, as shown in **Figure 6**. [*Note: Clear tape can develop static charges that can easily destroy sensitive devices, such as certain types of ICs. For those applications, standard electrical tape may be best. — Ed.*] Pulled tightly over the components and lapped over the ends of the board, the tape holds everything in place when I flip the board. For shorter components, I put a second layer of tape across them. It's so simple and neat that I don't know why I never thought of it before. — *73, Bruce MacAlister, W4BRU,* **w4bru@arrl.net**

Transformer Covers from IKEA Toothbrush Holders

There is a segment of audio enthusiasts who prefer the harmonic richness of vacuum-tube amplifiers over that of statistically superior solid-state amplifiers. The simple circuitry is popular with audio homebrewers who often like to display the tubes and transformers on an uncovered chassis. However, less expensive audio output transformers are not particularly display-worthy and are typically hidden under transformer covers. The problem is that these covers can cost as much or more than the transformers they are hiding.

The homebrew solution, of course, is to find an alternative item and repurpose it (i.e., "hack" it) to fit the new application. I found that IKEA's unpronounceable MJÖSA stainless-steel toothbrush holders (**www.ikea.com/us/en/catalog/products/90284952/**) make workable transformer covers. At $4.99 each, the price is hard to beat. The problem I faced was figuring out how to attach them to the chassis (see **Figure 7**).

My somewhat over-engineered solution was to put two threaded brass inserts into a plywood disc and cement it to the bottom of the holder with construction adhesive (see **Figure 8**). Threaded rods screw into the in-

Figure 7 — Transformer covers fashioned from IKEA toothbrush holders. [Barry Shackleford, W6YE, photo]

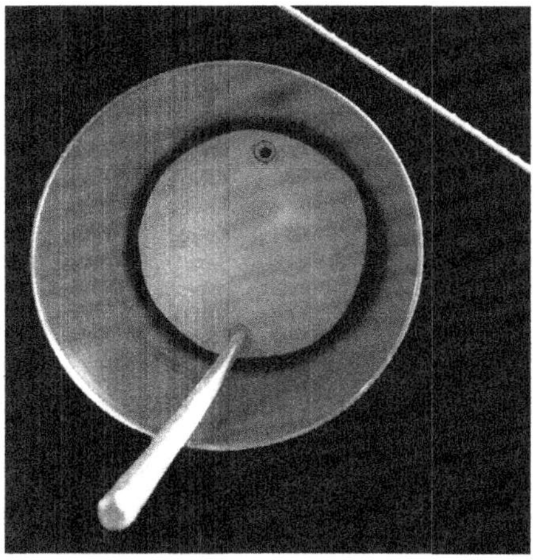

Figure 8 — A cemented plywood disc with threaded brass inserts provides a mounting means. [Barry Shackleford, W6YE, photo]

serts and the assembly is pulled tight to the chassis by nuts underneath. Next time, I think I will simply use epoxy to fix either two decapitated screws or two threaded spacers to the inside lip of the holder. If you use this simpler method, take care not to over-tighten the chassis connection, which may cause the epoxy bond to shear. — *73, Barry Shackleford, W6YE,* **w6ye@arrl.net**

Desoldering Tips

After listening to an episode of the "ARRL The Doctor is In" podcast about radio restoration, which discussed removing multiple solid-state devices, I wanted to share a tip. First, if you're working on a surface-mount device, it's a lot easier. You can use a standard, cheap heat gun on low, with a small metal funnel attached to the front to act as a nozzle and focus the hot air. Heat the component and then pick it up with tweezers. This can also work with through-hole components. Make sure to heat from the component side, not the underside of the printed circuit board (PCB). Heating from the "top" will protect the PCB from the heat (older boards are not nearly as fire retardant as newer materials), and the substrate inside the chip will act as a heatsink to evenly heat up its pins. Older semiconductors are not as likely to survive this process as new ones.

The other option for removing large dual in-line packages (DIPs), especially if you don't have a heat gun or are nervous about using one, is to use fine side cutters or snips to cut each leg of the DIP where it meets the body of the package. Then you can simply heat up each lead on its own and remove it from the board with tweezers. Finally, and I cannot stress this enough, leave your solder sucker in the drawer, or give it an extreme cleaning before using it on a dense PCB. The spring action tends to rain a fine layer of solder balls all over whatever is under it with every use, and these will cause short circuits. Instead, invest in some quality desoldering braid (get two or three sizes), and you'll never look back. — *73, Chris Parker, VA7PK,* **parkerca@ieee.org**

Cocoa Tin Enclosures

One of the challenges of building is finding suitable enclosures. A metal cough drop box is often used for small semiconductor circuits and even, occasionally, for projects involving one tube.

The idea of using small metal containers can be extended to larger ones, like those that hold fancy chocolates, teas, and similar products, or even seasonal cookie tins. These can often be obtained for free from coffee shops and other businesses, or you can buy the products yourself.

I've found that rectangular tins are best (see **Figure 9**). It's difficult to fit circuits into tins that are round, and round prototyping boards are very hard to find. A tin's original decoration can be used for effect, or the tin can be painted.

Figure 9 — A homebrew digital clock housed in a tin that once held hot chocolate mix. [Bryant Julstrom, KCØZNG, photo]

In general, it's best to use a tin's lid as the unit's mounting surface and/or front panel, and to minimize the number of holes in the tin's body. Implement the circuit itself with sockets and plugs so that it can be taken apart for debugging and maintenance.

The main issue with using such tins as enclosures is that the metal is very thin and deforms easily when worked. A backing plate of more substantial metal can be attached to any surface you need to work and become a permanent part of the finished unit. — *73, Bryant Julstrom, KCØZNG,* **kc0zng@arrl.net**

Easy Power Resistor Clip

While working on a Yaesu FL-2100B, I needed to hold a 10 W resistor in place while soldering it. I usually use glue to hold these in the right spot, but I was in a hurry and didn't want to wait for the glue to set.

I use binder clips for lots of things around the shack, and I grabbed one to hold the resistor (see **Figure 10**). I removed the handles off the clip and had an instant resistor clip. It's inexpensive, and it really held the resistor in place. If necessary, a bit of thermal paste would add some extra dissipation for the resistor.

Other uses for the clip could be for holding batteries, large capacitors, and so on. If a panel lip isn't available, a screw or epoxy could hold the clip in place. There are lots of different 5-minute epoxies that'll do the job. The clip is easily removable and no holes had to be drilled. — *73, Clinton Millett, VE3CMQ (formerly VE6CMM),* **ve6cmm@secrs.com**

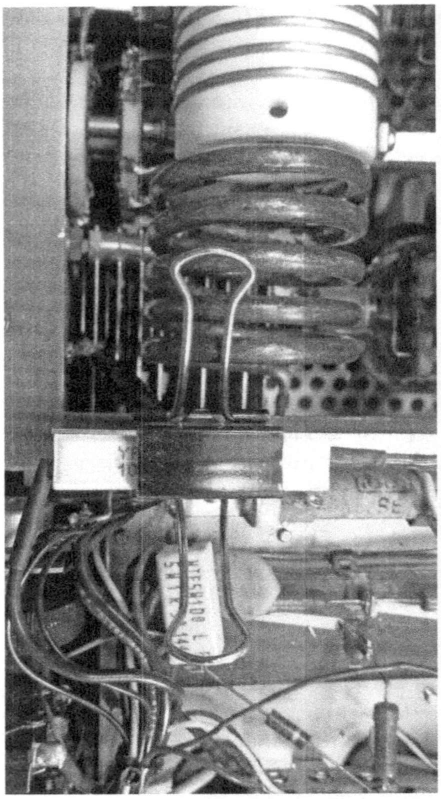

Figure 10 — The clip was installed with the wings still on, to hold the resistor in place. [Clinton Millett, VE6CMM, photo]

Figure 11 — A small magnetic dish is handy for keeping track of screws and metal parts. [George Taylor, K6RWI, photo]

Magnetic Dish

I recently bought a small magnetic dish that's great for holding screws and other small metal parts so they don't get lost or roll away (see **Figure 11**). At about $4 each from Harbor Freight or other online sources, it's worth buying a few to have around. I don't know what I'd do without them. — *73, George Taylor, K6RWI,* **georgejtaylor@yahoo.com**

Figure 12 — A 1/2-inch square neodymium magnet enables return of the drill press chuck key to its storage position with a minimum of fuss. [Barry Shackleford, W6YE, photo]

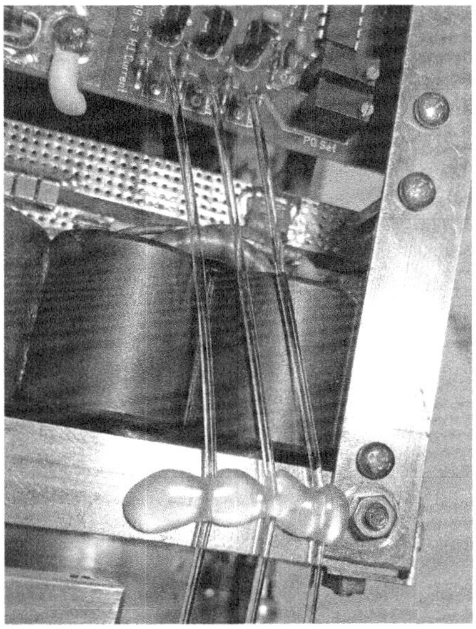

Figure 13 — Clear acrylic rods secured to the chassis with glue, after being glued to the LEDs. [Chuck MacCluer, W8MQW, photo]

Figure 14 — The fault indicators, now easily readable from the front panel. [Chuck MacCluer, W8MQW, photo]

Magnet Keeps Drill Press Chuck Key Handy

A small neodymium magnet lets you store your drill press chuck key where you want it (see **Figure 12**). The greater magnetic mass of the drill press housing prevents the key from pulling off the magnet. — *73, Barry Shackleford, W6YE,* **w6ye@arrl.net**

LED Light Pipes

In the process of building a case for a solid-state amplifier, I was faced with removing the three fault LED indicators from the PC board and re-mounting them to the front panel. Instead, I piped their light. I routed three 1/16-inch clear acrylic rods from the LEDs to the front panel, superglued the rods to the LEDs, and secured the rods to the chassis with hot glue (see **Figure 13**) — no PC board modification required. The fault indicators can now be easily read from the front panel (see **Figure 14**). — *73, Chuck MacCluer, W8MQW,* **maccluer13@gmail.com**

Mounting an SO-239 Connector Without Nuts

It is sometimes necessary to mount an SO-239 UHF connector in a tight location, which can make assembly with screws and nuts difficult. Fortunately, the SO-239 mounting holes are just the right size to accept an 8-32 tap. A few twists of the tap (see **Figure 15**) and the need for mounting nuts is eliminated and the SO-239 can easily be assembled into a tight location (see **Figure 16**). An additional advantage is that the tapped hole provides a positive contact to the connector body.

If you don't have an 8-32 tap, inexpensive light-duty sets are available online or at your local hardware store. You'll wonder how you ever got along without them. — *73, Barry Shackleford, W6YE,* **w6ye@arrl.net**

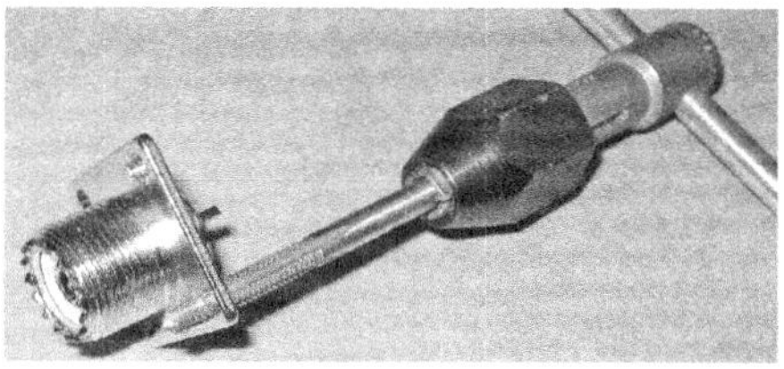

Figure 15 — SO-239 mounting holes are just the right size for tapping with an 8-32 tap. [Barry Shackleford, W6YE, photo]

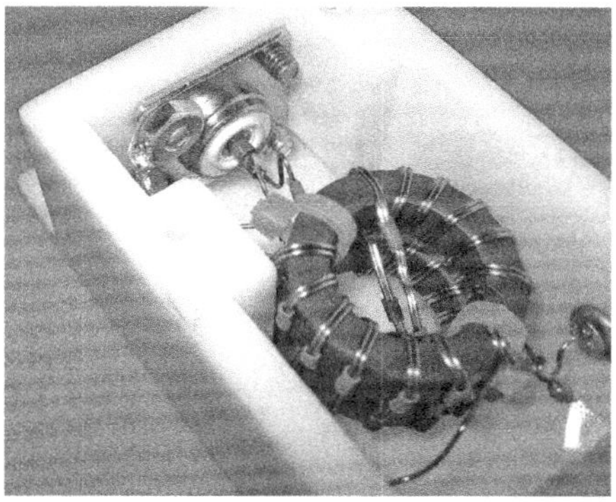

Figure 16 — Tapping the mounting holes of an SO-239 allows installation in tight spaces without nuts and provides a positive connection to the connector body. [Barry Shackleford, W6YE, photo]

Cutting Light-Gauge Aluminum Sheet

Occasionally, my homebrew projects involve cutting light-gauge (16 AG) aluminum sheet for building enclosures or enclosure panels. Because the cost of a metal shear is too expensive for my limited use, I developed a simple and inexpensive cutting fixture using a few pieces of wood, an aluminum angle, and a hacksaw blade. The fixture consists of two parts: a base and a cutting tool (see **Figure 17**).

The base consists of a base plate and two parallel guides that form a 3/4-inch-wide channel for the cutting tool. The aluminum sheet to be cut is inserted between the base

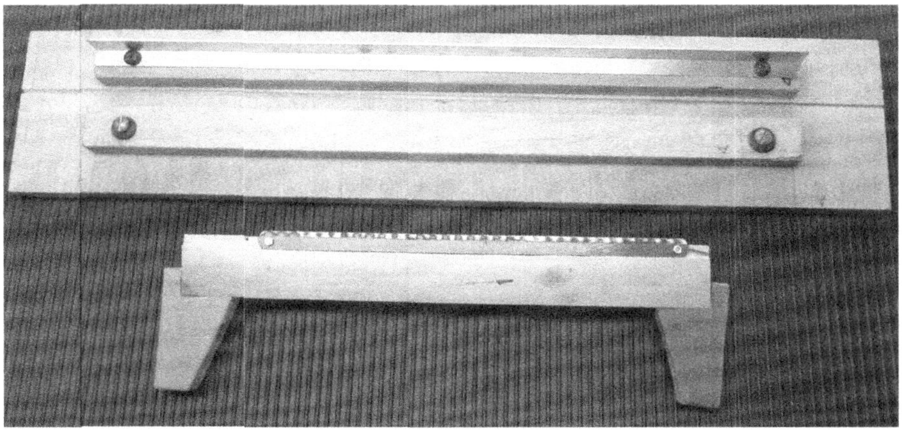

Figure 17 — The cutting fixture base (top) and the hacksaw cutting tool (bottom). The arrow on the cutting tool indicates the direction of the cut. [Stan Ekiert, K3KKH, photo]

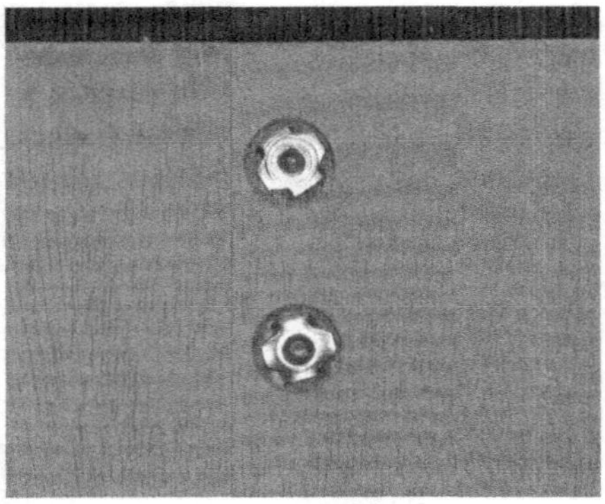

Figure 18 — The T nuts were mounted on the underside of the base plate in 1-inch diameter, 1/4-inch deep counterbore to make them flush with the base plate. [Stan Ekiert, K3KKH, photo]

plate and the guides, then bolted the assembly together using 1½-inch-long ¼-20 screws. I secured the screws to the underside of the base plate with ¼-20, three-prong **T** nuts. On the underside of the base plate, I counterbore holes for the mounting screws to a depth of ¼ inch to mount the **T** nuts flush with the base plate (see **Figure 18**).

I made the base plate and channel guides from ¾-inch plywood. The base plate was 4½ × 23 × ¾ inches, while the guides were 18 × 1 × ¾ inches. The center-to-center spacing of the screws was 16 inches and could accommodate a 15¾-inch-wide sheet. A ½ × ¾ inch aluminum angle bracket was added to the left channel guide to minimize bowing of the guide when cutting narrow sheets.

The cutting tool consists of a 10-inch, 32 teeth-per-inch (TPI) hardened carbon steel hacksaw blade mounted to a 12¾ × 1½ × ¾ inch pine support. A 32 TPI blade was selected because its small kerf produces a narrow cut approximately 3⁄32 inches wide. I cut a 5⁄16 × ⅛ inch deep notch along one side of the support to accommodate the hacksaw blade and the 4-40 mounting screws and nuts. The hacksaw blade teeth extended approximately ⅛ inch beyond the support. I screwed and glued two handles fashioned from scrap wood to each end of the support, so the cutting tool could be pushed in a forward and backward motion within the channel to generate a cut.

I used the cutting tool to cut a 1⁄16-inch-wide slot in the top side of the base plate when the fixture was assembled. The slot can be used to align a scribe line (the cut line) on the aluminum sheet to the slot. If the piece to be used is to the right of the slot, then the scribe line should be aligned with the right side of the slot; otherwise, it should be aligned to the left side.

After the sheet was aligned and bolted to the fixture, I **C**-clamped the entire assembly

Figure 19 — Enlarge the diameter of the plug's clamping jaws. [Barry Shackleford, W6YE, photo]

to a bench top and began to cut. Because the cutting occurs on a horizontal surface, debris accumulates along the cut site and needs to be manually cleared every 10 to 12 cutting cycles (more frequently for long cuts). I used an old toothbrush to remove the debris.

I have also used this fixture to cut circuit boards. When cutting circuit boards and brushing away debris, take care to minimize any airborne dust and wear a surgical-type face mask. — *73, Stan Ekiert, K3KKH,* **roadspyder@msn.com**

Weatherproofing Wood Components

I often use wood to fabricate various components because it's readily available, inexpensive, and easily machined. The only problem is that it doesn't stand up to the weather very well. I eventually came up with a weatherproofing technique for my wooden components.

I started by mixing 60-minute epoxy in equal parts, as I normally would. I then thinned the epoxy mixture 50/50 with acetone. The resulting mixture was very thin and watery. Next, I laid my parts out on wax paper and generously brushed the mixture over them. I allowed them to dry overnight, as the acetone slowed the curing process. I then flipped the parts over and repeated the process on the other side. Once dry, the parts were lightly sanded to smooth the wood. The part could then be left as is or painted.

The acetone allows the epoxy to wick into the fibers of the wood, allowing it to penetrate more deeply. Once the acetone evaporates, the epoxy cures normally. I find this protects wooden components better than fi-

Figure 20 — Use a vise and C-clamp to close the clamshell while screwing the halves together. [Barry Shackleford, W6YE, photo]

berglass resin because it does not trap moisture in the wood. I wouldn't recommend mixing amounts less than a few ounces at a time, as it becomes more difficult to get the proportions right.

I have used this weatherproofing technique on my wooden components for years in a harsh, salt-water marine environment with good results. — *73, Steve Appleton, K4PH,* **k4ph@windstream.net**

Installing Heavy-Duty Power Plugs

Heavy-duty power plugs, such as the Leviton 5256-VY, typically feature a clamping piece or clamshell closure to effect a strong grip on the attached power cord. The halves of the closure are held together by two screws threaded into the opposing piece, which is typically nylon or plastic. When installing a heavy-gauge power cord, achieving a complete closure by means of the two screws alone is likely to be problematic.

I have found a procedure that ensures an easy and neat installation of heavy-gauge power cords. First, measure the diameter of the power cord and select a drill bit of the next smallest diameter. Next, enlarge the diameter of the plug's clamping jaws with the selected bit. It's best to hold the plug closed in a vise while drilling, with caution not to drill past the clamping jaws (see **Figure 19**). After connecting the power cord wires to the plug blades, use a vise and possibly an additional clamp to close the clamshell tightly around the power cord sleeve while screwing the two halves together (see **Figure 20**). The result is a snug and secure fit to the power cord (see **Figure 21**). — *73, Barry Shackleford, W6YE,* **w6ye@arrl.net**

Figure 21 — The completed installation, snug and secure. [Barry Shackleford, W6YE, photo]

Getting a Grip

I often run into rusty screws and nuts that need to be removed. I've found it useful to keep a tube of valve-grinding compound handy. The compound is a paste with a type of sand mixed in, available from most auto parts stores. If you have a corroded Phillips-head screw, first use a fine pick to clean out as much rust as you can before you attempt to remove it. Fill the head of the screw with the compound (or put a dab on the tip of the screwdriver; see **Figure 22**), wiggle the tool into the head, and carefully attempt to remove it. The compound should give you a better grip and makes things much easier.

This also works for just about any type of fastener. Or, if you have a mast that keeps rotating in the wind, apply some of the compound to the mast and clamp. — *73, Bob Bastone, WC3O,* **radiofreebob@gmail.com**

Figure 22 — The author's mic hooked to a car vent using a bent binder clip. [Bill Kerber, WA2ILY, photo]

Figure 23 — Plastic takeout containers can hold small parts while working on projects. [Phil Karras, KE3FL, photo]

Handy Parts Holders

Takeout food often comes in small, clear plastic containers. I thought these would make great small parts holders for working on electronics repair projects. I used these containers in a repair/redesign of a power supply. One container held the screws and some other parts when I re-moved the case, while another held the parts taken off to remove the small circuit card (see **Figure 23**). By placing clear tops on the containers, I made sure the small parts wouldn't be knocked over and scattered by accident while working on the project. — *73, Phil Karras, KE3FL,* **ke3fl@yahoo.com**

Solderless Breadboards

You can purchase complete solderless breadboards starting at around $25 (see **Figure 24**), which go up in price as you add features like built-in power supplies, function generators, interfaces, displays, etc.

However, you can save a few dollars if you're willing to do a little construction on your own by buying and mounting solderless breadboard modules on a board of your choosing. My own homebrew version was mounted on a cutting board (see **Figure 25**). Although it doesn't have the dedicated dc supply connections that my other board does, it's easy to provide power to the supply rails directly. The solderless breadboard modules simply snap together and you can mount them to your board using screws, glue, or double-sided tape. You choose to make it as small or big as you want. One of the many sources for these modules is **amazon.com**. Just search for "solderless breadboard" and you'll find plenty to choose from.

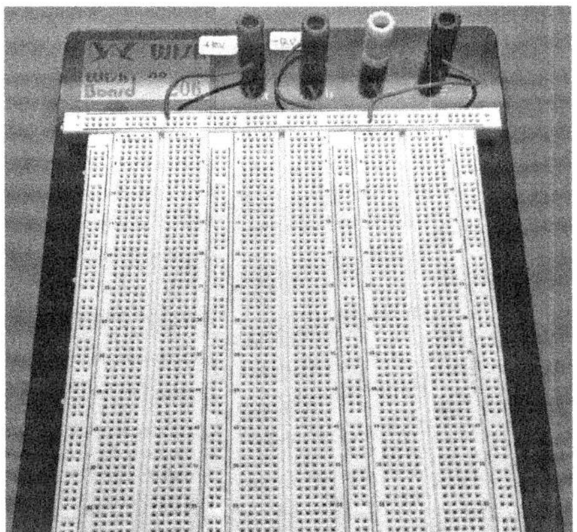

Figure 24 — One of the author's simple solderless breadboards. [Dino Papas, KLØS, photo]

Figure 25 — The author's homebrew version, made up of several breadboard modules mounted on a cutting board. [Dino Papas, KLØS, photo]

Along with mounting components such as transistors, integrated circuits, resistors, and so on, you can make circuit interconnections with your own wire or purchase commercial jumpers made especially for this purpose. Just make sure your homebrew wire jumpers fit snuggly into the holes.

External connections to your breadboard can be made in a few different ways. One method is to use pre-made breadboard connector adapters. They're available for lots of connector styles — including BNC and computer types — and plug directly into the breadboard. A few of the adapters I've used are shown in **Figure 26**. One place to purchase solderless breadboard adapter modules is **www.daqstuff.com/breadboards.htm**. RadioShack also sells several solderless breadboards and associated supplies. — *73, Dino Papas, KLØS,* **kl0s@cox.net**

Figure 26 — Different adapters used by the author. The connector shown in the middle of the top is a complete RS232 to TTL level adapter built right onto the small PC board, making circuit interfacing quick and easy. [Dino Papas, KLØS, photo]

Figure 27 — DuPont Nomex paper works as a chimney for Russian GI-7BT tubes. [Roger Gibson, K4KLK, photo]

Figure 28 — The new front plate mated with an old panel as a substrate. The handles are mounted through the back of the old panel, ensuring that both plates are electrically connected. [Robert Miller, KE6F, photo]

Figure 29 — A rear view of the high-voltage power supply panel showing the old holes. [Robert Miller, KE6F, photo]

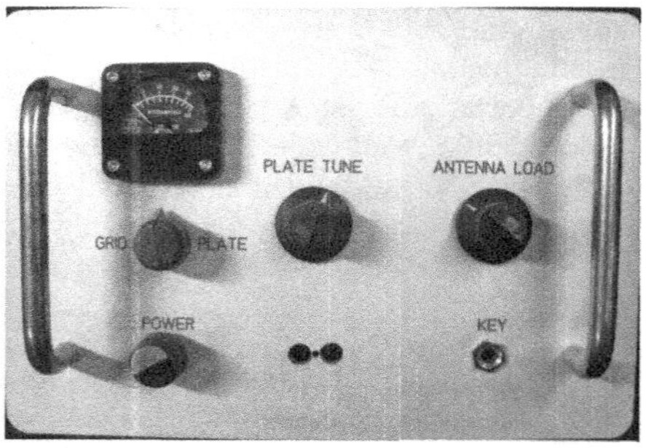

Figure 30 — A transmitter in a Heathkit electronic switch cabinet with a Front Panel Express face plate. [Robert Miller, KE6F, photo]

Russian Tube Chimneys

DuPont's Nomex paper is safe up to 200° C and makes excellent chimneys for Russian GI-7BTs and other tubes. I bought my paper at a motor shop, but you can also find it online. Your best bet is to get the thickest paper you can find. The paper I used was 10 mils.

I made a paper template before cutting the Nomex, then secured the edges with red RTV (room temperature vulcanizing) silicone. Small brass bolts work fine too. Sized properly, the ears on the GI-7BT heatsink keep the chimney in place. I've used the Nomex paper chimneys on four tubes for several years without a problem (see **Figure 27**). — *73, Roger Gibson, K4KLK,* **rogerk4klk@aol.com**

Reused Homes for Radio Projects

Ham radio experimenters often need cases for their projects. I like to use obsolete Heathkit test equipment cabinets. Color bar generators, TV sweep generators, TV alignment generators, and other pieces of vintage Heathkit items are perfect for hosting homebrew projects.

Most of these Heathkit items were housed in simple aluminum boxes with stamped front-panel assemblies. I've used them to hold several homebrew projects, including using a former Heathkit electronic switch to house a general-purpose high- and low-voltage bench supply; a 1,500 V dc 250 mA metered bench supply; a rubidium standard and power supply, and a 6AG7/6DQ5 CW transmitter.

Once I've removed the internal components from the Heathkit item, I fabricate a face plate to cover the holes left by the retired item's controls and connectors. Once I cut the plate material to size, I glue it to the original panel using E6000 industrial adhesive (see **Figure 28**). If you're worried about the aesthetics of holes being visible on the inner side of the original panel, the same treatment can be applied there (see **Figure 29**). I used 0.060-inch aluminum for this panel restoration.

I always include handles with mounting screws to ensure the new plate is electrically connected to the original panel. I only apply the adhesive around the edge of the panel to keep it in place during the initial fitting. If the small gap around the edge caused by the radius of the original panel is bothersome, a small treatment of body filler putty will take care of that. Just scrape and sand, then once you have drilled and punched your holes, coat the panel with spray paint.

The Heathkit cabinet scheme used the internal chassis as a fastening point to hold the front panel in place. If this scheme is not used in your application, you can apply #6 sheet metal tapping screws to the lip of the original panel to hold it in place.

Heavy projects — such as my 1,500 V bench supply — need structural reinforcement, because the flimsy metal of the case will distort with the heavy transformer load. To that end, I added an aluminum plate to the inside of the bottom of the case and another aluminum plate under the inside top of the cabinet, using the handle screws to fasten it. I have also housed a project in a Heathkit electronic switch cabinet, using a panel I had laid out and submitted to Front Panel Express (**www.frontpanelexpress.com**) for fabrication (see **Figure 30**). — *73, Robert Miller, KE6F,* **millerke6f@aol.com**

Quick Magnetic Pickup

It's always a nuisance when a tiny screw disappears that you know dropped onto the floor. Magnets are useful to help find and pick up these stray screws, and I've found that a small handheld magnetic pickup with release serves as a useful tool that can save a lot of effort in these instances. The one I use is item number 150152, available from Northern Tool + Equipment (**www.northerntool.com**). — *73, William Couloufacos, KC2RIM, 4233 192nd St., Flushing, NY 11358*

Mini Ground Plane for Solderless Breadboard HF Circuits

Contrary to popular opinion, solderless breadboards can be used at high frequencies — even at 28 MHz. **Figure 31** shows a solderless breadboard ½ W 10-meter beacon band transmitter. It is an updated version of a 25-year-old design (which I wrote about in the November 1994 "Hints & Kinks" column), and uses a single 74AC240 integrated circuit (IC) in a crystal oscillator and power amplifier configuration. I found that a four-legged mini ground plane made the difference between success and failure of this project.

The reason solderless breadboards have a bad reputation for high-frequency circuits is that the stray inductance caused by long lead lengths often results in VHF parasitic oscillation, low RF output, and crystal oscillator issues. A ground plane is the usual cure for these problems. Although a full-sized ground plane is impractical for a solderless breadboard, I found that a four-legged mini ground plane was easy to make, easy to use, and worked nearly as well as a full-sized one.

I used a 200-mil-wide, 10-mil-thick piece of tin-plated steel to make the mini ground plane. I soldered wire legs to each end. The four-legged mini ground plane sits in the middle of the solderless breadboard and underneath the RF-generating IC.

The four legs allow short ground connections to the rows next to the corners of the IC. A short 1⁄10-inch jumper connects the lower left leg to the IC's ground at Pin 10. The output filter is grounded on the rows connected to the two upper legs. The critical bypass capacitor is plugged in next to the upper-right-hand leg and the hole next to the IC's supply at Pin 20. To minimize output to oscillator coupling, I connected the oscillator grounds to the lower-right leg.

The result is a very robust ½ W, 10-meter CW transmitter. I have also used similar four-legged ground planes on successful 1 W solderless breadboards for 40, 30, and 20 meters. The only serious problem I encountered was a VHF parasitic oscillation in a 40-meter transmitter. That was traced to a sloppy 2-inch-long lead in series with the input of the logic gate used for the crystal oscillator. (A series 100 Ω resistor killed that parasite.) Otherwise, these solderless breadboard projects were trouble-free.

If you like to experiment or are looking for a simple club or school RF ham project, consider using a solderless breadboard with a four-legged mini ground plane. — *73, Lew Smith, N7KSB,* **evieandlewsmith@gmail.com**

Helpful Wooden Spacers

By setting my vise on a short piece of wood, the wood serves as a spacer and allows me to tighten or loosen the vise without having to slip the handle every time, so it won't hit the table or desktop (see **Figure 32**). Now I can continually spin the screw to tighten or loosen the vise. I've found that 7-inch-long, 2 × 3 inch spacers work well.

Also, when I was working on a power supply, I turned it on its side and found that my solder fume extractor was too low and would not suck in the solder fumes. I found a longer piece of 2 × 3 and cut two more of these spacers to hold the fume extractor at a better height to capture the solder fumes (see **Figure 33**). These blocks of wood continue tocome in handy. — *73, Phil Karras, KE3FL,* **ke3fl@yahoo.com**

Figure 32 — The vise with the spacer, which allows for adjustment without hitting the table. [Phil Karras, KE3FL, photo]

Figure 33 — The extractor at the new height for removing fumes from the circuit card soldering. [Phil Karras, KE3FL, photo]

The Machinist Approach to Surface-Mount Soldering

I combined a couple of my hobbies (or at least the tools from them) to simplify the task of soldering surface-mount components. I use machinist double-sided tape (it holds tight, but doesn't leave any residue when removed) and a small machinist surface gauge (see **Figure 34**) to hold the circuit board and surface-mount components in place for soldering.

The double-sided tape holds the project board in place while the gauge gives very precise adjustment and just the right amount of force to hold the components to the board while the solder is applied.

I position the tool over the part using the gross adjustment (side screw) and then use the fine adjustment (rear screw) to contact the part and just barely lift the front of the tool. The weight of the tool holds the part in place without bending any leads. The fine tip allows the soldering iron to get to all the leads on the part. — *73, Dave Penrose, K1DHP,* **k1dhp@arrl.net**

Caution with Homebrew Inductors

Like many who enjoy building their own equipment, I often need to wind various inductors. I usually wind coils with enameled wire from surplus stores or hamfests.

On one occasion, I had just completed winding a coil for a project and securing the windings with cyanoacrylate cement. But upon measuring the coil's inductance, I discovered that it was not at the value I had expected — not even close.

As I was removing the wire, I noticed the insulation was coming off in several places. I put some cement on a cotton swab and rubbed the swab on the wire. Turns out, the cement was dissolving the insulation on the wire. By using this particular cement, I was creating a coil with shorted turns.

If you're planning on using cement to hold coil windings, be sure to test it first with your chosen wire and make sure it will not dissolve the insulation. — *73, Chris Maukonen, WA4CM,* **wa4cm@arrl.net**.

Figure 31 — A spare four-legged mini ground plane (left) and a 28 MHz transmitter (right). One end of a four-legged mini ground plane can be seen. [Lew Smith, N7KSB, photo]

Figure 34 — A small machinist's surface gauge is great for holding down surface-mount devices while soldering. [Dave Penrose, K1DHP, photo]

Figure 35 — The author's project before applying the lettering sheet. [Roy Hansen, KL7GQ, photo]

Figure 36 — The project with the lettering sheet applied. [Roy Hansen, KL7GQ, photo]

Photo Paper for Front Panels

I used my computer to create lettering for the front panel of my latest homebrew project. Since my panel was already drilled (see **Figure 35**), I measured carefully and made several lettering sheets. Each lettering sheet was the size of the entire front panel, with the lettering correctly positioned on the paper according to my measurements. I used Microsoft *Visio* software to create the layout on the paper, but there is nearly identical, free software called *Draw* available in OpenOffice.

I sprayed the panel with a thick layer of lacquer and let it set. While it was setting, I printed the front panel lettering on glossy photo paper. Glossy photo paper doesn't soak through like regular printer paper (I tried regular paper first and it looked awful).

When the lacquer was sticky, I set the paper on the lacquered panel and tweaked it just a bit until it was properly aligned. I then let it set and trimmed it with an X-ACTO™ knife.

The glossy photo paper makes for a nice front panel (see **Figure 36**). I've struggled with press-on lettering in the past, so this was much better. — *73, Roy Hansen, KL7GQ,* **wizard@cvinternet.net**

Mounting Terminal Strips with a Unique Adhesive System

I have a 12 V dc power supply that powers my transceiver as well as several smaller accessories. This supply has two high-current binding posts dedicated to the transceiver and a pair of spring-loaded terminals for use with stripped wire meant to power accessories. Unfortunately, the spring-loaded terminals will hold only two pairs of wires, and because I have additional accessories that required dc power, I needed to provide additional 12 V dc connections.

My preference for an application like this is to use terminal strips and small crimped lugs. My plan was to run a pair of medium-gauge wires from the spring-loaded terminals on the power supply to a multi-position terminal strip that would provide additional dc feed points. It was important that the terminal strip be easily accessible without the need to drill mounting holes in equipment or resort to something unsightly or awkward. I also wanted to be able to easily remove or relocate the terminal strip as my needs changed.

My solution was to use one of the unique two-part adhesive strips supplied with the popular 3M "Damage-Free Poster Mounting Systems" sold in hardware stores. These fasteners are designed to be easily removed without damaging finished surfaces like painted or wallpapered walls. Key to the clever design is an integrated pull tab that, when stretched, will safely and cleanly release the adhesive's impressive grip. While normally sold packaged with mounting hooks, these adhesive strips can also be purchased separately.

As seen in **Figure 37**, I took an eight-section barrier-type terminal strip and jumpered four terminals of the strip together for the positive feed and the other four terminals together for the negative. I then mounted the terminal strip vertically to the side of the power supply using one of the adhesive strips and ran a pair of wires back to the spring-loaded terminals. This gave me a neat, accessible, and flexible way to connect up to four additional accessories.

Figure 37 — An eight-section barrier-type terminal strip mounted vertically to the side of the power supply using an adhesive strip. [Barry Green, W1JFK, photo]

Note that these strips come with one side labeled "wall side." To use, strip the release paper off the unmarked side and press it firmly against the back of the terminal strip, making sure that at least ½ inch of the release tab sticks out beyond the end of the terminal strip. Next, pull the release paper off the wall side of the adhesive strip and press it and the terminal strip securely against the mounting surface. Take care to position the terminal strip exactly where you want it, because once pressed into place, the only way to reposition it is by slowly pulling the release tab to free the terminal strip and starting over. Also, be sure to use good practices to ensure that appropriate wire sizes are used and that the overall current capacity of the power supply isn't exceeded.

Obviously, this system can be used to mount many other types of items such as small SDR modules, rotary antenna switches, and so on. Two adhesive strips can be used for larger accessories, even though the holding power of just one strip is pretty impressive. — *73, Barry Green, W1JFK,* **w1jfk@arrl.net**

Conquering the DIN Plug

Many hams want to use their transceiver accessory jacks for digital operating and other activities, but they find the prospect of wiring a DIN connector intimidating. My transceiver accessory port requires a 13-pin plug, and these are among the more challenging connectors you'll encounter. I started by ordering a few plugs from Digi-Key (**www.digikey.com/product-detail/en/cui-devices/SD-130/CP-1013-ND**) and immediately encountered an issue — trying to connect wires to a DIN connector that lacks solder cups at the ends of the pins.

In an attempt to achieve a secure connection while avoiding shorted pins, I decided to try attaching female connector sockets to the wires then pushing the sockets over the pins.

For the connector sockets, I used Molex

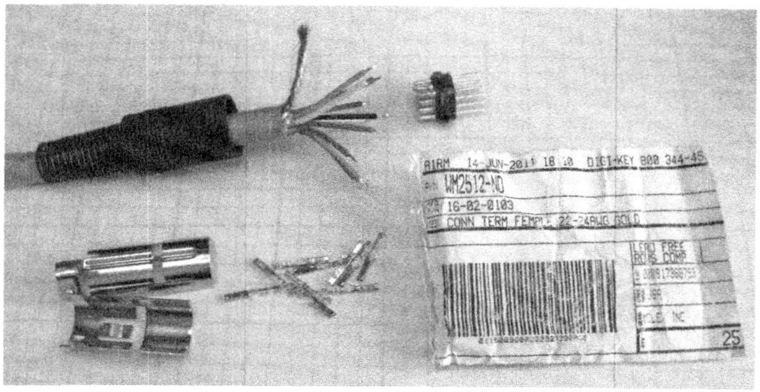

Figure 38 — The preparation of the cable. [Ken Williams, KC6PUQ, photo]

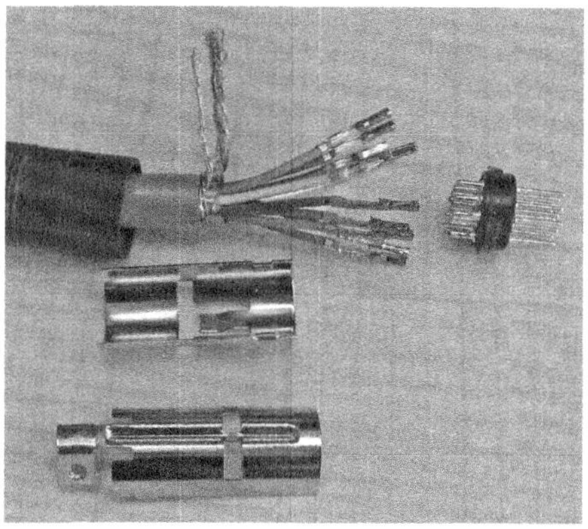

Figure 39 — The sockets attached to the wires. [Ken Williams, KC6PUQ, photo]

16-02-0103. These sockets are meant for smaller pins, but they can be made to go over the pins on the DIN connector. Sockets like these are very inexpensive, but specialty crimpers like the ones used for this project (**www.digikey.com/products/en?keywords=WM9017-ND**) are often more costly than one might want to spend (you can always try a different socket type that doesn't require this crimper).

The first step is to prepare the cable. Remove about an inch of the outer jacket, remove the foil (if any), and strip the individual wires in preparation for the sockets (see **Figure 38**). If the cable has a braid or a drain wire, make sure to leave it intact, as the metal shell of the connector often serves as a chassis ground and the braid or drain wire can be connected to the shell in order to preserve this ground connection. The amount of insulation that needs to be removed for the sockets is typically specified by the manufacturer (mine specified a strip length of 0.100 to 0.125 inches). Additionally, don't forget to put the shroud on the cable at this point in the process, especially if the other end of the cable is not available for later assembly.

After preparing the cable, attach the sockets to the stripped wires using an appropriate crimper (see **Figure 39**). This is probably the most important step in the process. Double-check your work when finished.

The next step is to put some heat-shrink tubing over each of the pins, as shown in

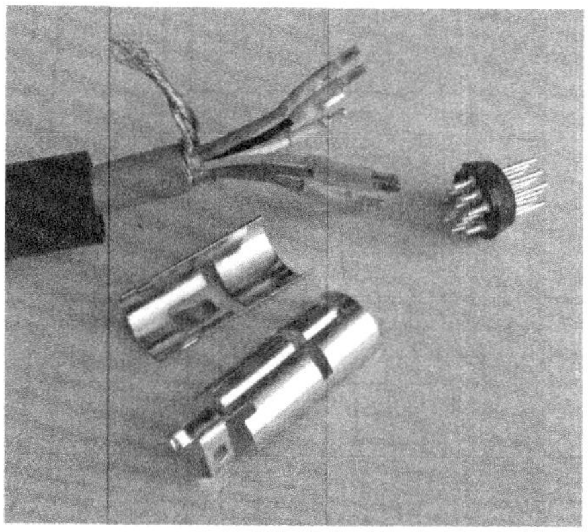

Figure 40 — Heat-shrink tubing added to the pins. [Ken Williams, KC6PUQ, photo]

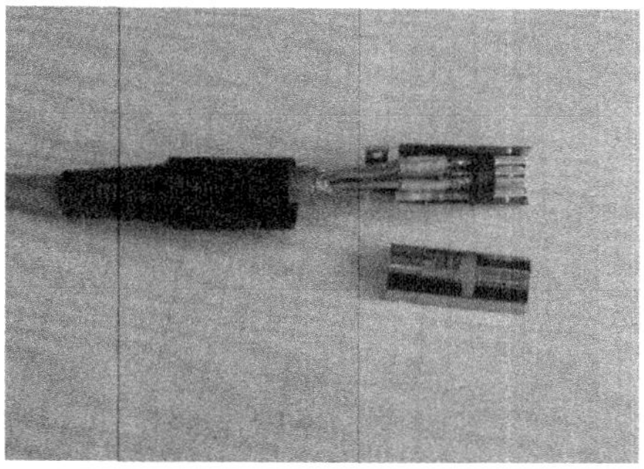

Figure 41 — The sockets attached to the DIN connector and ready for closure. [Ken Williams, KC6PUQ, photo]

Figure 40. The tubing should completely cover the ends of the sockets, where the wire is attached. The heat-shrink tubing will add some mechanical strength to the socket-wire connection. Also, these sockets are not meant for this application, so the sockets will touch each other when the connector is all wrapped up and the tubing will prevent short circuits.

Finally, push the sockets over the ends of the DIN pins and solder the pins. The solder is primarily for mech-anical strength, so don't worry if you're unable to access the inner pins of the connector. The final cable-connector assembly ready for closure is shown in **Figure 41**. — *73, Ken Williams, KC6PUQ,* **ken@williamsclan.us**

Polarity Protection

If you plug in and unplug equipment often enough from a power supply (like when transferring base station equipment to the car, Field Day, or a vacation setup), eventually some piece of equipment will get plugged in with the wrong (aka reverse) polarity.

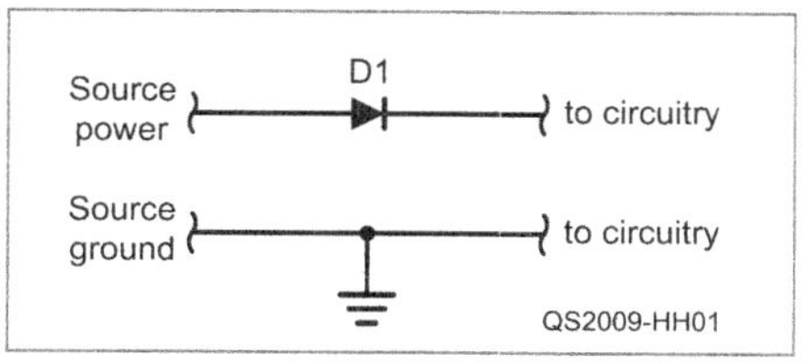

Figure 42 — Some reverse-polarity protection circuits use a diode in series with the power line.

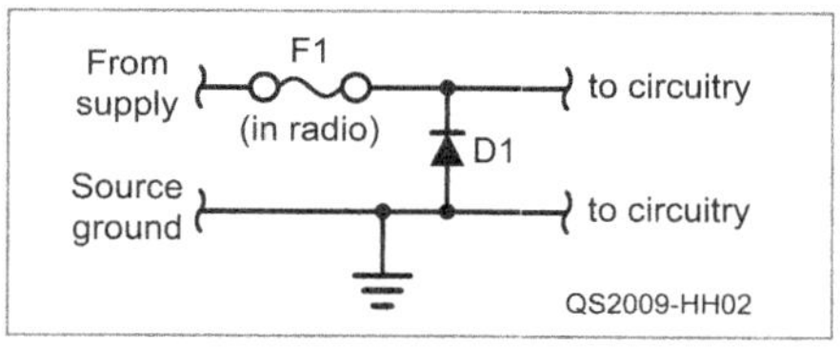

Figure 43 — Another protection approach uses a diode across the supply lines, along with a fuse.

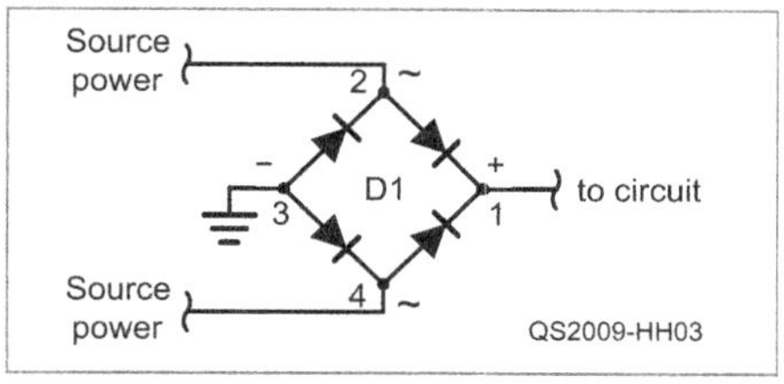

Figure 44 — By using a bridge rectifier, a polarity reversal at the power supply still results in the correct voltage polarity applied to the radio.

In some cases, the manufacturer has added a bit of protection right inside the equipment with a diode in series with the power line (see **Figure 42**) or across the supply lines (as shown in **Figure 43**) to keep it from going up in smoke. The circuit in Figure 42 will make the radio appear "dead" with no further explanation, while the circuit in Figure 43 will cause the radio's fuse to blow. Neither of these techniques are done universally, and if a schematic from the radio is not available, you're pretty much relying on luck that protective circuitry was built into your equipment.

While the standard approaches are the easiest and least expensive ways to add protection, they fall a bit short. If a device doesn't power up due to power cables being plugged into the supply with reverse polarity and we don't realize what happened, we lose valuable time looking for the problem.

If we build our own protection circuit, we can go a step further. By using a bridge rectifier wired as shown in **Figure 44**, a polarity reversal at the power supply still results in the correct voltage polarity applied to the radio. The trick here is that this bridge rectifier should be installed directly and permanently into a radio or circuit, so if wires get mixed up it won't be between the protection circuit and downstream circuitry.

You also need to consider the current draw of what you're powering. For example, what's the maximum current that will pass through the protection diodes? In the case of a 100 W transceiver, we could be looking at as much as 25 A in full-output, key-down mode. Even with low-power circuits, there is still the inrush current to consider and this current can be higher than the operating current. So whatever approach you use, make sure the diodes are selected accordingly, and even attached to heatsinks if necessary. — *73, Klaus Spies, WB9YBM,* **wb9ybm1@yahoo.com**

A Relocatable Power Tap (RPT)

I've been thinking about ways to manage ac power distribution for my desktop amplifiers. Until recently, I selected an amplifier by manually switching power plugs under the operating table. One branch circuit was used, which consisted of a 30 A circuit breaker and a run of four-conductor #10 AWG electrical cable to a 30 A twist-lock wall-mounted receptacle. I use four-wire cabling because my older Alpha amplifiers use a neutral for the blower. The #10 AWG cabling to my shack helps to maintain stiff power supply regulation and reduce voltage drop losses over an 80-foot run. To start simultaneously feeding ac power into three amplifiers, I decided to construct a relocatable power tap (RPT), which is a 240/120 V version of a multiple outlet strip (see **Figure 45**).

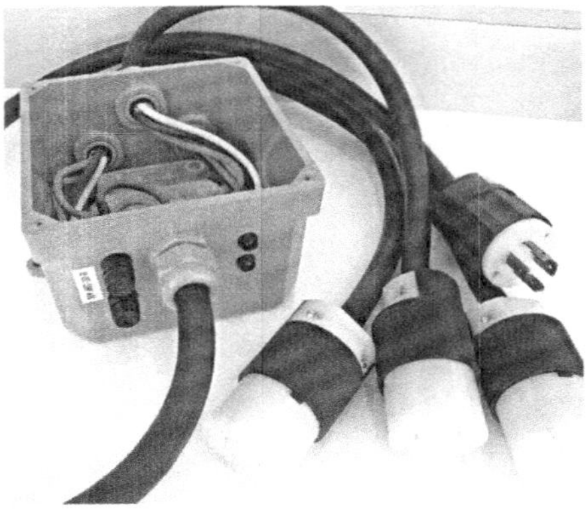

Figure 45 — This remote power tap, like an outlet strip, can provide 240 V power hookups to several amplifiers. [Paul Christensen, W9AC, photo]

[Note: Safety first! This project is intended for use as part of a high-voltage, high-current power circuit. While the wiring is simple and straightforward, if you haven't had much experience with high-voltage wiring, find a mentor to help, or better yet, hire a professional electrician. — *Ed.*]

I wanted to come as close to being 100% compliant as possible for the relevant UL standard. UL 1363 had a significant limitation: the RPT was limited to 20 A service and all device connectors must be of the same type and current rating. Because the RPT was limited to 20 A, then so were the connectors. Consulting NEC guidelines, the plug was limited in size to 20 A, so the mating premise receptacle must be 20 A. Finally, NEC 210.21(B) in this application stated that the serving circuit breaker must match the receptacle rating. Use of the 20 A RPT forced a redesign of the branch circuit from 30 A to 20 A components, although the wiring could remain. UL 1363 limited the length of the supply-side cord to no more than 25 feet and no less than 1½ feet.

Figure 46 — A close-up view of one of the bus bars with the cover removed. The copper bus bars are rated at 130 A and provide a safe connection point for the four conductors. [Paul Christensen, W9AC, photo]

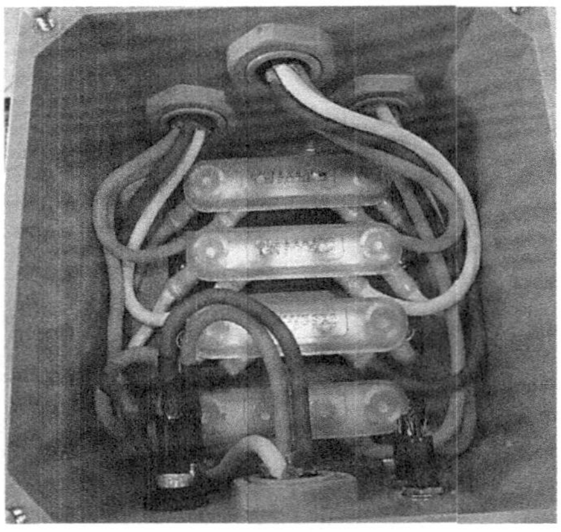

Figure 47 — The interior of the RPT box. The ac mains come in at the bottom, pass through the fuses on the left, and are then connected to the bus bars for distribution to the three output connectors exiting at the top. [Paul Christensen, W9AC, photo]

I started my design with 130 A copper bus bars that distribute service into three outputs (see **Figures 46** and **47**). I chose to use pigtails to install panel receptacles, because they allowed the use of different styles of 20 A plugs/receptacles, provided they remained the same type.

A summary of the RPT features includes: 1 × 3 device, designed for either three or four wire (240/120 V); #10 AWG wiring and hardware; Heyco compression cable entry/exit connectors; nickel-plated copper bus bars (with protective covers) rated at 300 V ac and 130 A; dual fuse holders with 20 A, ABC ceramic fuses; dual neon light ac power indicators, and 20 A Hubbell twist-lock connectors used on all cable ends.

If you want to add secondary-type surge protection devices (SPD) inside the RPT, their installation and type must comply with UL 1449, *Surge Protective Devices*. Place SPDs only between Line 1 and Line 2, and one each from the line to neutral — with no SPDs connected to the grounding conductor. Even then, I wouldn't use secondary SPD protection inside the RPT unless "whole-house" Type 1 or Type 2 protection is first used at the service entrance as the primary SPD. This is another reason why adding a neutral may make sense even if your present amplifier only requires three-wire service and no neutral. Adding the neutral also ensures that you can use older Alpha and Henry amplifiers in the future. — *73, Paul Christensen, W9AC,* **w9ac@arrl.net**

Crimping Caution

When I was on active duty in the US Air Force many years ago, we did an analysis of a failed power circuit. Our conclusion substantiated published information available at the time, which was: don't crimp a solder-covered wire to a terminal. Instead, use the correct type and size crimper, the correct size terminal for the wire, and crimp properly. Next, do a pull-out test and *then* solder the wire, as crimped, to the terminal.

If you solder before you crimp, the solder may heat under high load currents, causing the connection to loosen. That will lead to failure. Adding the solder as the last step reduces corrosion possibilities and adds mechanical strength to the connection. Clean terminals and wire are required, of course. — *73, Gary Charniga, ADØPE,* **gcharniga@centurytl.net**

A Breadboard for Vacuum Tube Projects

Vacuum tube technology may be more than 100 years old, but designing and building with tubes is still fun. Tube circuits can accomplish a lot with a small number of parts, and they produce that satisfying orange glow.

A major step in any project is breadboarding: knocking together a temporary

Figure 48 — The vacuum tube breadboard. [Bryant Julstrom, KCØZNG, photo]

implementation that's easy to change and adjust. Breadboarding is easy with semiconductor circuits when you use breadboards designed for that purpose. Breadboarding tube projects is more difficult. It generally involves soldering parts. In this project, I adapted a semiconductor breadboard to developing tube-based circuits.

The unit is shown in **Figure 48.** It holds several tube sockets and a semiconductor breadboard. Its back panel holds a connector (a four-pin Jones Plug in this case) for filament voltage, B+, and ground. The latter two connect directly to the breadboard. Connections from the sockets' pins are brought to SIP sockets on a length of protoboard. These enable connections from tubes to components on the breadboard.

Exceptions are the pins that connect to tubes' filaments. This unit is de-signed for seven- and nine-pin miniature tubes with 6.3 V filaments. In seven-pin tubes, Pins 3 and 4 connect to the filament, and in nine-pin tubes, Pins 4 and 5. These pins are connected to the power plug on the back panel.

Additionally, the unit supports nine-pin dual triodes like the 12AU7, 12AT7, and 12AX7. In these tubes, the 6.3 V filaments of the two triodes are connected in series between Pins 4 and 5, for a total of 12.6 V, and Pin 9 is connected between the two filaments, so that they can be operated in parallel at 6.3 V.

Associated with each nine-pin socket, then, is a DPDT switch that toggles between two arrangements. In one position, Pins 4 and 5 lie across the filament lines, for nine-pin tubes that require 6.3 V, like the 6U8 and 6T9. The other position connects Pins 4 and 5 together and Pin 9 to the other filament line, to power 12.6 V dual triodes. **Figure 49** shows the filament circuit. The three bolts at the unit's front allow temporary panels to hold controls and larger components.

Tube B+ voltages are high compared with the semiconductor voltages we're more accustomed to, so exercise caution. In addition, the semiconductor breadboard is not designed for high voltages, so keep the B+ low. — *73, Bryant Julstrom, KCØZNG,* **kc0zng@arrl.net**

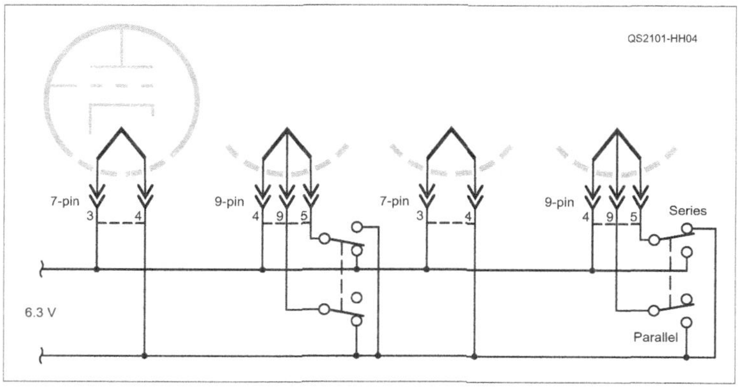

Figure 49 — A schematic diagram of the breadboard's filament connections.

A Spade Bolt Alternative

I was recently working on a restoration project that required some spade bolts to attach a coil shield to a chassis. Some people will remember them as the attachment bolts for IF and RF cans in old radios. These once common attachments are virtually unobtainable, and the price is through the roof for the few available, so I decided to make some myself.

I used a ¾-inch copper pipe cap for the shield and made the spade bolts using a wire terminal connector and a number-6 machine bolt. The head of the machine bolt needs to be filed down on one side to slip inside the terminal connector and lay flat on the side that's adjacent to the pipe cap (see **Figure 50**). I used a small sheet metal screw to attach them to the shield can, but they can be soldered or riveted. They function exactly like the old spade bolts. — *73, Michael Janis, KE3OQ,* **mjanis@ptd.net**

Figure 50 — A ¾-inch copper pipe cap serves as the shield and the spade bolts are made from wire terminal connectors and number-6 machine bolts. [Michael Janis, KE3OQ, photo]

Easier PL-259s

When installing PL-259 connectors on coaxial cable, it's often difficult to screw the shell onto the outer jacket. Next time, try wrapping Teflon tape on the cable first. Not only does this make putting the connector on easier, it also helps seal the connector against moisture. I find that it takes less than half the effort to install the connectors this way. — *73, Michael Melnyk, Jr., KD4KWP (SK)*

Center Pin Stabilizer

Pre-tinning the center pins of coaxial connectors can be a hassle. Instead, next time you open a bottle of wine, save the cork and cut it to a stable size. Insert the center pin and you have a stable, and somewhat insulated, cork base to perform any pre-soldering activities (see **Figure 51**). When attaching to the coax itself, the cork does not pull heat away from the pin like metallic pliers, so soldering takes less heat, which can also prevent melting of the heat-sensitive dielectric material. — *73, Al Rabassa, NW2M,* **nw2m@arrl.net**

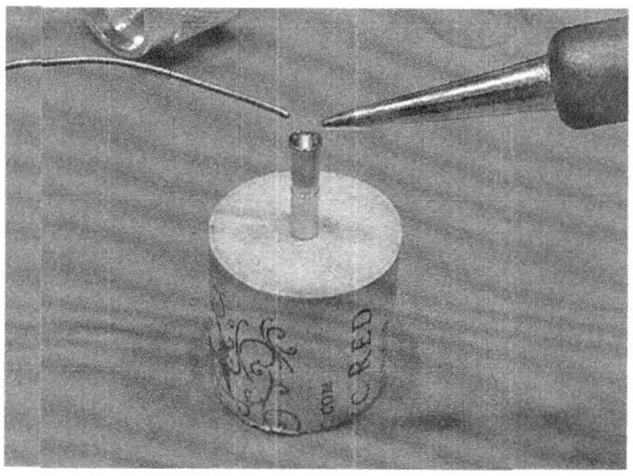

Figure 51 — Insert the center pin and you have a stable cork base to perform any pre-soldering activities. [Al Rabassa, NW2M, photo]

Make Tool Markings Easier to Read

Sometimes I need to use a metal tool with stamped-in titles, numbers, or settings that are difficult to read. I found a neat way to make those characters much easier to see.

Wite Out® is a correction fluid product that has been used for many years to cover errors on printed material, especially those made on manual or electric typewriters. As seen in **Figure 52**, the stamped-in numbers on my wire-stripping tool are hard to read on the tool's left arm. I tried to bring them into better view with the correction fluid on the tool's right arm.

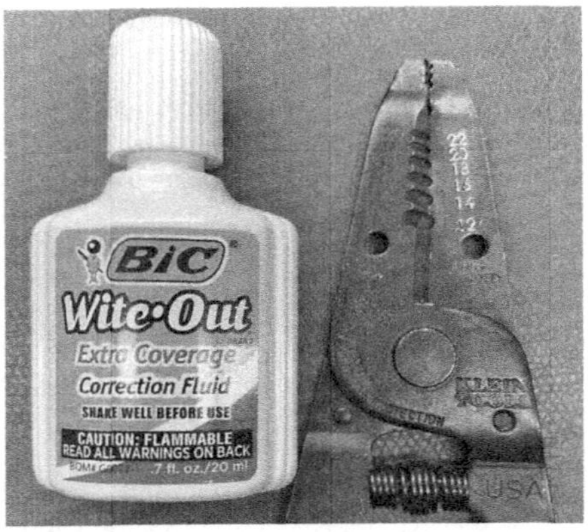

Figure 52 — The stamped-in numbers on this wire-stripping tool are difficult to read on the tool's left arm. The author attempted to enhance them with correction fluid on the tool's right arm. [Dino Papas, KLØS, photo]

Figure 52 — An application of correction fluid really makes the numbers "pop" on both arms of the tool. [Dino Papas, KLØS, photo]

To do this, I applied a small amount of the liquid across the characters with the bottle's brush and immediately wiped the area with a lint-free tissue to remove the excess. The material dries very quickly when you do this, so the result may not be perfectly clean around the imprinting, but you'll be able to read the text. You can remove most of the excess around the characters with a cotton swab dipped in isopropyl alcohol or with a sharp pick.

Now all the wire gauge markings are much easier to see (see **Figure 53**). — *73, Dino Papas, KLØS,* **kl0s@cox.net**

Magnetic Parts Keeper

When working on radios, there are many covers and enclosures that must be removed to gain access to various circuits. Some are secured with machine screws, while others use sheet metal screws. Still others use various nuts, bolts, and washers. When you start to reassemble the radio, perhaps months later, it's difficult to remember which screw, nut, or bolt goes where.

I have used low-cost, low-strength magnets with great success in this regard. A single magnet is used to hold the fasteners to a specific internal cover, while non-magnetic screws (stainless or brass) can be placed in a metal jar lid with a magnet. The magnet is strong enough to hold the lid to the cover, so there's no question as to what parts go where. The magnets shown in **Figure 54** were sold in units of 10 for $1 at a local bargain store. — *73, Al Rabassa, NW2M,* **arabassa@comcast.net**

Beware of 2N2222 Variants

I recently learned a surprising fact about the popular 2N2222 transistors: their pinouts are not standardized the way most other transistor pinouts are. I discovered this when a simple amplifier I built refused to amplify, but the transistor got very hot. A check of the device with a transistor tester and some online research provided an explanation.

It appears there are two variants of the 2N2222 in TO-92 small plastic packages. One is marked "PN2222" and has the common E-B-C pinout. The other is marked "P2N2222" and has its pins in reverse order: C-B-E.

Not all suppliers are aware of this issue, so you may find yourself buying a package of P2N2222 transistors while the seller's website shows the PN2222 datasheet, which will give you the wrong pinout. — *73, Paul Koning, NI1D,* **ni1d@arrl.net**

Reusing Hard Drive Magnets

I acquire old computers, refurbish them, and then donate them to various local charities. I often come across computers that are too old to be of use. However, every computer hard drive, whether it's a laptop or desktop, has two very strong magnets operating the actuator and head arms. I take the obsolete computer's hard drive, remove the

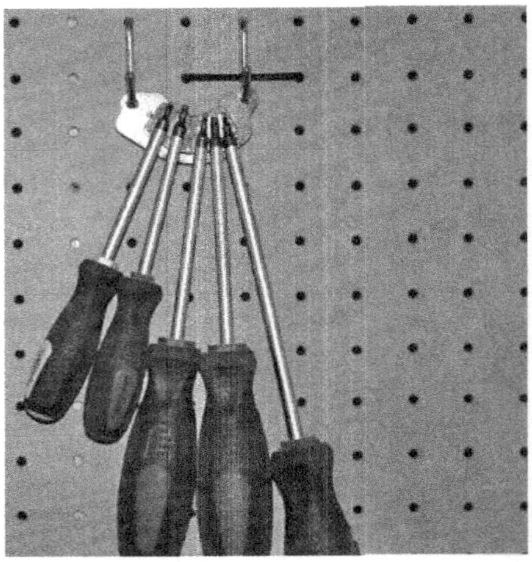

Figure 55 — Hard drive magnets repurposed as screwdriver hangers. [Mike Joyce, VE3LTN, photo]

Figure 54 — During disassembly, an inexpensive magnet can keep fasteners attached to the pieces they belong to. [Al Rabassa, NW2M, photo]

magnets, and mount them around my work area so I can place my tools on them for ease of access while working on projects, thus eliminating clutter on the bench. For example, I have mounted some on the pegboard where I keep my screwdrivers (see **Figure 55**).

To retrieve these magnets, keep in mind that every hard drive case opens differently. Torx, Phillips, or Robertson fasteners are all used. In some cases, you may have to drill out punched rivets. Once opened, you will see the magnets, with the actuator head hidden beneath the upper magnet. Simply remove the screws holding them and remove the magnets.

The next time you come across an obsolete hard drive, do some at-home recycling and get a neater workshop in return. — *73, Mike Joyce, VE3LTN,* **ve3ltn@gmail.com**

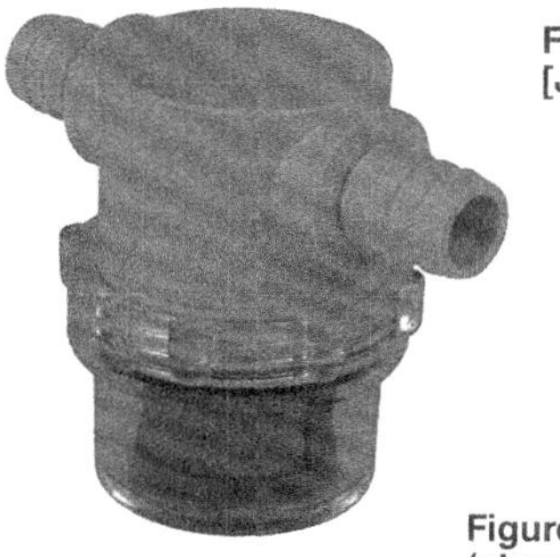

Figure 57 — A $10 solder trap. [Jeff Whalin, KC9JOB, photo]

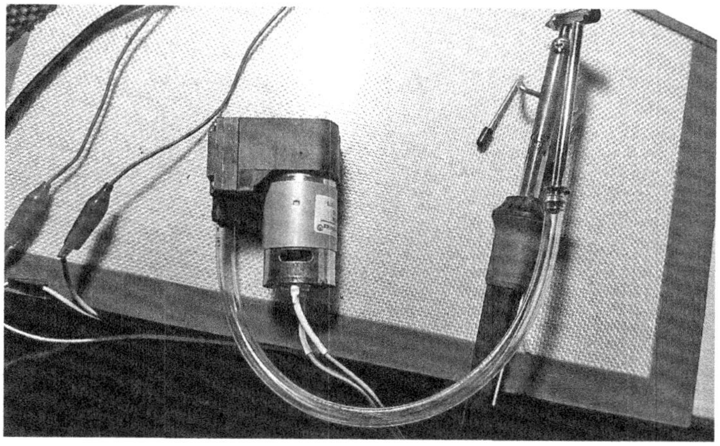

Figure 58 — The finished desoldering tool (shown without the solder trap). [Jeff Whalin, KC9JOB, photo]

Upgrading a Simple Desoldering Tool

Most inexpensive desoldering tools that use squeeze bulb suction are rather anemic, not to mention a bit awkward to use. The old, larger solder pumps are usually too big for modern printed circuit board work, and professional desoldering stations cost hundreds of dollars.

Here's a simple $35 modification you can make to an inexpensive combination tool or bulb-type solder sucker to remove solder like a pro. It's as easy as replacing the bulb with a small but powerful 12 V vacuum pump operated by a footswitch. The pump is available from sources like Amazon for about $25 (see **Figure 56**). You'll also want a solder trap, which can be purchased on Amazon for about $10 (see **Figure 57**).

To make this modification, wire the pump's 12 V dc line to a footswitch (used models can be found for less than $20), or any type of switch that you prefer (see **Figure 58**). Remove the bulb and attach tubing between its attachment point and the solder trap. Then use another short tube between the solder trap and the pump, and you're done. — *73, Jeff Whalin, KC9JOB (submitted by Fred Glenn, K9SO,* ***wa9mvz@yahoo.com****)*

Figure 56 — A 12 V vacuum pump available online for $25. [Jeff Whalin, KC9JOB, photo]

An Ad-Hoc Heatsink

I was trying to test parts for an analog adjustable-voltage 5 A power supply using an LM338 voltage regulator. Unfortunately, I discovered that the two different types of LM338 packages — TO-3 and TO-220 — did not work as they should. The TO-3 case-type adjusted voltage just fine, but wouldn't deliver more than 100 mA of current, even when using a proper heatsink. The TO-220 package would run at 7 A, but the voltage could not be adjusted.

I needed to quickly test both components without the laborious process of attaching and removing them from sizeable heatsinks. The solution was as simple as clamping a pair of locking pliers onto the regulators. The pliers have enough surface area to safely dissipate heat for brief tests.

Figure 59 shows two LM338 TO-220 regulators, one in the locking pliers, connected for testing. The other LM338 is there to show how the leads were bent to make it easy and safer for clip-on connections. Bending the pins in this way makes it less likely for any clip-on connection to short out with another. — *73, Phil Karras, KE3FL,* **ke3fl@arrl.net**

Budget Spacers

I needed 16 spacers to mount four small circuit boards onto a sub-chassis. At the local home improvement store, I found aluminum spacers, about 8 millimeters long, but the price of a single spacer was nearly $1.

Because part of being a ham is always looking for a cheaper alternative, I visited a hobby store. There I discovered all sorts of inexpensive plastic beads. The ones

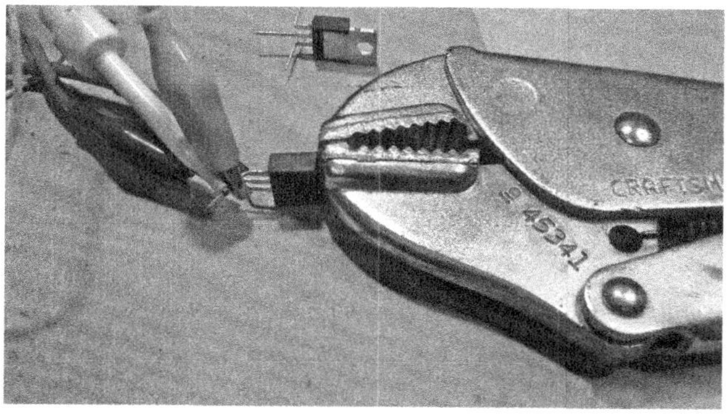

Figure 59 — A pair of locking pliers can be pressed into service as a heatsink. [Phil Karras, KE3FL, photo]

Figure 60 — An inexpensive assortment of colorful plastic beads makes a great substitute for circuit board spacers. [Ron Rausch, K5HZ, photo]

Figure 62 — An old PC case repurposed as an enclosure for the author's homebrew amplifier. [Don Opdycke, KT3V, photo]

Figure 63 — PC power supply cases put to use as enclosures for a remote control for an antenna switch, an antenna rotator control, and workbench power supplies. [Don Opdycke, KT3V, photo]

I picked out measure approximately 6 × 7 millimeters and have a hole size that permits a 6-32 machine screw to pass through it with ease (see **Figure 60**). The best part, a bag of 500 beads only costs $1.99, or less than 0.004 cents each. I now have enough spacers to last a lifetime. — *73, Ron Rausch, K5HZ,* **wn0mqx@gmail.com**

A Zero-Dollar Insulated Shaft

If you're ever in need of a plastic shaft for a homebrew project such as a receiver, transceiver, or antenna tuner, you can make one in short order from a plastic clothes hanger (see **Figure 61**) for almost no cost. A single hanger can provide two shorter lengths and one longer. Simply cut the necessary pieces to length and then grind the ends to fit whatever shaft couplers you are using. — *73, Edward Barbacow, K3ZCY,* **k3zcy@zoominternet.net**.

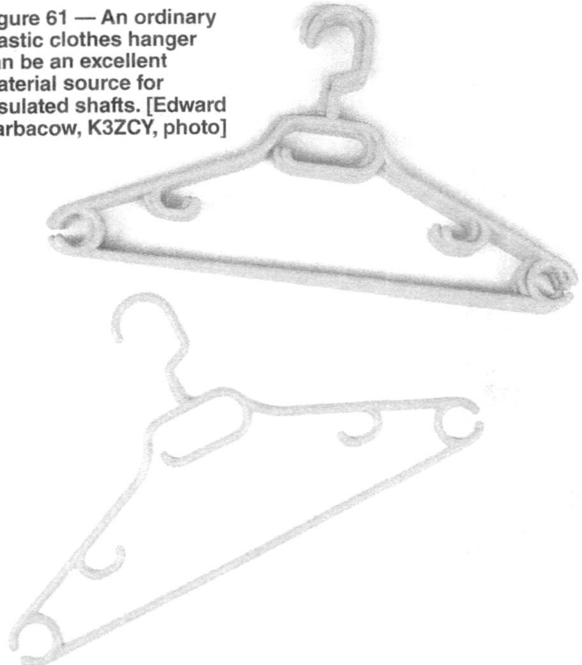

Figure 61 — An ordinary plastic clothes hanger can be an excellent material source for insulated shafts. [Edward Barbacow, K3ZCY, photo]

Old PC Cases Revisited

I recall a past *QST* article regarding the use of old computer cases for project boxes. It showed a linear amplifier configured in a PC tower or mini-tower case. Years ago, I put a pair of 4X150s with a Pi-El output, power supplies (2 kV, 300 V, –50 V), fans and blowers, in an old ITT XTRA PC desktop case. It still serves as my station's backup and medium power amp, delivering 400 – 500 W. **Figure 62** shows the interior of the amplifier. It was a very tight fit, a mechanical challenge, and a rewarding result.

I would like to extend the idea of using various PC parts as chassis boxes in building projects. PC power supply cases make excellent containers for many projects. **Figure 63** shows several examples, including a remote control for an antenna switch, antenna rotator control, and two bench power supplies.

The power supply cooling fan and power connector may be left in place if needed. The line voltage select switch can be replaced with an on/off slide switch. It is sometimes necessary to cut out grill openings and re-panel the front and rear of the chassis with aluminum sheet or Lucite® plastic for mounting of connectors and controls. Rubber feet can be added to the bottom of the case to keep it from sliding around and scratching a desk surface.

If the PC power supply is still in work-ing order, it can make a useful bench power supply. It is usually necessary to provide a light load of about 10% on the +12 and +5 supplies to maintain proper regulation of the outputs.

CD and DVD cases can also become shielded containers for items such as circuit boards. **Figure 64** shows one such CD case with the CD components removed. The front and rear of the case have been closed with aluminum panels so that connectors and controls can be mounted.

Figure 64 — A CD/DVD case with the components removed. [Don Opdycke, KT3V, photo]

That old PC that you were about to toss can live again with good metal (or plastic) working skills, imagination, and patience. — *73, Don Opdycke, KT3V,* **don.opdycke@gmail.com**

Painless Project Boxes

A quick and painless way to box up projects, especially projects involving LCD displays, is to squeeze four small dabs of Loctite® Super Glue on the corners, and glue the LCD display onto the inside of a transparent acrylic box.

In **Figure 65**, from left to right, there is a 0.1 to 500 MHz power meter (using an AD8307 module), a 0.2 to 40 MHz signal source (using an AD9850 module), and a position-readout/motor-driver control box for a 2.4-meter moonbounce dish. Each project incorporates a I2C 16 × 2 LCD panel

Figure 65 — From left to right: A 0.1 to 500 MHz power meter, a 0.2 to 40 MHz signal source, and a position-readout/motor-driver control box for a 2.4-meter moonbounce dish. [Chuck MacCluer, W8MQW, photo]

Figure 66 — An oversized hole in the aluminum box. [Michael A. Czuhajewski, WA8MCQ, photo]

Figure 67 — The oversized hole modified with an aluminum sheet. The sheet is drilled to fit the BNC connector. [Michael A. Czuhajewski, WA8MCQ, photo]

driven by an Arduino Nano.

The high-quality, thick-walled boxes (found at outlets such as The Container Store® at **www.containerstore.com**) are effortless to drill. Batteries and Arduinos can be fastened to the inside with hook-and-loop strips, while other parts can be glued in. Electrical interconnections are made with 4-inch breadboard jumper wires. This way, 10-hour projects become 1-hour projects! *— 73, Chuck MacCluer, W8MQW;* **w8mqw@arrl.net**

Mounting BNC Connectors in Oversized Holes

I wanted to put a pair of single-hole-mount female BNC connectors in a recycled die-cast aluminum box. There were holes in the desired spots already, but they were too large (see **Figure 66**). I cut pieces of aluminum sheet to cover them and drilled holes of the proper size for the connectors (see **Figure 67**). Then I mounted the plates on the outside. If mounted on the inside, the connector would be recessed and mating to it could be difficult. I purchased the sheet at a local home improvement store. Thickness is not critical. *— 73, Michael A. Czuhajewski, WA8MCQ,* **wa8mcq@verizon.net**

Accurate Pipe Drilling

A time may come when you'll need to drill precise, aligned holes in PVC or metal pipe; you might, for example, need them positioned at 90 degrees (quadrature). This can be a challenging exercise, but I've found a simple, low-cost method.

First, cut a 3-inch strip of paper and wrap it around the pipe. With a writing utensil, mark the paper where it overlaps, and cut it again so that its ends meet perfectly when wrapped around the pipe. For simple, straight-across holes, fold the paper in half to make a crease. Mark the top edge of the crease with a felt-tipped pen (see **Figure 68**). Next, wrap the marked paper around the pipe again, and mark the center of the first hole at the butted ends (see **Figure 69**). Then, mark the opposing hole at the center of the crease on the other side.

If you require four holes at quadrature, simply fold the paper in half again and mark the new creases. For other angles, divide all 360 degrees of a single wrap of paper by the needed marks. I find the fine divisions on a metric millimeter ruler to be helpful with this.

Center punch the marks and drill small guide holes to center the drill bit during the final drilling stage (see **Figure 70**). After drilling the guide holes, you can use a centering guide block for pipe drilling to make perpendicular holes. Inexpensive pipe drilling blocks are available on the internet and in the tool section of most hardware stores. *— 73, John Portune, W6NBC,* **w6nbc-mail@gmail.com**

Figure 68 — Marks have been drawn along the creases in the paper strip. [John Portune, W6NBC, photo]

Figure 69 — After using the paper creases as a guide, the pipe has been marked for 90-degree holes. [John Portune, W6NBC, photo]

Figure 70 — Four 6-32 screws drilled into the pipe at 90 degrees (quadrature). [John Portune, W6NBC, photo]

CHAPTER 5

Equipment Tips and Mods

GPS Interface for the ID-880H

The Icom ID-880H is a popular mobile rig for D-STAR digital voice mode operation. With an external GPS module (a combination of the GPS antenna and receiver in one package) connected, the ID-880H radio can also provide GPS reporting. This item describes a relatively simple GPS interface you can make in a few hours, which is powered by a USB outlet and doesn't require any special cables. Note that I chose to hardwire the GPS cable to the adapter and use interface connectors and plugs for the other two cables.

GPS Requirements and Connection

Icom states that any external GPS module used must have an RS-232C output, in the NMEA data format, and be connected to the radio's data jack. An "optional" data communication cable, OPC-1529R, is available from Icom to make the connection, with the proviso that a null-modem adapter also be used. As noted in the *QST* review of the ID-880H, newer USB-only GPS modules are not compatible with the ID-880H and cannot be used.[1]

GPS and Mounting

I used the Garmin GPS 18X high-sensitivity module. It has a magnetic base for easy mounting, and it comes with a 5-meter cable with six wire connections. I saw no reason to shorten the cable, but rolled up and tied off what I didn't need. For mounting using the magnetic base, I used a 1½-inch-diameter plated steel fender washer mounted with hook-and-loop fastener tape on the dash of my car, close to the front windshield, for a good view of the sky. The GPS 18X requires 4.0 – 5.5 V dc at less than 100 mA, which is easily supplied from the vehicle USB connector or a USB cigarette lighter plug.

Data Jack

The data jack is located on the rear panel of the ID-880H (J5 on the ID-880H schematic), and is an ordinary 2.5-millimeter stereo audio-type jack with three connections: ground, RX data (ring), and TX data (tip). (Note: the Icom schematic appears to have the RX and TX connections to J5 reversed, depending on your interpretation of the jack symbol.) The data jack is connected to a MAX3221 RS-232C transceiver IC, which requires RS-232C RX and TX signal levels (3 to 12 V space, –3 to –12 V mark) for the GPS icon to appear on the ID-880H display.

I used a 1 A fuse, as recommended by Garmin, which can be either on or off the circuit board, as best suits your assembly. Capacitors C1 and C2 are there as "good practice" to decouple any conducted noise on the +5 V supply line. The 3-millimeter

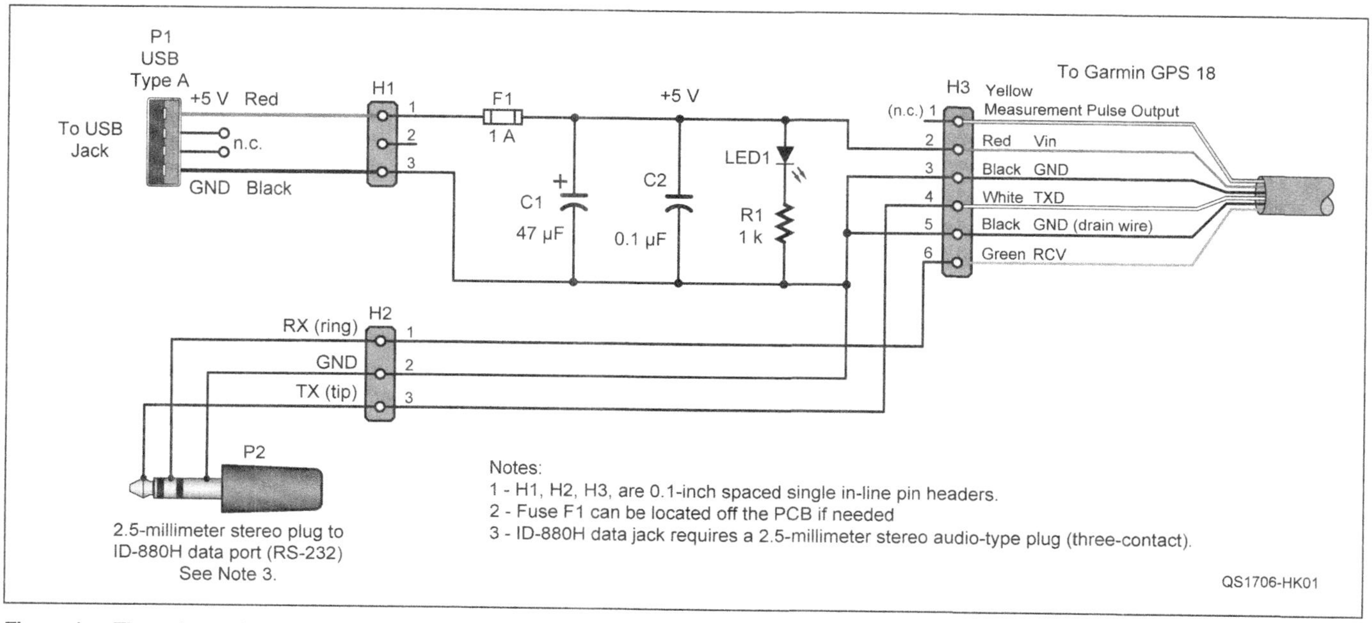

Figure 1 — The schematic diagram of the adapter.

Table 1
GPS Adapter Parts List

Reference	Part Description
C1	Capacitor, 47 uF, 25 V, radial electrolytic
C2	Capacitor, 0.1 uF, 50 V, ceramic
R1	Resistor, 1 kΩ, 1/4 W, 5%
pcb	0.062-inch perforated copper stripboard

green LED is a "power on" indicator and is mounted 1/4 inch above the board by a nylon spacer. This brings it closer to the top of the box, and also lessens the possibility of LED damage when soldering. You may have to adjust the value of R1 to accommodate the LED you use.

USB Cable Preparation

The USB plug cable (Jameco 229682) can be adapted from almost any unused or extra USB cable you have around. Some of the cheaper ones use very thin wire and may not be very robust. A good-quality cable will be marked as using two #28 AWG wires and two #24 AWG wires. The #24 AWG wires are usually the red and black, used for the +5 V supply.

I used an old printer USB cable that had the USB A on one end and the USB B on the other. I simply cut off the USB B end and separated the wires. Keep the red and black, and cut off the others, which are usually green and white. Note: there are some badly made cables that don't adhere to the color code, therefore, verify that the red and black are indeed the +5 V lines by using a voltmeter while plugging the USB A connector into a known good USB outlet.

2.5-Millimeter Plug and Cable

I used the Digi-Key CP-2203-ND cable for my project. Pre-made stereo audio cables can *sometimes* be found with a 3.5-millimeter plug on one end, and a 2.5-millimeter plug on the other. Simply clip off the 3.5-millimeter end, unless you want to have the cable detachable, then use it with a 3.5-millimeter jack wired to H2.

Assembly and Options

Referring to **Figure 1**, I mounted the few parts needed (see Table 1) onto a small piece of stripboard [Perforated board with copper traces running horizontally on one side. — *Ed.*] about 1 × 2 inches (see **Figures 2** and **3**), which fit nicely into a Jameco 18922 plastic box. I installed straight, single-row male pin headers for H1, H2, and H3 (Jameco 68339 cut to size). These made easy attachment points for the cables that make up the USB, radio, and GPS connections. You can solder the wires directly to them or use mating inline female headers.

You could also solder cable wires directly to the underside of the stripboard. Another possibility would be to trim the stripboard down to about 1 × 1.5 inches and use large diameter heat-shrink tubing in place of the box.

In the stripboard layout (see **Figure 2**), the four double crosses mark four points where the copper strip on the reverse should be cut. Mounting hole locations are shown by the crosshairs. There are also five wire jumpers shown in addition to the components.

This simple adapter is all that's needed to make your position data available for your D-STAR operations. — *73, Don Dorward, VA3DDN,* **va3ddn@arrl.net**

Fused Plug for Johnson Equipment

E.F. Johnson equipment often used a power plug with two fuses instead of a chassis-mounted fuse. These plugs are often hard to find; however, an almost identical replacement is used in electric fence applications, and is available from Kencove Farm Fence Supplies (**www.kencove.com**) as item MFP for $4.25. — *73, Pat Griffin, AA4PG,* **aa4pg@arrl.net**

SB-200 80-Meter Loading Fix

I wasn't able to load my Heathkit SB-200 on 80 meters after installing a W7RY QSK board. I found that the problem was RF getting into the grid meter shunt, possibly getting rectified by the meter protection diode. I added a small RF choke between the bottom of the supply and the meter components — three turns of 24 gauge on a medium ferrite bead (see **Figure 4**). The power went from 400 W to 600 W. The choke is similar to L401 in an SB-1000. I know I still have some tweaking left to do. I'm sharing this information to help others who may be having the same problem.

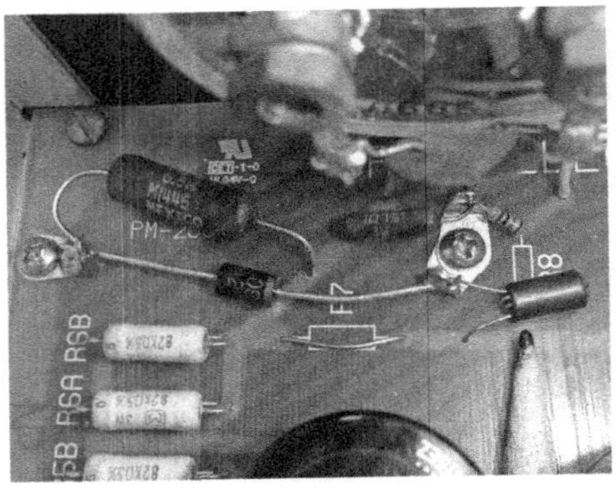

Figure 4 — The pencil tip points to the small RF choke between the bottom of the supply and the meter components. [Ed Daignault, WA1LJY, photo]

Note: this is only for an SB-200 *with* the W7RY QSK modification. The unit also has the Harbach Electronics power supply board, although that shouldn't make a difference.

The choke I used was a Mouser part 623-2944666671LF, making a three-turn winding from their existing two and a half turns by running the spare lead length through the remaining hole in the core. The choke is electrically positioned between the bottom of the power supply and the metering components.

A 9/64-inch hole was drilled in the power supply board for a screw, and solder lugs to accommodate the metering components. In a nutshell, the 0.1 μF cap stays on the supply side of the choke, and R12, R14, and the meter protection diode are on the metering side. — *73, Ed Daignault, WA1LJY,* **edaignault@netzero.net**

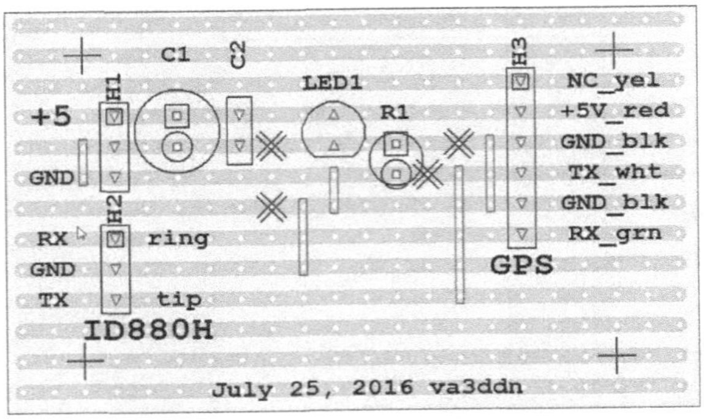

Figure 2 — The parts layout used for the adapter. The layout isn't critical and can be modified to fit your needs. The material is 0.062-inch perforated copper stripboard, cut to 1.2 x 2.1 inches (www.veroboard.com). [Don Dorward, VA3DDN, photo]

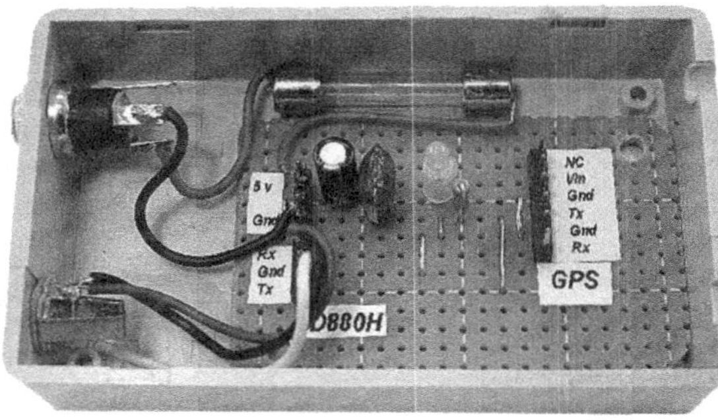

Figure 3 — The assembled adapter mounted in its case. [Don Dorward, VA3DDN, photo]

Figure 5 — The disassembled flashlight, ready for modification. [Joe Birsa, N3TTE, photo]

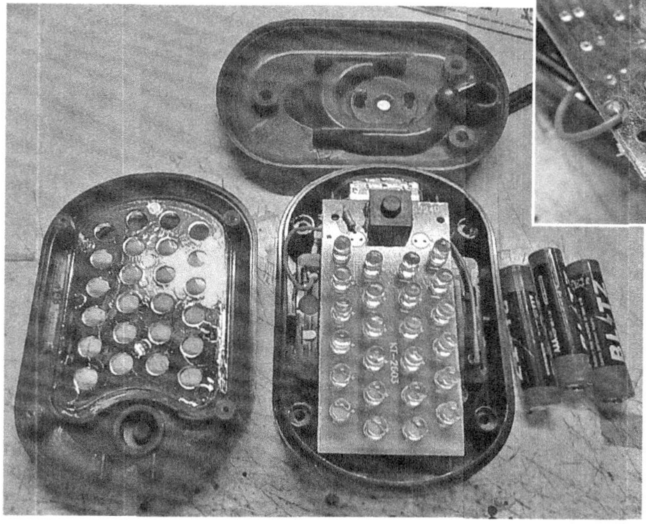

Figure 6 — The LED must be gently pried up in order to jumper from the LED/switch terminal to the other switch terminal using a small soldering iron. [Joe Birsa, N3TTE, photo]

Modifying a 27-LED Flashlight

I love the 27-LED flashlights sold by Harbor Freight (**https://www.harborfreight.com**). But I don't like that the flashlight alternates between the 3 LED and 24 LED modes (off, 3 LED, off, 24 LED), because it is never ready for the mode I happen to want at any given time. To fix this, I came up with a simple modification to make the flashlight turn on all LEDs at the same time. Here's how you can modify a similar flashlight yourself.

Remove the back by unscrewing the three screws and take out the batteries. Then, disassemble the flashlight by unscrewing the four screws at the corners of the flashlight (see **Figure 5**). At this point, notice that the **ON/OFF** switch has two terminals on one side and one terminal on the other side. To modify the flashlight, we want to short the two terminals on the underside of the circuit board.

Carefully turn the circuit board over. Note where the switch terminals will be shorted. Gently pry up the end LED (see **Figure 6**). Tin a piece of bare wire (I used 20-gauge wire from a craft store) and jumper from the LED/switch terminal to the other switch terminal using a small soldering iron. I soldered the end of a long piece to the LED first, then cut it off and soldered the small terminal.

At this point, put the batteries in and verify the operation is off, then all LEDs on. It may be necessary to resolder the small terminal if the solder didn't take at first. When it works as intended, reassemble in reverse order.

It's worth noting that I always use a hot soldering iron with the temperature set for lead-free solder whenever I modify a circuit board. Also, it may be necessary to unsolder a wire between the batteries and the circuit board to flip the circuit board. Unsolder the wire at the circuit board. — *73, Joe Birsa, N3TTE,* **jjbirsa@yahoo.com**

MFJ-452 Keyer Modifications

Several years ago, I purchased an MFJ-452 keyboard keyer, attracted by the keyboard feature; the intuitive, high-capacity memory banks; the ability to use a paddle as well as a keyboard, and the option to key either positive (most solid-state transceivers) or negative (grid-block tube type) transmitters. The unit, which is still in production today, was reliable and easy to use.

For use with only one rig and keyer paddle, no problems should be experienced. However, I use a multitude of different rigs, both modern and vintage, and like to use many paddles interchangeably, so I encountered a bit of inconvenience. Fortunately, this was easily remedied.

Most keyers — both standalone and those built into modern transceivers — are standardized, with the ground connection wired to the shell of the stereo plug, the *dah* connected to the ring, and the *dit* to the tip. This keyer is wired in reverse. The socket is a printed circuit type, so the wires cannot simply be reversed without cutting traces on the board. Many keyers include a "paddle reverse" feature in the software, but this one does not. Of course, if only one paddle and one keyer are used, it's a simple matter to wire the paddle to match. However, I needed to have standardized paddles that could be quickly transferred from keyer to keyer. Lucky for me, it's easy to assemble a "pigtail," with a stereo 3½-millimeter plug on one end and a 3½-millimeter or ¼-inch stereo socket, depending on your preference, wired in reverse. I leave this pigtail plugged into the MFJ-452.

Another obstacle was the changeover between positive and negative keying. Making this change involved removing nine tiny screws, removing the cover, repositioning a jumper, and then reassembling. I drilled a suitable hole on the rear panel above the output socket, and mounted an SPDT miniature switch in the hole. Thin wires were soldered between the pins and appropriate switch contacts (see **Figure 7**). Now changing between the two types of keying is as easy as flipping a switch. Please note, however, that making these modifications will void your warranty.

I find that I use this device much more often after making these small modifications. — *73, Ron Pollack, K2RP,* **k2rp@arrl.net**

Figure 7 — Detail of wiring the keying polarity switch. [Ron Pollack, K2RP, photo]

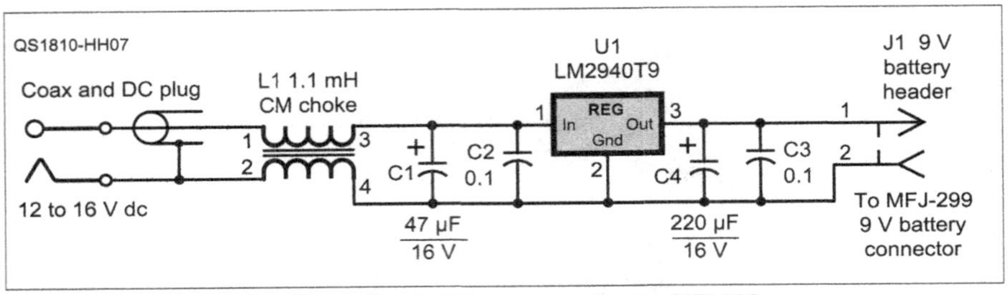

Figure 8 — A schematic of the author's 9 V dc source for the MFJ-299.

L1 — 1.1 mH common-mode choke
C1 — 47 µF, 25 V electrolytic capacitor
C2, C3 — 0.1 µF, 50 V ceramic capacitor
C4 — 220 µF, 25 V electrolytic capacitor
U1 — LM2940-T9 LDO voltage regulator
J1 — snap terminals taken from the top of a dead 9 V battery

Stripboard: 18 × 5 holes, approximately 1.8 × ½ inches (Veroboard epoxy/fiber or CEM-1 stripboard or equivalent)

Coax and plug: 24 inches of RG174 mini coax, dc power plug of your choice (the author used 5.5 × 2.1-millimeter center)

Grommet: (To protect the RG174 coax where it exits the corner of the battery compartment on the MFJ-299.)

Heat-shrink tubing, clear, ¾-inch diameter, 2¼-inch length (optional)

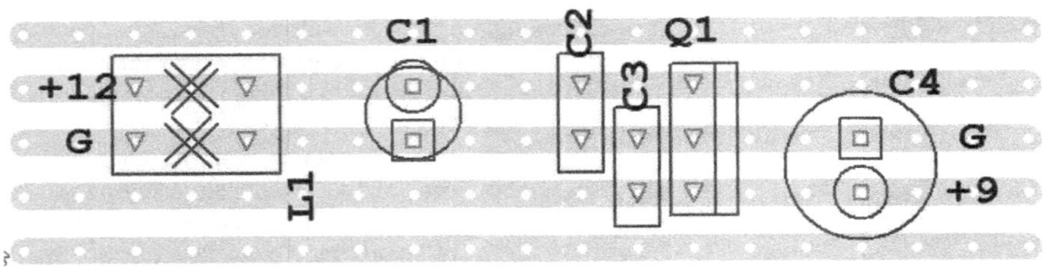

Figure 9 — The circuit on a small piece of perforated stripboard. Note: Cut the copper strips in the locations indicated by the double X marks under L1.

A 12 to 9 V DC Adapter for the MFJ-299 Desk Mic

The MFJ-299 desk microphone is my usual choice for HF use with my Kenwood TS-570S. I especially like the mic's volume unit (VU) meter, which allows me to easily monitor the level of outgoing audio. Of course, the other features such as the built-in PTT transmit indicator LED, selectable compression, and graphic equalizer make it a powerful unit indeed. The only negative aspect is that it's designed to be powered only from a 9 V battery, with no other option.

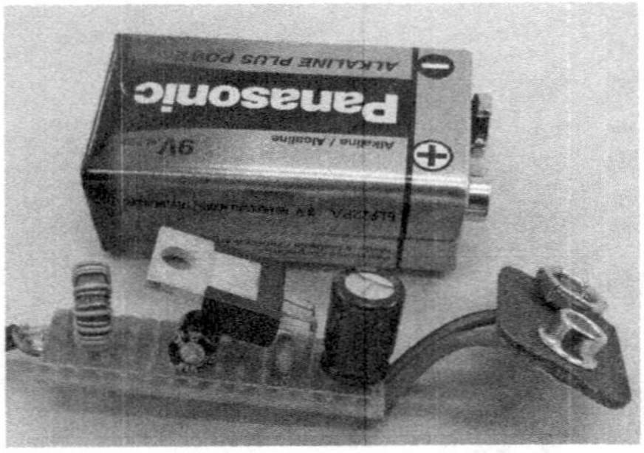

Figure 10 — A comparison of the stripboard power supply to the 9 V battery it's replacing. The stripboard circuit is enclosed in a piece of clear heat-shrink tubing for protection. [Don Dorward, VA3DDN, photo]

The problem is if you forget to switch the microphone off after use, then approximately 11 mA is continuously taken from the battery, which will necessitate its replacement several times per year.

I decided to build a simple 9 V dc source and install it in the MFJ-299 in place of the battery. Ideally, it would be no larger than a conventional 9 V battery and would take its power from the same 13.8 V dc bus I use to power my rigs, meter illumination, shack lighting, and so on.

Figure 8 shows the simple schematic, using the low-dropout voltage regulator LM2940T9, which will maintain the 9 V output as long as the input is 10 V or higher. I included the common-mode choke L1 as a precaution to prevent RF from coming in on the 12 V dc input. For L1, anything from 1 –10 mH will work fine. In my case, I found some low-cost 1.1 mH common-mode chokes on eBay. However, if you want to make your own, the Laird LFB127079 ferrite (DigiKey 240-2278-ND) with 10 bifilar turns on it will work as well.

I decided to build the circuit on a small piece of perforated stripboard (as shown in **Figure 9**). This is no larger than the standard 9 V battery it is replacing and easily fits into the battery compartment of the MFJ-299 (see **Figure 10**). — *73, Don Dorward, VA3DDN,* **va3ddn@arrl.net**

Using Dynamic Microphones with Electret Mic Transceivers

Many popular HF transceivers were designed for use with electret microphones that have 10 – 20 dB higher output levels than typical dynamic microphones. The microphone gain adjustment range on many of those radios is frequently insufficient to accommodate many popular dynamic microphones.

After purchasing one of these radios, I found that the Audio Technica dynamic desk microphone, which performed well with several of my previous transceivers, was unable to provide sufficient output for use with my new transceiver, even with the microphone gain maxed out.

The problem was solved by constructing a simple single-transistor gain-adjustable preamp that fits nicely inside the microphone body, powered by the 8 V present on the microphone positive lead from the transceiver's microphone connector. I used parts I had on hand to build the preamp, all inexpensive and commonly available for less than $5.

Since installing this preamp, I've received numerous compliments on my transmitted audio and I'm able to run the transceiver mic gain between 25 – 30% with proper ALC action. Prior to adding the preamp, I had to speak loudly into the microphone and run the transceiver mic gain at 100% with the compression set at +3 to get close to normal power output.

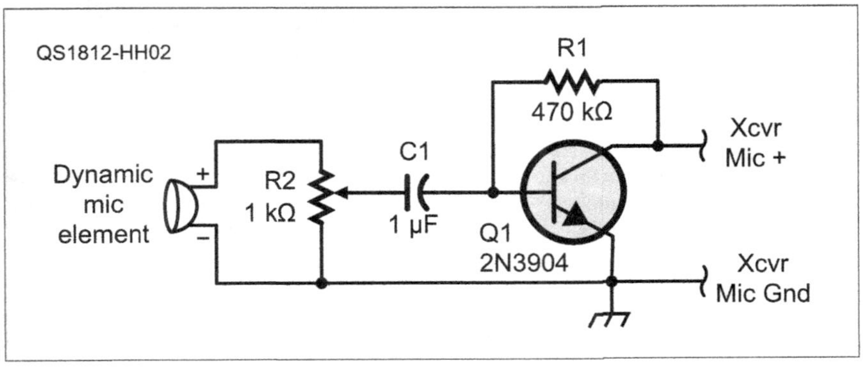

Figure 11 — A schematic with phantom power, for use when the transceiver powers the microphone element on the positive audio lead.

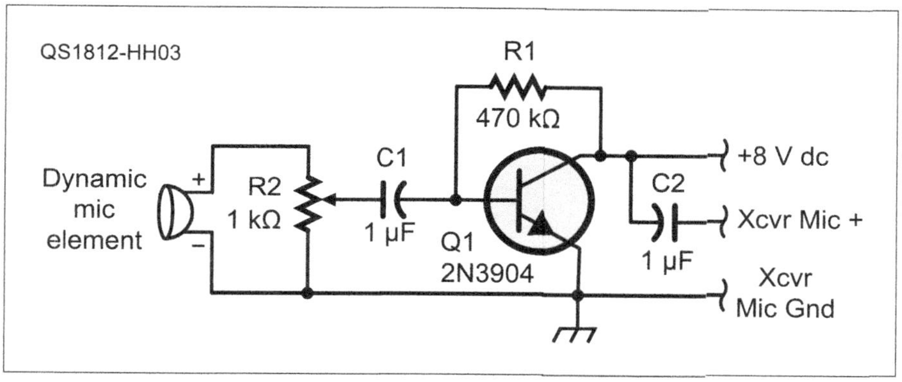

Figure 12 — The schematic without phantom power, for when the microphone is powered on a separate lead.

Figure 13 — The sky-wired assembly before heat shrink. [Paul Dobosz, K8PD, photo]

Figure 14 — The assembly enclosed in heat-shrink tubing. [Paul Dobosz, K8PD, photo]

There is one thing to be aware of when using dynamic microphones with transceivers designed for electret or condenser microphones. If the transceiver powers the microphone element on the positive audio lead, this preamplifier must be connected as shown in **Figure 11**. If the microphone is powered on a separate lead, it will be necessary to connect to the preamplifier as shown in **Figure 12**. Most modern transceivers have 8 – 12 V available on the microphone positive pin or a separate pin of the microphone connector, so this circuit should work with pretty much any radio needing extra microphone output from dynamic microphones.

Each physical installation will be unique to the actual microphone. It will depend on the shape and size of internal space available, and your ability to disassemble and reassemble the microphone without damaging it. In my case, I opted to "sky wire" components together with fly leads (see **Figure 13**) and enclose the assembled circuit in shrink tubing (see **Figure 14**). I placed a miniature gain adjustment trim potentiometer under the microphone windscreen to allow adjustment without fully disassembling the microphone (see **Figure 15**). If placement of the circuitry internal to the microphone is not possible, it can also be enclosed in a small inline enclosure or in the base of a microphone stand, or combined with a PTT switch. — *73, Paul Dobosz, K8PD,* **pjdobosz@gmail.com**

Figure 15 — The gain potentiometer installation. [Paul Dobosz, K8PD, photo]

Prevent Transceiver from Keying Amp During Auto Tune Cycle

Note: This hack only works in the semi-automatic tuning mode. Do not attempt to use this in the fully automatic tuning mode. Also, this circuit has been tested with the Icom IC-7300/Ameritron ALS-600/LDG AT-600 Pro II automatic antenna tuner configuration and may or may not be compatible with other configurations.

Overview

If the ALS600 amplifier switch is in the **OPERATE** position and **TUNE** is requested, the amplifier will be keyed on by the IC-7300 transceiver, and seeing an initial high SWR, the ALS600 will trip out. This transient has the potential to cause damage to the ALS600 and/or the LDG tuner, as well as causing the ALS600 to need a manual reset.

This circuit prevents the ALS600 from operating during the **TUNE** cycle by opening the **RELAY** line between the IC-7300 and the ALS600 during the LDG tuning cycle. After the **TUNE** cycle ends, the **RELAY** line is restored and the ALS600 will operate normally.

The circuit uses the **KEY** line in the tuner to control its operation. The Tuner Interface described the **KEY** line as follows: "The **KEY** line is pulled low to signal to the radio that a CW tuning carrier is requested. The **KEY** line will remain low for the duration of the tuning cycle." The notes also explained why this cannot be used with fully automatic tuning, saying, "Tuning cycles that are initiated because of fully automatic tuning (SWR exceeded preset threshold) will not activate the tune request (**KEY**) line."

Construction

I constructed this circuit in a small aluminum box for RF shielding using an experimenter-type pre-drilled printed circuit board.

Originally, the LDG IC-PAC-6 power and signal cable went directly from the **TUNER** jack on the back of the IC-7300 to the LDG tuner. The dc 2 ½-millimeter coaxial jacks and the stereo 3½-millimeter signal jacks route the dc power and control signals from the LDG IC-PAC-6 cable through the box to the LDG tuner. The dc power and **KEY** signal for the keying preventer circuit were obtained by tapping onto the power and signal lines as they passed through the box. The RCA type **XCVR** jack connects to the **SEND** terminal on the IC-7300 and the RCA type amp jack connects to the **KEY** terminal on the ALS600. The two plugs (dc and signal) from the LDG IC-PAC-6 adapter cable from the IC-7300 **TUNER** jack connect to the box **XCVR** jacks. The dc and stereo cable jumpers from the box **TUNER** jacks connect to the LDG tuner.

through the top and bottom. Going through the side would have meant grinding off a corner of the nuts to make room for them between each other when installed. On the smaller gun, you could use a large nut so you have more thread to work with, and ground the side of the nut that faces the other electrode so there's enough space between the electrodes when both nuts are installed. I haven't needed to do this on my small gun yet, but I will eventually.

Also, I know that new tips can be hard to find when the shops are closed. In a pinch, I have used #14 and #12 solid wire successfully as tips in the smaller gun. #10 or #8 should work for the larger gun, but I have never had to replace this tip. To do this, form a loop of wire and tightly twist the curved end of the loop to form the tip. Only two or three turns are needed.

Both of these Weller guns have been trusty tools of mine for over 40 years, and I am sure that my grandchildren will appreciate them for many more. — *73, Clint Millett, VE3CMQ,* **cmillett@prairiemobile.com**

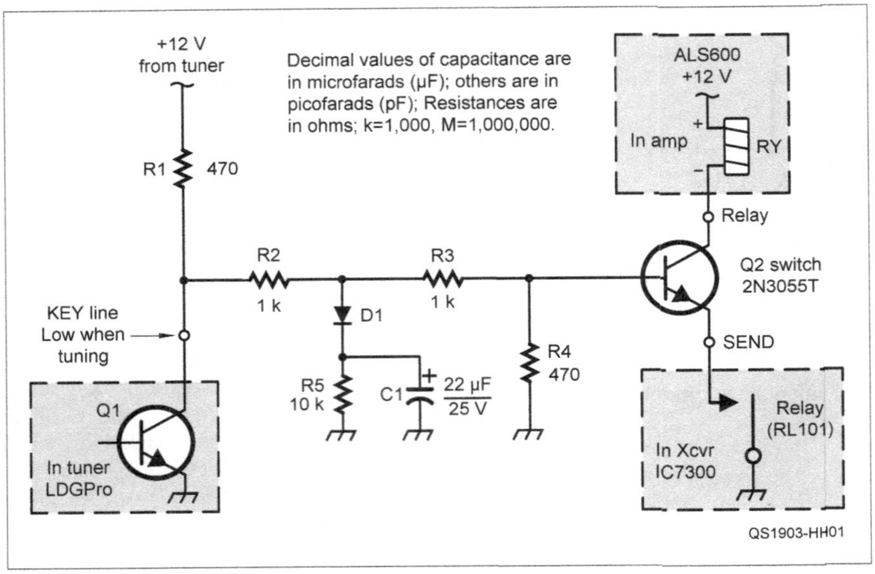

Figure 16 — A schematic illustrating the circuit's use.

C1 — 22 μF at 15 V or greater
Q2 — NPN 2N3055T, TIP31, etc. type with a T0220 case
Resistors (R1 – R5) — ¼ W
Diode (D1) — 1N4001 or similar
Power jacks — Type N 2½-millimeter coaxial
Signal jacks — 3½-millimeter (⅛-inch) stereo type

Operation

Q2 acts as a series switch in the **RELAY** line between the IC-7300 and the ALS600 (see **Figure 16**). When Q2 is on, it acts like a closed switch and the ALS600 relay RY is switched on or off by the IC-7300 relay **SEND** terminal. When Q2 is off, it acts like an open switch and the ALS600 relay RY cannot operate.

During normal operation, while not tuning, Q1 is turned off and R1 – R4 form a voltage divider providing the needed voltage and current to the base of Q2 for it to turn on, but it does not turn on until the **SEND** relay in the IC-7300 pulls the emitter of Q2 to ground. The **SEND** terminal on the IC-7300 going to ground turns Q2 on (closed switch), pulling the **KEY** line on the ALS600 low, activating the ALS600 relay RY. During this time, the capacitor C1 is charged to about 5 V through D1.

When **TUNE** is requested, Q1 turns on and the **KEY** line in the LDG tuner goes low, pulling the base of Q2 to nearly 0 V, turning Q2 off (open switch), in turn preventing the ALS600 relay RY from turning on. Also, during the tuning, D1 is reverse biased (off) and C1 discharges through R5 to 0 V.

At the end of the tune cycle, Q1 turns off and the LDG tuner **KEY** line goes back to high and D1 is forward biased (on) and C1 begins to charge. Because of the R3 and R4 voltage divider, C1 must charge to approximately 0.8 V through R1, R2, and D1 before Q2 can turn on again. This delay keeps Q2 off long enough to prevent the ALS600 from coming on until the IC-7300 RF output has stopped after the end of the tuning cycle.

Now, with the ALS600 amp set to the **OPERATE** position, the amplifier remains off during a semi-automatic **TUNE** cycle. — *73, Wayne Waller, W4TZB,* **w4tzb@comcast.net**

New Life for Weller Soldering Guns

I have two Weller soldering guns, a 100 W and a 300 W. I use them often and started having problems with both, especially the larger one. I decided to repair the issue, and hopefully make the guns better than before.

The problem was the screws holding the tips in place were always getting loose. The smaller gun was easy to fix. I just enlarged the hole for the screw and re-tapped it, using a larger screw. This worked quite well. The larger screw now has better bite into the tip and can apply more force, producing better contact.

The larger gun had aluminum electrodes, and the contact screws stripped out the threads quickly. I first tried the drilling/tapping trick, but the stripped threads returned. I knew there had to be a better way.

I found two steel nuts large enough that I could drill out the threads and fit both of them onto the electrodes, still leaving enough of a gap so there wasn't a possibility of them shorting together. I then drilled and tapped a hole in the side of the bolt, so a 10-32 steel screw could be used as the new tip contact. I drilled the original hole in the electrode large enough that the 10-32 screw would slip through it.

This modification turned out great, and it can be done on both models of gun. The new screws could even be put through the side of the electrodes, but I was in a hurry and went

ARI-500 Mod for Restoring MFJ-998 Auto-Tuner's Amplifier Bypass

The MFJ-998 Auto-Tuner prevents the Ameritron ALS-600 amplifier from keying during the tuning cycle, avoiding potential damage to the tuner and/or amplifier by controlling the relay/key line between the amplifier and the transceiver. MFJ calls this feature "Amplifier Bypass." The ARI-500 accessory adds automatic band-switching to the ALS-600 amplifier, but as supplied, the MFJ-998 Amplifier Bypass feature is deactivated because the ARI-500 controls the ALS-600 keying and bypasses the usual **RELAY** or **KEY** cable between the transceiver and the tuner.

Circuit

The jumper in the back left corner of the ARI-500 is JMP2, and the front panel **OPR/STBY** switch S1 connects between Pins 1 and 2 and controls power to the ARI circuitry. RC_IN on Pin 3 is from the optional ALS-500RC remote control head connector REM A or jack J1.

RC_IN connects to Pin 7 on the REM A jack J1 and Pins 3 – 4 provide +12 V source. By moving the jumper JMP2 to position 2 – 3, the ARI circuitry gets +12 V from Pin 4 after it is passed through the MFJ-998 Amplifier Bypass contacts. The associated ARI circuitry gets no power while the MFJ contacts are open during tuning and the ALS-600 amplifier remains off (see **Figure 17**).

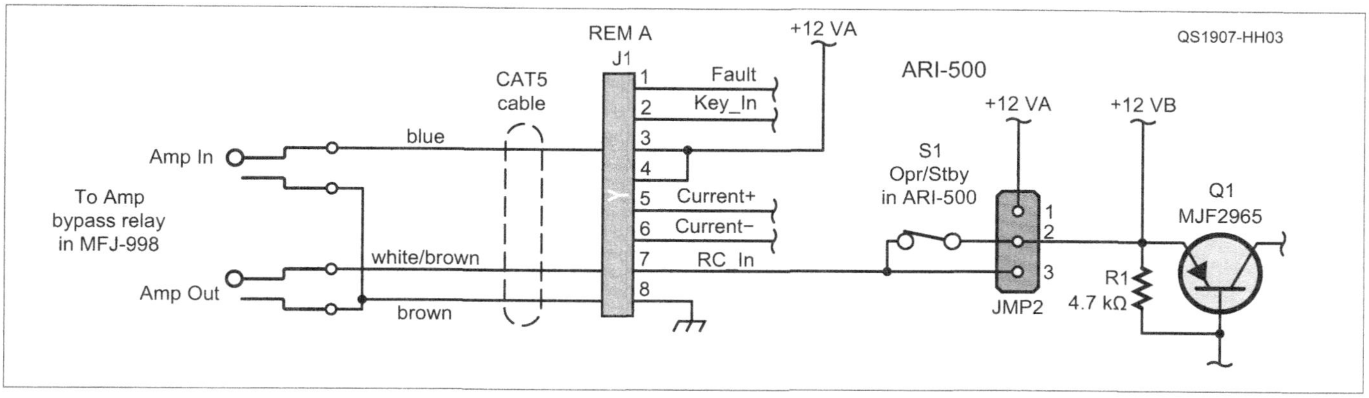

Figure 17 — A schematic of the author's circuit.

Construction

I cut a CAT5 cable and a dual RCA cable, then spliced them together and covered the splice with heat-shrink tubing. The braid shields were tied together to ground.

To install the modification, I changed the jumper JMP2 from 1 – 2 to 2 – 3, then connected the CAT5 end of the cable to REM A on the ARI-500 and the two RCA plugs to **AMP IN** and **AMP OUT** RCA jacks on the MFJ-998.

Operation

To operate, I simply choose a band (the ALS-600 relays could be heard clicking changing bands), select a frequency, push **TUNE** (on the MFJ or the transceiver), and it's ready to go. I also remember to check the transceiver power level before or after tuning, because ALS-600 requires different drive levels for different bands, and 30 meters is legally limited to 200 W.

Additional Use

Another possibility, with or without an MFJ-998 tuner, is to put a switch in series with the circuit, creating a second **OPR/STBY** switch in any convenient location. Then, if desired, the ARI-500 can be moved out of sight. If you're not using an MFJ-998, simply wire a single-pole, single-throw switch between Pin 4 and Pin 7 (blue and white/brown). If you are using an MFJ-998, put the switch in series with the **AMP OUT** line. — *73, Wayne Waller, W4TZB*, **w4tzb@arrl.net**

CHAPTER 6

Interference (RFI/EMI) and Operating

Solar Power Optimizer RFI

I was very interested in Tony's, K1KP, article in the April 2016 issue on solar-power system RFI because I am personally annoyed by a significant noise problem coming from a solar-power array that uses optimizers.[3] In my case, unfortunately, the source of the disturbance is not in my own house, but on the roof of a neighbor, some 250 feet away.

As an active DXer, retired professor of photovoltaics, and author of more than 150 technical papers, I realized in the early '90s that solar-power systems might be a future problem for Amateur Radio. Depending on their dimensions, solar arrays can be quite efficient antennas on certain frequencies, especially in the HF range. Therefore, during inverter tests in our laboratory from the 1990s on, we always tested conducted emissions not only on the ac, but also on the dc side.

If standard dc/ac inverters (without optimizers) are used, it is possible to avoid generating RFI by using suitable filters on each Maximum Power Point Tracking (MPPT) input on the dc side (typically 1 to 4). Good inverters should have internal filters that are sufficient in nearly all cases. If additional filtering is required, only one external filter per MPPT input is required. By the way, there is a new international standard with limits for conducted disturbances on the dc side of inverters: CISPR 11 Ed.6: 2015.

Tony's, K1KP, article could be misinterpreted as a general incompatibility between solar power and ham radio. Fortunately, this is not the case. So far, such problems occur mainly if there are distributed switching devices, like optimizers, in the array, as described in K1KP's article. In my case, I have no RFI problems with my own stand-alone solar-power system, nor with another solar-power system on the house of another neighbor at about 120 feet. Both systems have no optimizers.

Hopefully, manufacturers of optimizers will be able to improve their products to avoid RFI problems in the future by adding suitable filters. — *73, Henry Haeberlin, HB9AZO,* **hb9azo@uska.ch**

A Different Kind of Mobile Interference

A week ago, I experienced a strong unmodulated carrier on 441.500 MHz, a UHF frequency I regularly monitor. My first inclination was to blame the vandals that have been deliberately interfering with local repeaters in the area. Observing the signal over the course of a week, however, I became convinced that it was a solar-charged, battery-operated device. The signal would routinely come on at dusk and disappear at dawn.

Then, last weekend, the pattern changed. The signal was present in broad daylight as well as at night. I fired up my customized portable Ramsey DDF-1 Doppler Direction Finder and proceeded to walk around the neighborhood. Oddly, all signs pointed to my neighbor's parked car. I dismissed that as unlikely and concluded that I must have the Ramsey 180° out of phase. The readings now led me to another neighbor's residence, but he appeared to be away. I noted that several houses had coach lights that were on, which led me to believe the signal was coming from a poorly filtered LED or CFL bulb.

One day around dusk, I was doing yard work with my handheld transceiver on my belt, tuned to 441.500 MHz. If that mystery signal came on, I was going to see whose house lights were lit and find the offending bulb. Sure enough, around dusk, the signal returned, but none of the suspect coach lights were on.

I tried to think of what else could be causing the signal. Soon, I had my answer. My neighbor pulled out of his driveway, and as he drove away, the signal got weaker and weaker. The transmitter must be in his car — my direction finder had been right after all.

When my neighbor returned later that evening, I told him his car was transmitting a signal. Jokingly, he suggested that someone had planted a tracking device in his SUV. He had no other explanation for what might be emitting a signal, especially while the car was parked and not running. Together, we searched for the device, combing the vehicle inside and out using my handheld with the antenna removed to decrease sensitivity. It didn't take long to figure out that his rear seat TV/DVD player was emitting the signal.

We checked the settings and determined that the rear seat after-market TV was sending sound to the vehicle's FM radio on 88.3 MHz. Changing the frequency to 88.5 MHz caused the interference to cease. Mystery solved!

What I observed now makes perfect sense. Five times the TV's 88.3 MHz audio signal is 441.5 MHz. This strong fifth harmonic signal (detectable a half-block away) was the source of my interference. Because my neighbor's car was only parked at home from evening until morning, it behaved like a dusk-to-dawn device, except on the weekends when he was home during the day.

Two things conspired to create this problem. First, the rear seat TV should have been wired to turn off when the car was not in use. Second, the manufacturer should have employed better filtering and shielding to prevent the high-frequency UHF harmonics from radiating.

Should you need to track down the source of an offending UHF signal, keep the following in mind:

- Look for simple explanations before assuming there is a sinister source.
- Remember that poorly made and/or improperly installed electronics of any type can be a source of interference.
- Believe what your equipment is telling you. My Ramsey DDF-1 took me to the offending vehicle right away, but instead of trusting it, I assumed the signal could not be coming from a parked car.
- Be friendly to your neighbors. You never know when you will need their help. — *73, Tony Jedlinski, WA9PEB,* **tony@konoso.com**

Revisiting Touch-Sensitive Lamp Circuit Interference

For more than 20 years, touch-sensitive lamps have limited my HF operations because when I transmit, the lamps cycle through the various states. I tried the methods on the ARRL website (**www.arrl.org/radio-frequency-interference-rfi**) and they helped somewhat. I added common-mode chokes to the ac cords, added inductors to the sense lead, and so on. These improved the radio-frequency interference (RFI), but did not fully cure the problem — until now.

I experimented and found that the key was to prevent the RF from getting into the touch control integrated circuit (IC) and, if it got there, to shunt the remaining RF to "ground." Keep in mind that the ground of the IC is actually the ac neutral, but it is still a ground reference for this IC. Here's what I found helps:

1) Use a module that has a UL mark. The module that I used was "UL Recognized." This is for safety, but shows that it has gone through a certification process and should be a more reputable design than some modules. The module that I used as a replacement for the lamps was purchased from **amazon.com**. The module used a TT6061A touch control IC (see **Figure 1**).

2) The inductor (choke) should be as close to the "sense" input of the IC as possible. The easiest solution is to remove the 1 kΩ resistor close to Pin 4 of the IC and replace it with a wire-wound inductor of the same length so you can use the same feedthrough holes that the resistor used. I tried both 2.2 and 2.7 mH inductors from Digi-Key and they worked just fine. With the inductor external to the module, the RFI rejection was nowhere near as good as when the inductor was mounted next to the IC. Location is very important.

3) The larger the series resistor, the better to reduce the RF signal input even more. I knew that 1 kΩ still allowed for normal operation, but 3.3 kΩ worked just as well, so I stuck with that value. I knew that if the series resistance was too high, the touch circuit would not work anymore (I did not find out what that upper resistance limit would be).

4) Adding an 82 pF (1 kV rating recommended) from Pin 4 (input) to Pin 7 (V_{SS} or ground) of the touch IC was necessary to get that extra isolation by shunting the remaining RF to ground.

5) I pretested each lamp, before and after modification, by using an incandescent light bulb as my dummy load. I located the bulb right next to the lamp's body to inject RF at close range. There was a clear difference between an unmodified lamp and a modified lamp.

The final test was putting the modified lamps back in their original locations and transmitting at 100 W on various frequencies and antennas. In all cases, the lamps remained in their original state during all transmissions. I used my phone to record a video of the lamp during the transmission just to make sure.

If the primary method for RFI reduction as outlined in the first part of this item is difficult due to limited soldering skills, or if your RFI problem still exists because you are running high power, try this brute-force method as a compromise solution:

1) Open the base of the lamp to expose the wiring.

2) Find the sense wire (this might be green or yellow depending on your lamp manufacturer) that runs from the lamp body to the touch module.

3) Cut the wire in the middle to add a single-pole single-throw (SPST) switch. You will likely also need some additional wire leads to go from the switch location (best to mount these on the lamp base) to the sense wires. Make sure the switch mechanism is insulated from the lamp housing, otherwise touching the switch to deactivate it will change the light levels. This can be done by using a switch with a plastic lever or adding heat-shrink tubing to the outside of the switch lever if it is metal.

This switch will disconnect the sense circuit to make the lamp "deaf" to RF. Simply set the light level (off, low, medium, high) to the desired state and turn the switch to the **OFF** position. Even in an RF field, the lamp should remain in its current state. When you are finished operating, just turn the switch back to the **ON** position for normal operation. — *73, Kevin Scott, K4GTR,* **k4gtr@att.net**

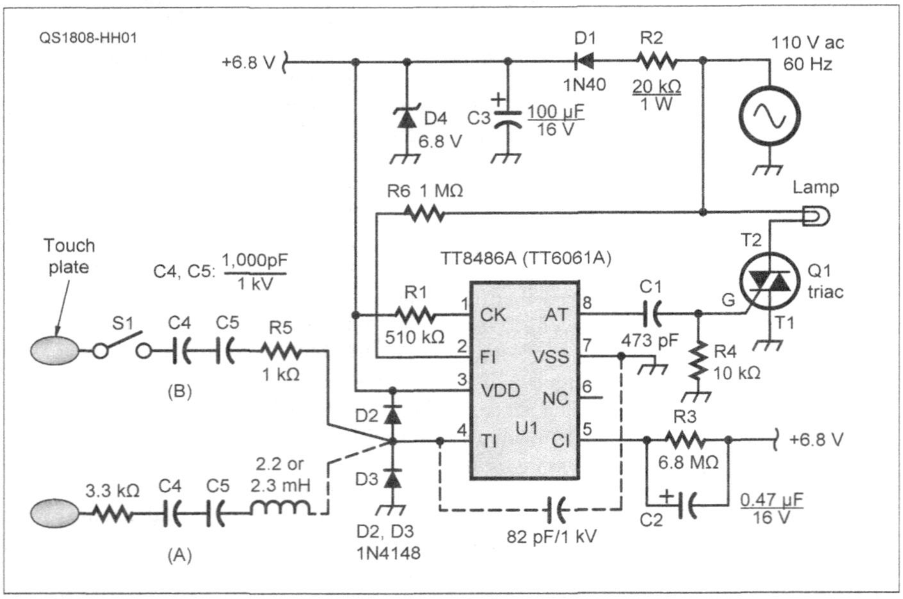

Figure 1 — A schematic of the author's modified circuit used to prevent RF from getting into the touch control integrated circuit and shunting the remaining RF to ground if needed (A). (B) shows how to disconnect the sense circuit to make the lamp "deaf" to RF. Any of the inductors listed below will work for this circuit.

Digi-Key Part Numbers

Value	Digi-Key Part Number	Description
2.7 mH	AIAP-01-272K-TCT-ND	Inductor 2,700 UH 10% 90 mA
2.2 mH	M8395-ND	Fixed inductor 2.2 mH 57 MA 45.6 Ω TH
2.2 mH	M8305-ND	Fixed inductor 2.2 mH 170 MA 5.3 W
82 pF	445-15979-ND	Cap cer 82 pF 1 kV SL radial
SPST switch	360-2826-ND	Switch toggle SPST 6 A 125 V with cap

*Touch lamp module from Amazon: **www.amazon.com/AmerTac-6503BC-Westek-3-Level-Replacement/dp/B00004XS95**

Oil Burner Woes

I've operated SSB for years on the HF bands without interfering with our household electronics, but when I started using FT8 — a 100% duty cycle mode — my average output power was much higher. As a result, each time I transmitted, our oil burner would immediately shut down.

After some trial and error, I found the RF was getting into the burner control through the wire from the thermostat. Fortunately, the solution was straightforward. I purchased a Type-43 ferrite core and wound about 14 turns of the thermostat line through the core (see **Figure 2**). Problem solved! — *Dan Bowman, K1FEV,* **k1fev@arrl.net**

Figure 2 — Assemble a stack of fliers and align them at the edge of a table, using a heavy weight (a power supply in this example) to compress the paper and hold it in place. Using your finger, apply Elmer's glue along the entire edge of the stack. Allow to dry completely. [Bob Bastone, WC3O, photo]

Wireless Speaker Interference

For about a year, I was plagued by a series of signal spikes every 33 kHz or so, including one right in the middle of a portion of 75 meters, from 3.877 to 3.883 MHz. The signals also appeared above and below the band. The spikes would generally peak at about an S7 or so, and were only about 1 kHz wide, but were just wide enough and loud enough to create an annoying buzz on weaker AM signals, as well as SSB signals.

The spikes were particularly apparent when using my 80/160-meter shunt-fed tower, which is located about 90 feet from our house and has an extensive radial field. After an exhaustive hunt around the house with a portable AM radio, I found a variety of noise sources from switching power supplies, computers, and more, but none proved to be generating interference beyond a few feet from the respective device. I was convinced that the noise was not coming from our house or property, but decided to turn off each circuit breaker in a last-ditch effort to see if I could find the culprit.

After going through our entire breaker panel and subpanel, I turned off one of our exterior ground fault circuit interrupter (GFCI) breakers and found that the spike was suddenly gone. That outdoor line extended about 75 feet or so through our backyard and fed several low-voltage transformers and electronic timers, but the line also picked up one outlet located on our back porch that powered a high-end wireless speaker. I never suspected the speaker, but it ended up being the culprit. Its Wi-Fi transceiver must have induced oscillator noise into the long outside ac line and various low-voltage cables running around the yard. I'm sure the proximity of an extensive radial field to those lines didn't help. **Figure 3** shows the RF spike at about S7 just below the passband of an AM signal on 3.885 MHz. **Figure 4** shows the result with the speaker disconnected.

Plugging the device into one of the interior house circuits, in combination with wrapping several turns of the line cord around a snap-on ferrite suppressor, has all but eliminated the noise. The speaker was made by Definitive Technology, Model W7.

The lesson learned here is that you should check your outdoor GFCI underground wiring for devices that might be coupling RF from wireless sources. That wiring serves as a nice antenna system to transfer the noise to your antenna farm. And you might also want to check your wireless speakers as sources of interference. — *73, Tom Daniels, N3CXP,* **tkdaniels@ptd.net**

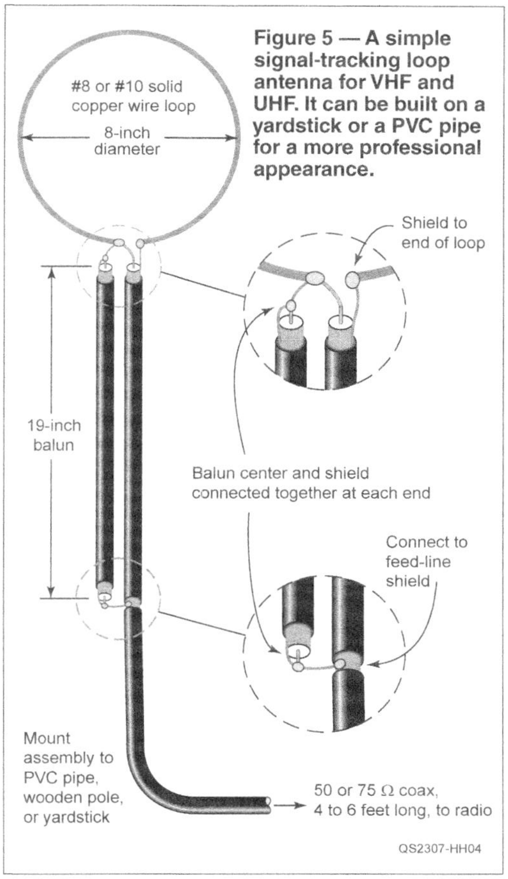

Figure 5 — A simple signal-tracking loop antenna for VHF and UHF. It can be built on a yardstick or a PVC pipe for a more professional appearance.

Interference Tracking Antenna

The Central Ohio Radio Club received a message from a non-ham who was looking for help because his garage door openers had suddenly failed to work. Several of his neighbors were experiencing the same problems.

Suspecting an interference issue, I contacted the individual and arranged for a visit. Upon arrival, I connected my handheld transceiver to a small loop antenna and searched between 310 and 315 MHz. The antenna I used was an easy-to-build homebrew design that functions well at VHF and UHF (see **Figure 5**).

It didn't take long to find an interfering signal that appeared to be strongest at the west side of their home. It turned out that they had a home office on the second floor and the wireless modem/router on the west side of the room was generating the signal. They contacted their internet provider, and a replacement unit solved the problem. — *73, John T. Perone, W8RXX,* **w8rxx@arrl.net**

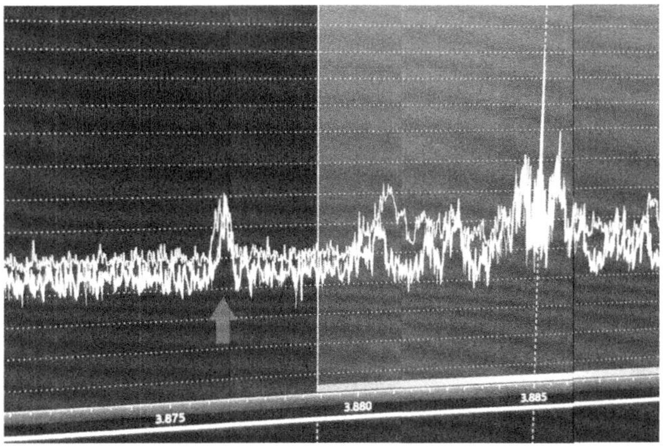

Figure 3 – Here you can see one of the offending signals (marked with a red arrow) just below an AM signal. [Tom Daniels, N3CXP, photo]

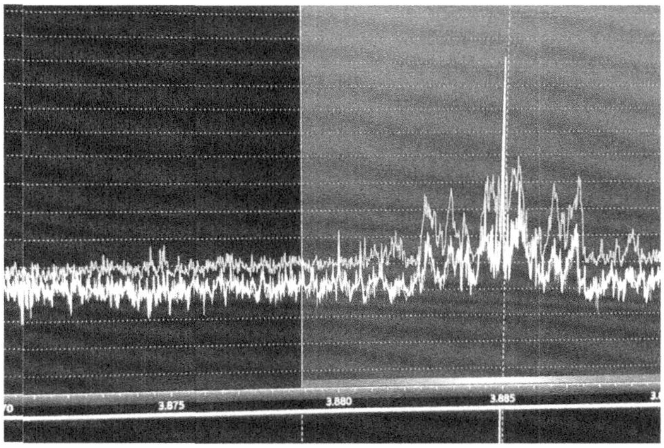

Figure 4 – When the wireless speaker is disconnected, the signal vanishes. [Tom Daniels, N3CXP, photo]

Managing Frequency Memories

In response to Pete's, N9SFX, suggestion for remembering repeaters, "Remembering Repeaters" in the July 2016 column, I'd like to suggest a system that has worked well for me, for both the radio in my car and the scanner and 2-meter rig I keep at home.

The dual-bander in my car has 180 memory channels, which is far more than I will ever need, so I group them in blocks of 10 channels. Memories 1 – 20 are for repeaters that are local to me. This area does not have 20 2-meter and 70-centimeter repeaters, but this method leaves room for expansion if needed. Frequencies for nearby cities are in 21 – 30, 31 – 40, and 41 – 60.

Frequencies for towns I visit on occasion are stored somewhere above memory 61. At the other end, memories 174 – 180 hold National Weather Service (NWS) frequencies at 162.xxx MHz. The ability to receive the NWS has proven valuable when I find myself traveling through inclement weather. For simplex channels, 146.52, 446.0, and so on, I start at memory 173 and work backwards.

The scanner also holds repeater and simplex frequencies for 10 meters, 6 meters, 1¼ meters, 33 centimeters, and 23 centimeters, grouped in the same fashion.

This arrangement has worked well to keep repeater frequencies grouped by locale, and leaves plenty of wiggle room to add or delete frequencies as the need arises. — *73, Todd Brandenburg, KØKAN,* **k0kan@arrl.net**

Offset Error Prevention

You can avoid a red face, an Official Observer notice, or even a notice from the FCC by setting the offset features of your transceiver to "000." My Icom IC-7300 has **RIT**, **DELTA TX**, and **SPLIT** options, but the icon showing on the screen is small and easily overlooked. Just set the value in all of those and any other features that give an offset to "000," so if they're accidentally pushed, there is nothing to worry about. — *73, Ken Shubert, KØKS,* **k0ks@k0ks.com**

Wind Chime Zero-Beating Aid

It's always been a task for me to zero beat the person on the other end when working CW. My Elecraft K2 transceiver has an automatic tone function, but my high-power rig does not. I have an app on my Android phone that produces any audio tone I want, but the volume on the phone is low, especially when I'm wearing headphones. I looked for a variable tone generator, but none that I found quite matched my needs and budget. (I prefer my sidetone set at 600 Hz, but other rigs may have a different frequency.)

I had an old set of homemade wind chimes lying around and found one of them was resonant at 622 Hz. However, I got tired of trying to finesse the 22 Hz difference in my head, so I set out to make a wind chime that would be much closer to 600 Hz (see **Figure 6**).

There is a website (**www.leehite.org**) with charts for making wind chimes from steel or copper pipe of various dimensions. I used ¾-inch copper plumbing pipe from my local hardware store. It has an outer diameter of 0.875 inch and an inner diameter of 0.811 inch.

In my experience, the tables from the website produce a chime that is slightly high in frequency. I determined the exact length needed for ¾-inch copper plumbing pipe at 600 Hz to be $15\frac{23}{32}$ inches, rather than the table-interpolated value of 15.671 inches. I drilled the suspension holes 3½ inches from one end.

I added a bit of monofilament line through the holes, used an old Tinker Toy stick as a spreader, and was all set. According to my Android tone generator, I'm right on the money. It also zero beats the 600 Hz tone from my transceiver. And the harder I hit it, the greater the amplitude — more than enough to compete with loud signals in my headphones. — *73, Randy Tompson, NC4RT,* **rtompson@triad.rr.com**

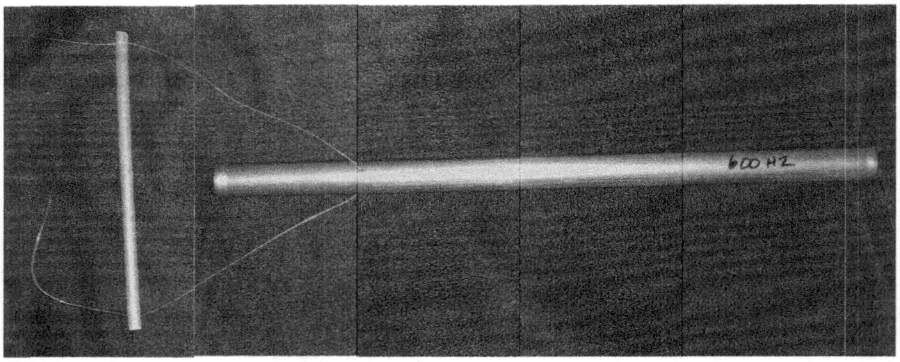

Figure 6 — The author's homemade wind chime, resonant at 600 Hz. [Randy Tompson, NC4RT, photo]

CHAPTER 7

Restoration

Improving SWR, a Little at a Time

If you run vintage gear with a separate receiver and transmitter, one thing you usually need is an outboard transmit/receive (TR) relay. With vintage equipment, this relay usually has to multitask between switching the antenna, muting the receiver, and keying an outboard amp or other ancillary device. Many folks use a commercial coaxial relay, originally known by the company that made them, Dow-Key. But, after having to maintain communication sites with 50 or 60 relays in this style, I came to appreciate the simplicity of a common open-frame relay.

The basic open-frame relay does the trick just fine, even if there is an impedance bump. That bump is not really an issue below 30 MHz, but can be a problem as you go up in frequency. This tip focuses on relays I have in service on 6 or 2 meters.

I use a standard plug-style 4PDT relay with an Ameco TX-62 transmitter and Hammarlund HQ-170A receiver. Two of the four sets of contacts are paralleled for antenna switching, and the remaining sets have one to mute the receiver and the other to key an amp. I do not use the socket to plug these into because those often have screw terminals and are nothing but signal-loss generators. These relays are hardwired to coax and control lines. I needed to redo my TR relay enclosure because the Altoids box it was housed in turned out to be too flimsy, so I took the time to see if I could improve the loss in the relay. With a good UHF dummy load on the SWR meter, I read 1:1. But with the TR relay inline and terminated, I got 1.5:1, with the SWR occurring within the relay. I thought eliminating the common contact from the terminals and going directly to the coax connector might help.

Because this relay was intended to plug into a socket, it has all the fixed contacts exiting via their eye terminal/plug-in contact directly. However, the movable contact has a small, flexible wire soldered to it, and this wire loops down to be soldered to the terminal in the relay base. I opened the relay case, unsoldered one of the flexible wires from the movable contact, moved it out of the way, unsoldered the other movable contact wire from the relay base lug, and added a jumper across the two paralleled moving contacts, because they are doubled up for the antenna. This essentially makes it a 3PDT relay.

I guesstimated location, then drilled a small hole in the plastic case opposite the point where the remaining attached common contact wire was affixed, and slid the plastic cover back on, fishing the wire out of the hole. You must have a flexible wire going to this style relay, because anything rigid will prevent free movement of the relay armature. I assembled the relay in its new metal enclosure (see **Figure 1**) and soldered the relay wire directly to the antenna connector. A re-check of the relay-only SWR revealed my effort brought it down to 1.2:1. Every little bit helps! — *73, Charlie Liberto, W4MEC,* **w4mec@arrl.net**

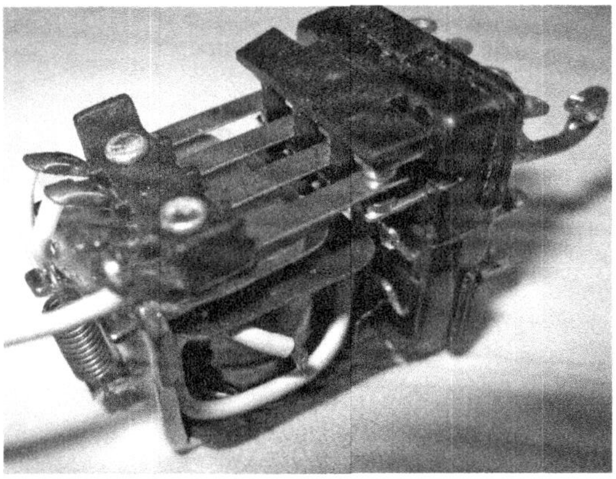

Figure 1 — The relay ready to go into its new metal enclosure, with a flexible wire soldered to the antenna connector. [Charlie Liberto, W4MEC, photo]

Microphone Coil Cord Recoating

After years of service, the outer covering of the cord for my Icom HM-133 microphone started to shed in multiple places. The microphone's keypad allowed programming, so I wanted to repair the cord. Icom no longer made my old microphone, and new microphones looked like they wouldn't program my IC-208H transceiver. Electrical tape proved bulky and made the cord unable to coil. The only option was to somehow recoat the coil cord.

I bought a can of Plasti Dip® — a thick, black plastic liquid used to provide an insulated coating for hand tools (see **Figure 2**). I rigged a support system consisting of two vertical sticks, each with a nail in the surface nearest to me. I spaced the sticks so the microphone cord could be stretched to open the coil turns and then hooked onto the nails. I dipped the coil cord into the liquid, coating the entire coil cord, the base of the micro-

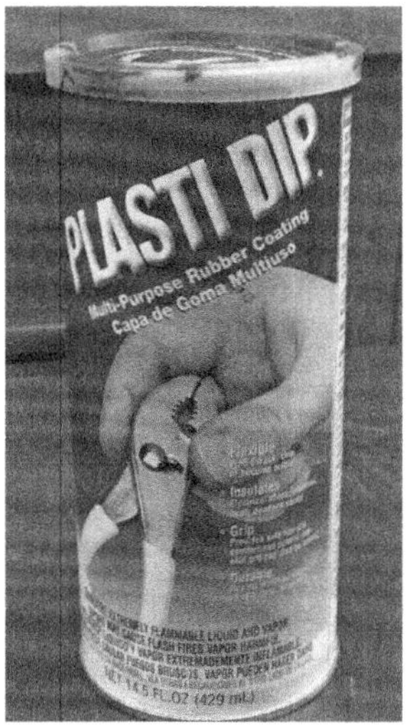

Figure 2 — Plasti Dip was used to recoat the author's old microphone coil cord. [Peyton Barnes, KE5ZDZ, photo]

Figure 3 — The recoated microphone cord: flexible, insulated, and ready for use. [Peyton Barnes, KE5ZDZ, photo]

Figure 5 — Three examples of different self-clinching nut extractors. [John M. Franke, WA4WDL, photo]

Figure 6 — A small magnet can correct a meter needle that rests above or below zero. [Peter Ravn, OZ8CTH, photo]

phone, and the base of the plug at the other end. Finally, I stretched the cord to open all of its coils (so they would not stick together) and then hooked the two ends of the cord onto the nails.

After drying overnight, the cord was completely insulated and flexible (see **Figure 3**). I don't know how it will hold up in the long term, but over the last month it has been entirely satisfactory. This was a handy solution for fixing up equipment in preparation for ARRL Field Day. — *73, Peyton Barnes, KE5ZDZ,* **skaevola@hotmail.com**

Simple Self-Clinching Nut Extractor

I recycle or repurpose aluminum from commercial and military surplus equipment. Many of the chassis have self-clinching threaded nuts, which can be difficult to remove. The fasteners need to be removed from panels being reused to return to a flat, smooth working surface.

I have resorted to drilling out the fasteners or pounding them out with a drift punch. I also tried threading a screw into the fastener from the side opposite the rim and then pounding on the screw head to drive out the fastener.

These techniques take a lot of time, are hard to do in limited spaces, and can bend, dent, scratch, or otherwise mar the surface.

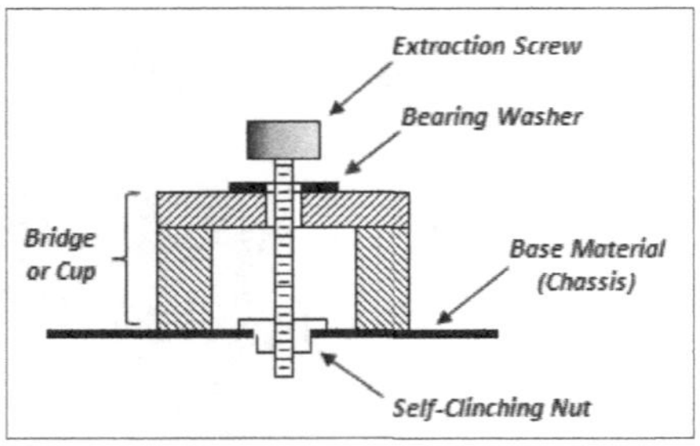

Figure 4 — A diagram of the author's self-clinching nut extractor. [John M. Franke, WA4WDL, image]

I did an online search for a better solution, but found there are plenty of videos on installing threaded fasteners, and very little on how to remove them.

My current idea, shown in **Figure 4**, is based on the general idea of a pulley puller. A cup or **U**-shaped structure with a central hole is placed over the rim side of the fastener and a screw is passed down through the hole and threaded into the fastener. Once the screw head reaches the top of the structure, further tightening will extract the fastener. A steel washer is used under the screw head as a bearing surface and allows the use of different sized extractor screws.

Figure 5 shows three designs I use. The first design (on the left) is a cup-shaped, hollow metal equipment foot and works well. The second (in the middle) is a bridge made from a salvaged aluminum bracket and improves on the first by allowing me to better see the fastener as I insert the extraction screw; the smaller footprint improves the chances I can insert the tool in narrow or cluttered areas. The third design (on the right) is a bridge made from a short bar and two metal standoffs, and improves on the second by being shorter and more compact.

There are several points to follow when making one of these designs. The base opening must be wider than the fastener rim. The extraction screw should be steel to reduce the chance of it breaking off in the fastener. The internal height should be higher than the thickness of the fastener rim to allow for full extraction.

The screw hole should have a diameter large enough to pass the largest thread size for the extraction screw. I typically make the hole large enough to pass an 8-32 screw. An appropriate washer is used under the extractor screw head to permit use with other sized extraction screws, such as 6-32 and 4-40.

To reduce chances of the screwdriver slipping, any screws used for the extraction screw should have Philips, Torx, or Allen heads. I can now remove most self-clinching nuts rapidly and without damaging the base surface. — *73, John M. Franke, WA4WDL,* **jmfranke@cox.net**

Meter Magnet Magic

There is no set screw on the Drake 4-line meters to zero the meter if the needle is somehow below or above the point it should rest while powered off. On the plate current meter, the point should read 0, and on the S-meter it should read 1. The out-of-place needle affects the accuracy of its readings. Here are some options for fixing this issue.

Production of these meters stopped years ago, but if you can find another on eBay, you can use the scale from a working S-meter in the plate current meter with the broken needle, or vice versa. In that case, carefully remove the plastic front, which is glued to the frame of the meter's metal case.

There is a simpler way, however. By moving a small magnet on the meter's case, you can pull the needle below or above where it should rest while powered off. When you've found the right spot, put a bit of glue on the magnet and attach it to the meter (see **Figure 6**). Few will notice the modification if they open the cabinet for visual inspection.

Don't worry that the magnet will affect the linearity and accuracy of the meter. I connected a repaired plate current meter with an original plate current meter in parallel and gradually increased voltages from zero. Both meters showed 1.4 V at full deflection and moved at 50 mA intervals at the same pace. — *73, Peter Ravn, OZ8CTH,* **fasan5@privat.dk**

Revamping an Analog Voltmeter's Power Supply

Even though I own several digital voltmeters, there's nothing like a big analog movement for fine tuning, which is why I often turn to my old-fashioned Heathkit IM-5228 vacuum-tube voltmeter (VTVM).

The Heathkit VTVM circuits were pretty

much the same over several incarnations and, like other VTVMs of the day, used a C or D cell battery to power the resistance function. The Heathkits specified a zinc battery (versus a modern alkaline) that was supposed to provide a fairly accurate voltage to use in calibrating the meter. It seems those old cells were pretty good at keeping a value right around 1.52 – 1.54 V, which would stay close even as the battery aged. For cells today, the current draw in the resistance mode can go well over 100 mA, which can draw a battery down with heavy use. As I didn't look forward to having to open the box and replace the battery, I decided to make an adjustment.

I had a *Popular Communications* article that I held onto so one day I could change the circuit from a battery to a powered circuit adjunct — one that would consistently provide a stable reference voltage. It is a simple full-wave bridge circuit driving a LM317 voltage regulator with the voltage set using a couple of precision resistors.

I built the circuit on a terminal strip (see **Figure 7**) and when I fed it 6.3 V ac from the bench supply, I got nothing out the other side. I was confused, because there didn't seem to be anything wrong with my wiring. I replaced the regulator next, just in case it was bad, but there was still no change.

I finally checked the resistor values and saw that one was defective. When I put a new resistor into the circuit, I got a very nice 1.54 V dc out the other end of the regulator. I then mounted it in the VTVM (see **Figure 8**).

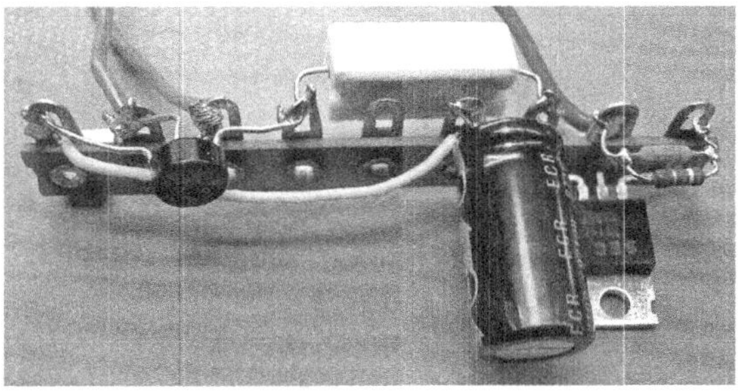
Figure 7 — The terminal strip with the supply circuitry. [Dino Papas, KLØS, photo]

After that, I added a pin jack to the front panel so I could access the voltage whenever I needed to double-check the calibration of the meter. The function knob on my VTVM was cracked, so I super-glued it back into place. Finally, I checked the calibration of the meter against the Agilent digital multimeter and it turned out very close, which is good enough for me. — *73, Dino Papas, KLØS,* **kl0s@cox.net**

Signal Generator VFO for Vintage Transmitters

I recently acquired a 1960s-era Heathkit DX-60 transmitter in good shape, but it had no matching VFO. The HG-10 VFO is the companion to this transmitter, but that VFO's price is more than I paid for the transmitter. I realized I had an Eico 324 Signal Generator on my bench, which I had found at an antique store for $10 (see **Figure 9**). I decided to modify the signal generator to use it as a VFO.

I lifted the cathode of the oscillator above ground by clipping the short lead from the 12AV7 cathode and routing it to a small switch on the front panel through RG-174 coax. The switch fit nicely through a hole that housed the binding post for the ground on the audio input to the signal generator. I also added a small trimmer capacitor of 15 pF on the front panel to fine-tune the frequency (see **Figure 10**).

In use, the output cable of the generator is connected directly to the VFO input jack of the DX-60. I flip the switch to turn the oscillator on and set the function switch on the DX-60 to the **TUNE** position and tune for maximum grid current, then reduce the drive to the recommended 2½ mA. I listen for the signal on my receiver with my audio reduced, then change to the CW position and tune up the transmitter with the tune and load controls. The signal generator is on the entire time I'm transmitting, and I turn it off when I switch to receive. The best settings I have found on this generator are maximum output on the fine and coarse output controls. If you try this, set the audio to internal, but reduce the level to zero. Failure to do so may result in a modulated CW (MCW) signal, which is not a legal form of CW transmission in Canada, where I live.

The results have been relatively good for equipment of this vintage. I get good results running about 50 W into a ground plane. There is the occasional report of a slight chirp, but that's not too bad for over 50-year-old equipment. The measured drift of the VFO is about 1½ kHz over a 2-hour period. I have no way of accurately measuring the output of the VFO, but it is higher than the few crystals I have installed in the DX-60. In fact, the grid reading goes off the scale.

The stability is good on 80, 40, and 20 meters, but not above that, because of frequency multiplication three and four times over. Although the added trimmer capacitor does offset the frequency somewhat, it is not critical, because I can measure the frequency with my Drake receiver or Icom IC-706, which I know to be accurate. Other signal generators may require the use of an amplifier to increase output if the drive into the DX-60 is not high enough. — *73, Art Horovitch, VE3AIH,* **suzanandart@gmail.com**

Figure 8 — The terminal strip mounted in the VTVM, after replacing the defective resistor. [Dino Papas, KLØS, photo]

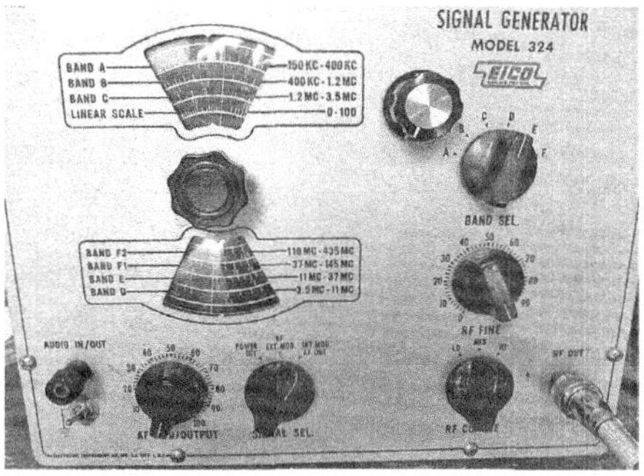

Figure 9 (top) — The EICO 324 Signal Generator, which the author modified to serve as a VFO. The switch fit through a hole in the signal generator. [Art Horovitch, VE3AIH, photo]

Figure 10 (bottom) — The location and wiring of the trimmer capacitor connected to one leg of the main tuning capacitor. [Art Horovitch, VE3AIH, photo]

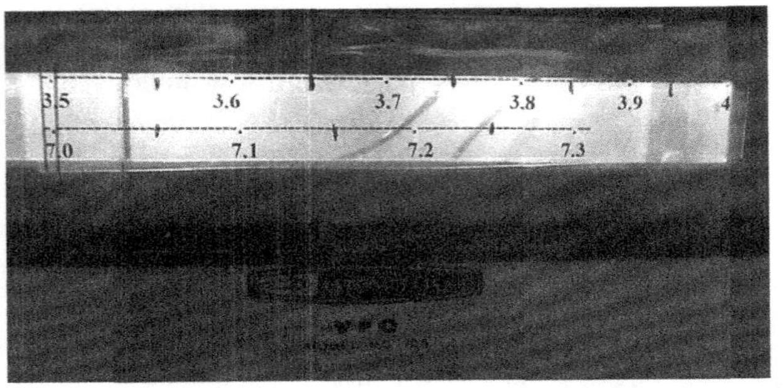

Figure 11 — The new slide rule dial, marked using Microsoft *Word*. [Arthur Horovitch, VE3AIH, photo]

An Improvised Heathkit Slide Rule Dial

I recently acquired a Heathkit HG-10 VFO for my 1960s vintage station. The original rotating band dial typical of most Heathkit rigs of that time was slipping and cracked. It actually started to fall apart as I changed bands, but the VFO was in good condition otherwise.

To fix this, I removed the rotating band dial and used Microsoft *Word* to print a slide rule dial on a transparent sheet, then taped it over the display window (see **Figure 11**). I first attempted to make a new slide rule dial using a permanent marker, but decided that creating one using *Word* would look more professional.

My new dial only covers 80 and 40 meters, but the lower band edge is quite accurate, so it isn't hard to extrapolate to 20 meters. I do frequently check the band edge using my Icom IC-706 MKIIG as a receiver. — *73, Arthur Horovitch, VE3AIH,* **suzanandart@gmail.com**

Easy Vintage Dial Face Replacement

In the process of restoring a Hallicrafters S-38, I changed and replaced various things, but I was stumped by how to fix the clear plastic window that protects the half-moon main tuning and bandspread dial. The window had darkened badly over the past 70 years since the radio was built, so it was almost impossible to read the frequency.

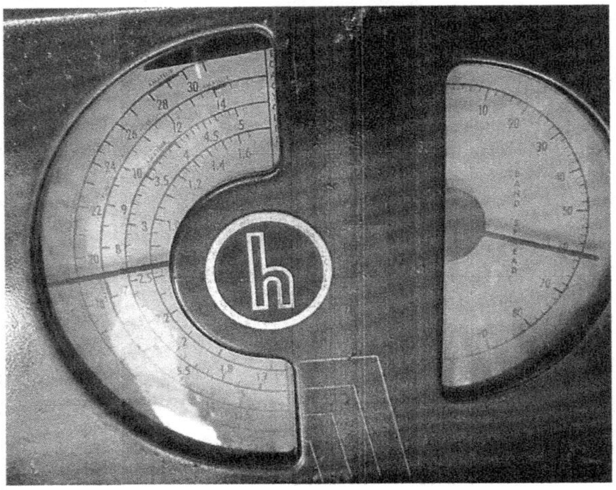

Figure 13 — The Hallicrafters S-38 with the new dial face installed. [Art Horovitch, VE3AIH, photo]

I decided to use a clear plastic "clamshell" container that had once contained peaches. I emptied the container and cut the top off with a pair of scissors. Then I removed the existing plastic window from the cabinet of the S-38, which was held in place with metal tabs and black tape. Once this was done, I traced out the oval shape of the plastic and marked the location of a small stud, which helped to center the plastic protector (see **Figure 12**). Finally, I replaced the tape with new electrical tape, not bothering with the metal tabs. I now have a clear view of the dial and the radio looks much better (see **Figure 13**). — *73, Art Horovitch, VE3AIH,* **suzanandart@gmail.com**

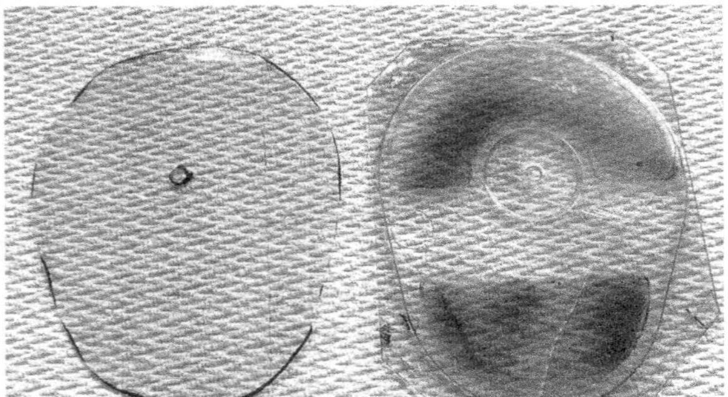

Figure 12 — The new dial face, made from a plastic fruit container (left), is much clearer than the original, 70-year-old dial face (right). [Art Horovitch, VE3AIH, photo]

Amplifier Tube Socket Repair

I noticed that as I was tuning up my Drake L-4B HF amplifier to chase a DX station, it was only putting out about half its normal power. I soon discovered that one of the two 3-500Z tubes was not "lit" (the filament was not on). After taking the amp apart, I found that the dark tube had a filament pin where the solder had slightly reflowed downward out of the pin — an indicator of excess heat at that pin, probably caused by a poor connection. I thought the tubes were just showing their age, so I replaced them with a pair of new units from RF Parts (**www.rfparts.com**).

The same half-power fault occurred again, and when I checked I found the tube in the same position as before that would not light. I disassembled the amp and found the same filament pin solder reflow issue had manifested itself on the new tube (see **Figure 14**).

I looked for solutions online and found a forum where this exact problem was being discussed. One contributor wrote that the spring clips that compress the small split-style tube holding connectors into which the tube pins are inserted can lose their "springiness" over time and that it was a simple affair to either replace the clip or attempt to squeeze it back into proper shape. This wouldn't be a great option for me, because it had already deformed previously and would be prone to doing so again. Fortunately, I had a few spare 3-500Z sockets and I found that it was very easy to remove one of the good spring clips.

I went back into the amp, replacing the spring clip in less than a minute. The difference between the old and new clip was substantial (see **Figure 15**).

I then "reflowed" the solder in the offending filament pin by turning the tube

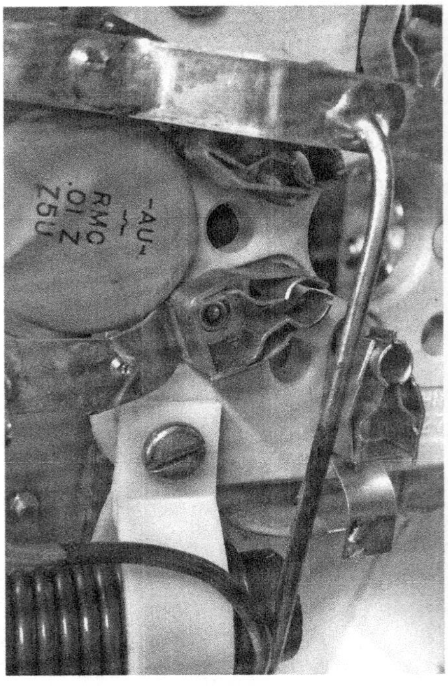

Figure 14 — The loose tube socket pin (compare the clip closure to the others). [Dino Papas, KLØS, photo]

upside down, carefully applying heat to the pin and adding a small amount of silver solder to the mix. After letting the pin cool, I checked the continuity between the two filament pins and it was still okay. I made the same repair to the original tube, since it's now part of the backup pair. With the tube reinserted into the socket, the difference was apparent (see **Figure 16**).

If you have an amplifier, you may want to check the tube socket pins. Be careful, however, because the voltages inside your amp are very dangerous and can prove lethal. Always make sure the power is disconnected, the high voltage is bled off properly, and that you never defeat any voltage interlocks. — *73, Dino Papas, KLØS,* **kl0s@cox.net**

Rejuvenating Tube Sockets

Dino Papas's, KLØS, "Amplifier Tube Socket Repair" item in the February 2020 "Hints & Hacks" column reminded me of a fix we had for seven- and nine-pin tube sockets when I worked in a shop that repaired TV tuners in the 1970s.

The procedure involves inserting a dental pick between both sides of the metal clip and

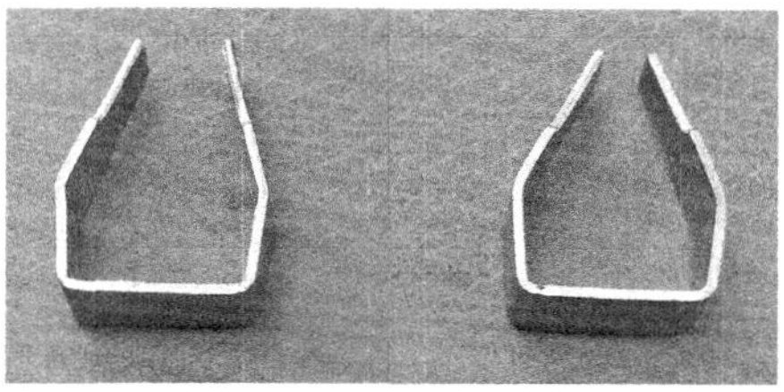

Figure 15 — The loose clip (left) compared to the replacement clip (right). [Dino Papas, KLØS, photo]

Figure 16 — The new spring clip compresses the pin holder tighter than before. [Dino Papas, KLØS, photo]

Figure 17 — The five 6 V batteries wired in series and installed in the old volt-ohmmeter. [Robert Miller, KE6F, photo]

the plastic socket housing, gently massaging the metal back into shape to retighten it, being careful not to crack the plastic. From there, squirt contact cleaner (non-lubricating) into the holes and run the tube in and out of the socket a few times to clean off any oxides or carbon that may have built up, and to ensure proper fit. The excess cleaner can be allowed to dry or be blown off with dust remover spray. Be careful not to over-tighten the clips, or damage to the tube can result when reinserting it. This method also works very well on the sockets of older, larger tubes. — *73, Joe Falletta, W6UDO,* **w6udo@arrl.net**

A New Battery for an Old VOM

Many hams still own and use classic volt-ohmmeters (VOMs), like the Simpson 260 or Tripplett 630-PL. The downside of owning a classic VOM is that many of these units require 30 V batteries for their high-range ohmmeters. These batteries have become expensive and difficult to find.

I found a solution in a collection of five "anti-bark dog collar batteries" that are available on Amazon for less than $6 for a set of five. Each 6 V battery is compact, and the terminals take solder easily. This makes it a simple process to wire them together in series to provide 30 V (see **Figure 17**). A few inches of double-sided tape mounts them inside the meter, and with two soldered connections to the existing wiring, the meter is ready for service. — *73, Robert Miller, KE6F,* **millerke6f@aol.com**

Making Vintage Gear Safer

To bring vintage equipment up to modern electrical standards, you can replace the dangling two-conductor ac cord with a three-conductor socket with the third conductor grounded to the chassis (see **Figure 18**). You can install either a standard socket or a filtered model to help reduce RFI.

Once you remove the old cord, you'll need to widen and reshape the opening to accommodate the new socket. Steel will resist nibbling tools, so you may need to use a power tool. With the socket installed, you can now power the equipment with a modern, three-wire cord. This simple change will ensure a safety ground on gear that may otherwise present a potential hazard. Before you perform this modification, however, check the equipment to make sure that one side of the ac line is not grounded to the chassis. Some older types of gear used this approach and, if so, this modification should not be attempted.— *73, Patrick Hamel, W5THT,* **w5tht@arrl.net**

Replacing an Old Battery Pack

I really like my old Kenwood TH-21AT 2-meter handheld, even if it's more than 30 years old. The radio continues to work well even after all these years, but the old NiCad

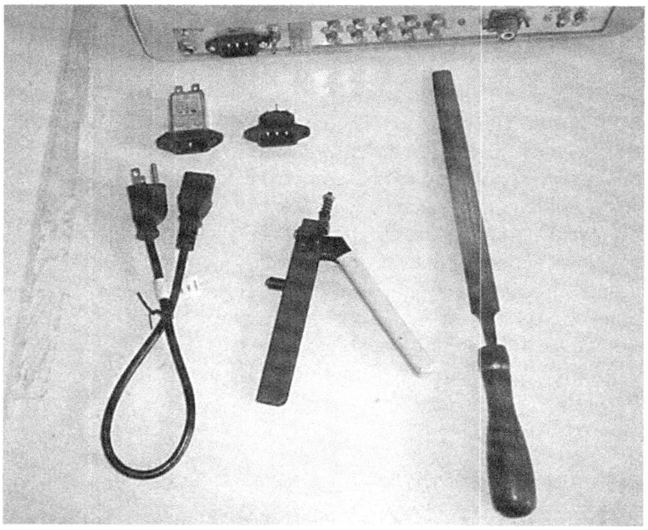

Figure 18 — You can enhance the electrical safety of older gear by replacing the two-wire ac cord with a socket to accommodate a three-wire cord. Both filtered and unfiltered sockets are shown here. [Patrick Hamel, W5THT, photo]

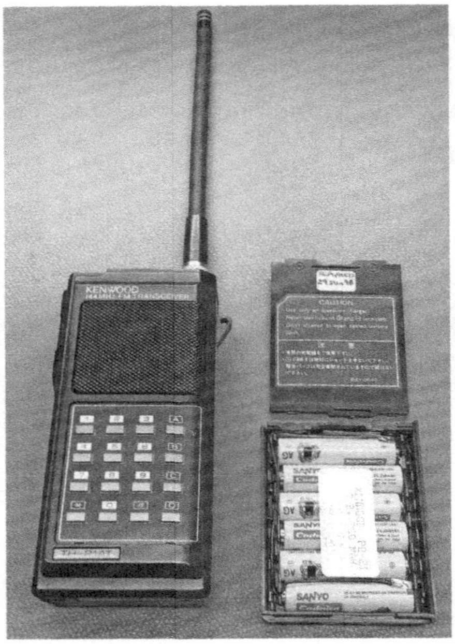

Figure 19 — The Kenwood TH-21AT and its battery pack with the cover surgically removed. The batteries are not the originals. [Dino Papas, KLØS, photo]

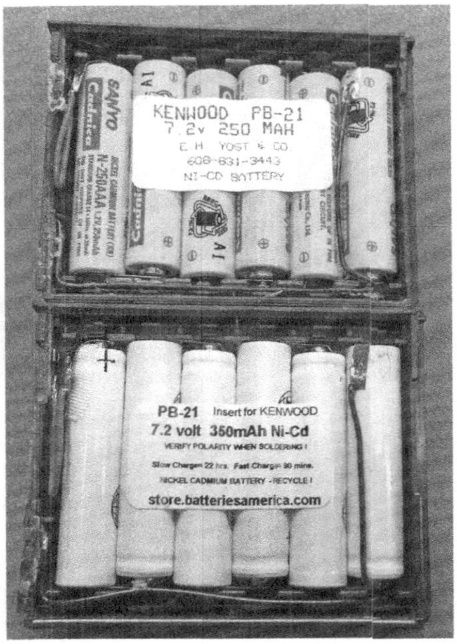

Figure 20 — The old batteries above and the new replacement batteries below. Use quick-bond glue to reattach the case cover. [Dino Papas, KLØS, photo]

batteries only last for so long (see **Figure 19**). When it's time to replace the battery pack in a radio this old, you might be fortunate enough to find that replacements are still available from the manufacturer. However, those replacements can be expensive. The alternative is to rebuild the battery pack yourself.

Most of the old battery packs rely on a collection of individual AA batteries. The packs aren't really designed to be opened; the idea is that manufacturer would sell you an entirely new pack. It takes a bit of work, sometimes with a blade or Dremel tool, but once the seal is broken around the battery cover's perimeter, it's easy to open (see **Figure 20**).

Companies like Batteries America (**www.batteriesamerica.com**) sell complete cell packs that will fit within the case. Just unsolder the two connecting wires, remove the old pack, solder the wires onto the new pack, then pop it into place. Reattach the pack cover with a small line of quick-bond glue around the edges and you're all done. Note that you may have to use a different charging system for a different battery chemistry (for example, replacing NiCD batteries with NiMH). — *73, Dino Papas, KLØS,* **kl0s@cox.net**

A Keypad Solution

I found a good way to recondition keypads that use conductive rubber contacts. I had an old Icom 24AT hand-held transceiver and its keypad button had become intermittent. I loved the old radio, so I wanted to see if could repair it myself. The procedure I used was a little unorthodox, but it worked.

I cut a thin strip from a foil gum wrapper and trimmed it into tiny squares. Next, using the tip of a fine-point knife, I put a bit of RTV silicone caulk on the back surface of each button and then attached the foil squares (see **Figure 21**), with the foil side up. I let it cure for several hours and was happy to discover that it is a permanent fix. Strange as it may seem, gum wrappers are ideal for the job because they offer a thin paper backing that adheres well to RTV.

The inside front needed a bit of repair to the four plastic rivets holding the shield (see **Figure 22**), but with some plastic cement it was soon ready to reassemble. The keypad works like new. — *73, John Gebuhr, WBØCMC,* **wb0cmc@arrl.net**

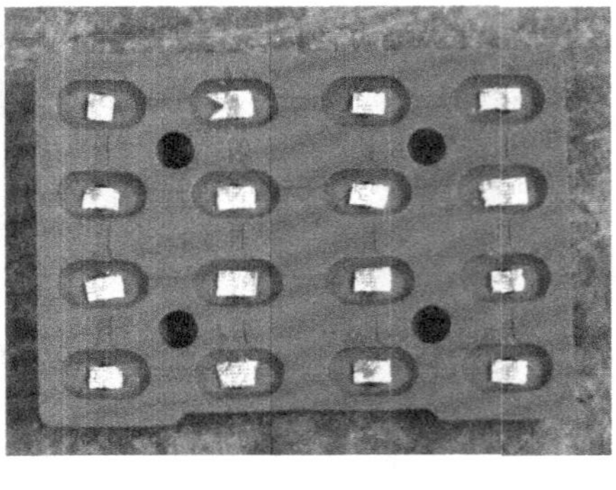

Figure 21 — Place the tiny foil wrapper squares onto the inside back surfaces of the buttons, with the foil side up. [John Gebuhr, WBØCMC, photo]

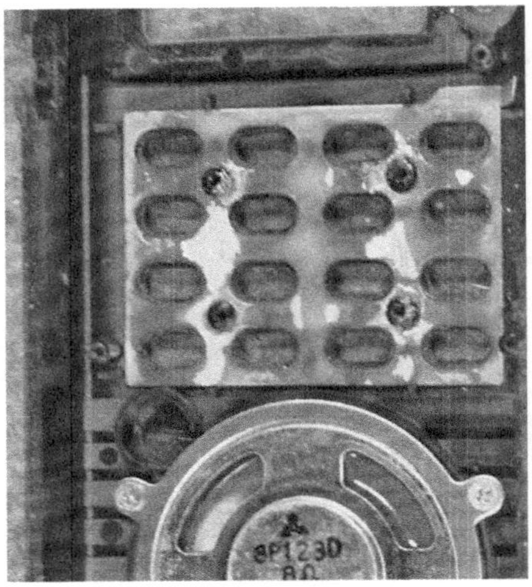

Figure 22 — The inside front needed a little help from some plastic cement, but once everything was reassembled, it worked perfectly. [John Gebuhr, WBØCMC, photo]

Isolating Old Radios

I recently acquired a Hallicrafters S-38 receiver at a hamfest for a bargain price. This model uses a two-wire power cord with no ground. There is no power transformer in the radio, so everything runs directly off the ac line. This can create a shock hazard, depending on how the plug is inserted into an ac receptacle, causing the metal case to be dangerously live.

My solution, at least for testing purposes, was a 60 W isolation transformer. You can purchase these new, but they are also common at fleamarkets and other outlets. The isolation transformers typically have multiple taps on the input (primary). Mine had taps for 120, 208, 230, and 460 V ac. Consequently, by

Figure 23 — A 60 W isolation transformer mounted onto a case with a neon light indicator. [Arthur Horovitch, VE3AIH, photo]

selecting the correct input, the output could be set at 120, 75, 65, or 45 V ac.

I selected taps to start at lower voltages to test the receiver's electrolytic capacitors. After replacing one capacitor and running for several hours at full power, the radio worked perfectly.

This testing method also worked on my vintage Heathkit HR-10B receiver and DX-60B transmitter. I decided to make the transformer a permanent part of my workbench.

I mounted the isolation transformer on a case (see **Figure 23)** and attached a neon indicator light to the case using a hot glue gun. It wasn't elegant, but it was quick, and it worked. I also installed a switch for the inputs, so that I could select either a 120 or 65 V ac output. A multiposition rotary switch would be even better for this purpose. Also consider a larger transformer for testing bigger radios, such as a HW-100 transceiver. — *73, Arthur Horovitch, VE3AIH,* **suzanandart@gmail.com**

Vintage VFO Connector

Years ago, many of the popular outboard VFOs were supplied with adapter cables that plugged directly into an FT-243 crystal socket. You could also use the Mosley 301 twin-lead connectors that were a close fit. Sadly, the supply of both of those connectors has vanished. You can sacrifice an old FT-243 crystal to make the connector, but even those aren't always available from the junkbox.

I realized that there are Molex connectors available with 0.093 pins and I wondered if those might be pressed into service. I picked up some Molex 02-09-2118 male crimp pins and put together the cable end shown in **Figure 24** for use with my Heathkit HG-10 VFO. It worked perfectly; the pins have good tension and stay securely in the socket. I just pinched the crimp ears with pliers and applied dabs of solder and a little shrink tubing. It took all of 10 minutes.

The 02-09-2118 pins are available from Digi-Key, Mouser Electronics, and many other vendors. Pricing is about $3 for 25 pins. — *73, Andrew Jarosik, K1APJ,* **k1apj@arrl.net**

Building Plate Chokes

Lately, I've been rebuilding an old Heathkit SB-200 amplifier. One problem I ran into was the discovery that the original plate choke was broken at the base. I knew getting a replacement was unlikely, so I decided to create my own substitute.

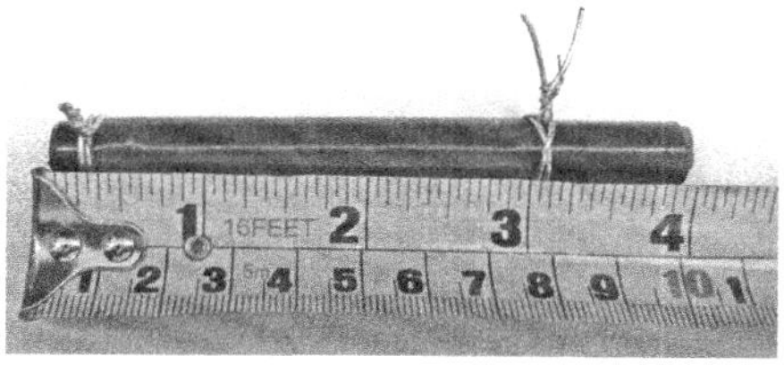

Figure 25 — The new plate choke was created from a 4-inch rod. [Clint Millett, VE3CMQ, photo]

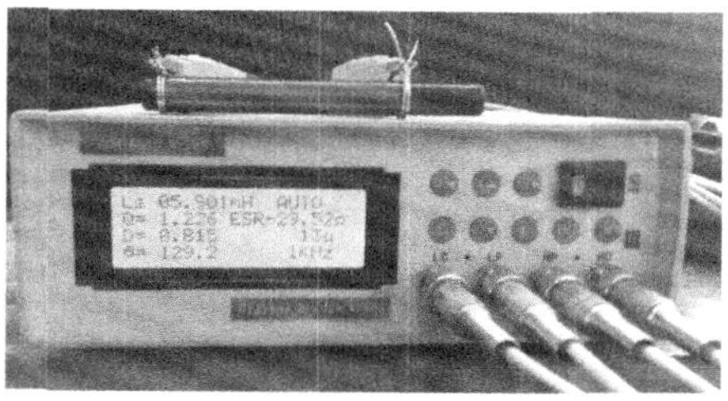

Figure 26 — Testing the 5.9 mH choke. [Clint Millett, VE3CMQ, photo]

I decided to use ⅜-inch fiberglass rod, because it's drillable and easy to work with. All I needed was a 4-inch-long rod, some enameled #28 wire-wrap wire, some heat-shrink tubing, and a variable-speed electric drill (see **Figure 25**).

I began by drilling a small hole ½ inch from one end of the rod along the shaft. At the other end, I drilled a small hole so I could screw in a ¼-inch 4-40 screw to hold the choke to the chassis.

I inserted one end of the fiberglass rod into the drill and an end of the wire into one of the holes. With a bit of tension on the wire, I started up the drill and within a minute I had 3 inches or so of perfectly wrapped wire on the rod. I put the remaining wire end through the other hole, then I covered the coil with the heat-shrink tubing. If you use the heat-shrink tubing with glue inside, the coil will never unwrap on you.

The result was a choke that was 5.9 mH — exactly the same as the original! That may have been a fluke, but I'll take it.

I then tried the same technique using a 4-inch ferrite rod. However, ferrite rod is difficult to drill, so I skipped that step. I first covered the rod with heat-shrink tubing, because it's pretty sharp and I didn't want to damage the wire. Three inches of coil later, I had another 5.9 mH choke (see **Figure 26**). It only took 5 minutes to build these.

I installed the 5.9 mH choke in the SB-200 and it worked very well. I'll probably use the other 5.9 mH choke in my next linear amplifier. Both chokes are good for at least 1A. — *73, Clint Millett, VE3CMQ,* **ve6cmm@secrs.com**

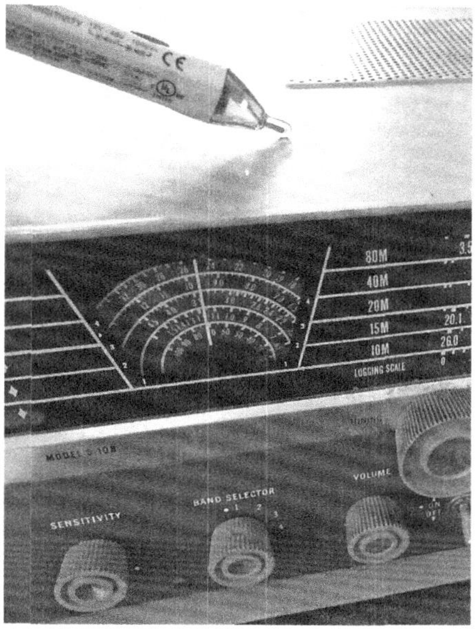

Figure 27 — This non-contact voltage tester is glowing red to warn of an electrically hot case. [Ruth Ring, KJ1V, photo]

Figure 28 — A polarized "cheater" plug offers a temporary solution for making sure the equipment can only be plugged into an ac outlet in the correct position. [Ruth Ring, KJ1V, photo]

The Hot Chassis Test

If you have older equipment with two-prong non-polarized ac cords, you could have a dangerous situation on your hands. A quick safety test is to hold a non-contact voltage tester (NCVT) near the metal case with the unit plugged into an active wall outlet. If the NCVT alarms (see **Figure 27**), the chassis is electrically hot. Try reversing the plug in the outlet and rechecking the chassis with the NCVT. It should indicate that the chassis is safe. You'll find NCVTs for sale online and at hardware outlets for less than $20 – 30.

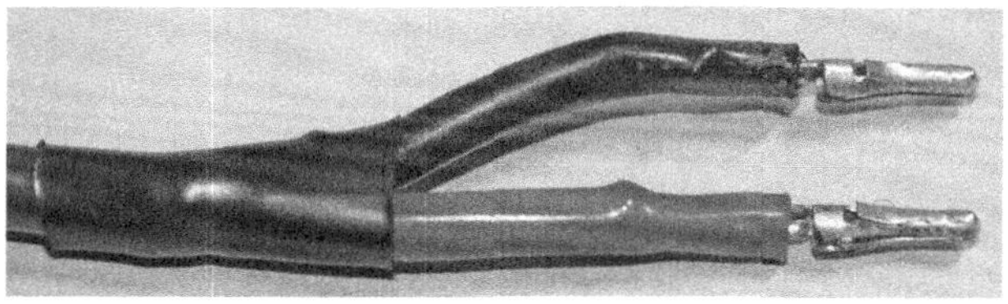

Figure 24 — A cable end for connecting a Heathkit HG-10 VFO made from Molex 02-09-2118 connector pins. [Andrew Jarosik, K1APJ, photo]

For a temporary fix, attach a polarized "cheater" plug (see **Figure 28**), so the unit can only be plugged into an outlet in the correct polarity. Plan on changing to a hardwired polarized cord (or better yet, a three-pronged grounded cord) soon. — *73, Ruth Ring, KJ1V,* **kj1v@arrl.net**

Figure 29 — Old crystals are available, but most aren't at frequencies within the amateur bands. [Joe Scoglio, KA4WJB, photo]

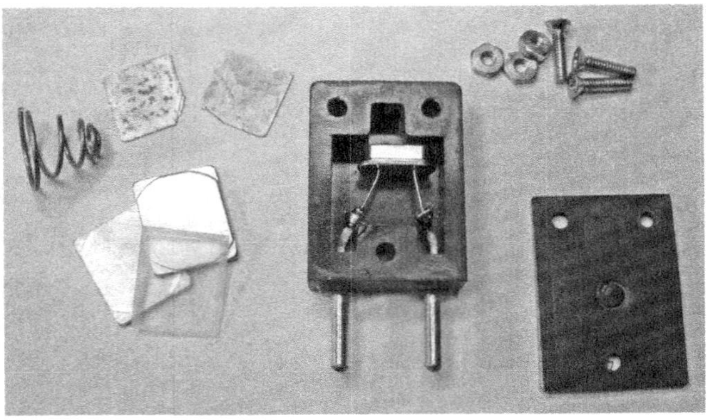

Figure 30 — After the components are removed from the old crystal case and replaced with the new crystal, solder the leads of the new crystal to the internal leads of the case. [Joe Scoglio, KA4WJB, photo]

Figure 31 — The old crystal case is re-labelled with the new frequency. [Joe Scoglio, KA4WJB, photo]

Repurposing Crystals

I enjoy operating several old ham transmitters that require FT-243 style crystals. However, I can never seem to find enough crystals with FT-243 holders at the frequencies I want. I did have some old crystals that weren't at ham frequencies (see **Figure 29**), so I decided to retrofit their holders with modern crystals at the frequencies I needed.

The first step is to remove everything from the interior of the old case (see **Figure 30**). Next, solder the new crystal onto the two internal pins and replace the cover. Finally, relabel the case with the new frequency (see **Figure 31**). This retains the vintage crystal look while providing crystals at frequencies I need. This approach also saves the trouble of building socket adapters. — *73, Joe Scoglio, KA4WJB,* **ka4wjb@gmail.com**.

Replacing a Boatanchor Backlight

I've owned two Realistic (RadioShack) PRO-2006 scanners since the mid-1990s and they're still top-quality analog receivers, covering the frequency range from 25 – 1300 MHz and capable of demodulating AM and narrow as well as wideband FM signals.

One of the historical problems with this series of scanners, starting with the PRO-2004, the PRO-2005 and finally the PRO-2006, is that the electroluminescent (EL) backlight that illuminates the LCD display weakens over time. Both of my displays had become almost unreadable, even under good ambient lighting. Fortunately, it is not only possible but surprisingly easy to replace these backlights, and this fix can be applied to other radios as well.

After disassembling the scanner and removing the front panel, the original backlight is easily unsoldered from the printed circuit board. The backlight then slips out from underneath the LCD display (see **Figure 32**).

The original parts are no longer available, but new EL panels are plentiful and even if they are oversized, they can be trimmed to size to create a suitable replacement. I found new panels on eBay for less than $10 each. You can see the difference in size between the original and the untrimmed replacement in **Figure 33**. I was apprehensive about cutting the large panel, as I wasn't sure whether the panel's brightness level was in some way related to the surface area of the electrical connection made on the top of the panel (you can see the metal connection from the top pin as it runs up along the left-hand side and across the top of the panel). It turns out, though, that as long as a connection is maintained at some point on the top of the panel it will in fact illuminate fully (the second pin connects to the bottom side of the layered panel).

To successfully trim the EL panel, you need a very sharp pair of scissors or a box cutter with a fresh blade; a dull cutting edge may potentially delaminate the panel's layers, which could cause dark spots or even failure. I taped the replacement panel down to my cutting surface and traced the outline of the old panel on it (**Figure 34**). I taped the straightedge down along the lines before making the cuts; that kept the straightedge and blade from wandering as I made the cuts.

After completing the cuts, I applied a small amount of Super Glue® (cyanoacrylate) along the edges to seal them and keep any moisture from migrating into the panel.

With the new EL panel in place, the improvement is remarkable, as you can see in **Figure 35**.

EL panels use an ac power source, usually a simple inverter circuit producing up to 100 V at a frequency of several hundred hertz or more. Although the current needed to power one an EL panel is low, the relatively high voltage posed a potential shock hazard — another reason to seal the edges.

This inexpensive repair should extend the service life of the scanner by many more years. Knowing now that the procedure works, I even purchased a couple of additional replacement panels to salt away for future needs. — *73, Dino Papas,* **KLØS, kl0s@cox.net**

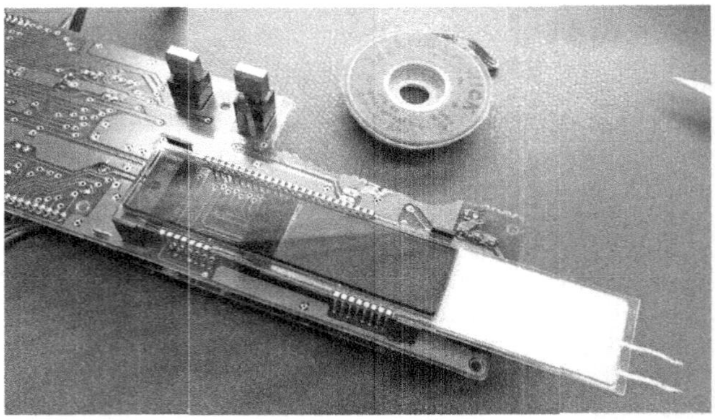

Figure 32 — After unplugging the display board and removing the metal shield, you simply unsolder two leads and slide the original backlight panel out. [Dino Papas, KLØS, photo]

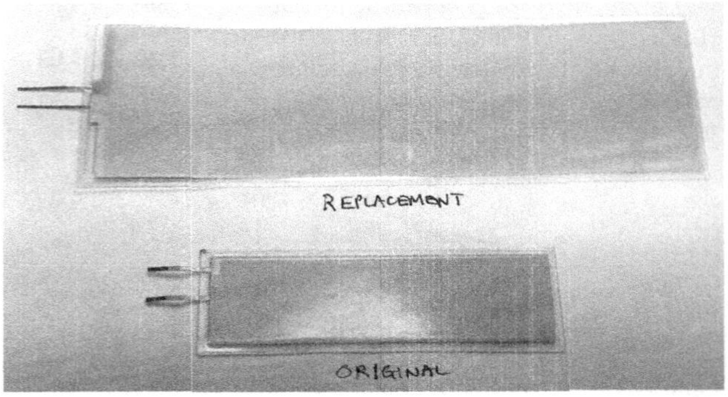

Figure 33 —The size difference between the original panel and the untrimmed replacement is not a problem! [Dino Papas, KLØS, photo]

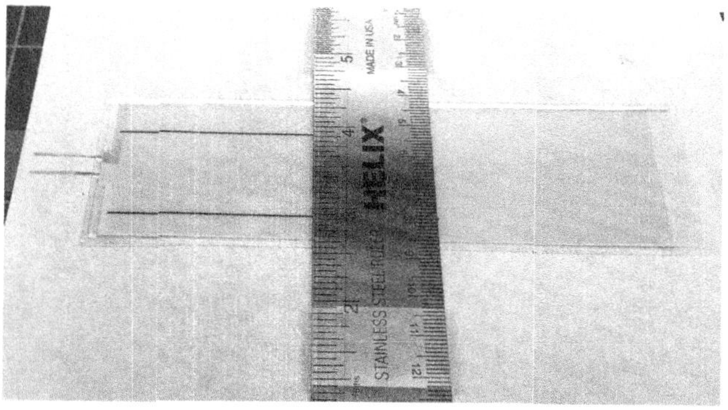

Figure 34 — With a marker and the original panel as a template, trace its outline on the replacement panel prior to cutting. Use a straightedge to guide your blade for clean, accurate cuts. [Dino Papas, KLØS, photo]

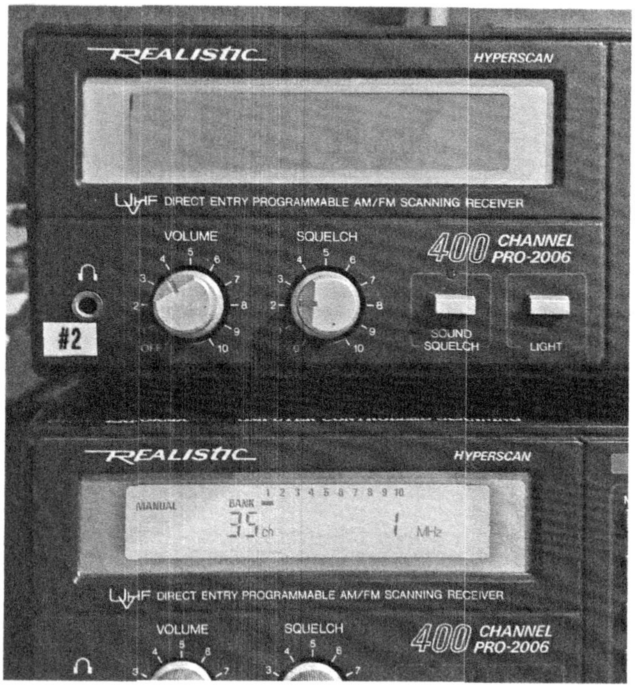

Figure 35 — It's obvious which scanner has the replacement backlight panel! [Dino Papas, KLØS, photo]

Removing Paint or Other Surface Finishes to Improve Electrical Bonding

Improving the electrical bonding connections on your radio or test equipment usually necessitates removing the existing paint or other surface coatings around the screw hole to ensure a quality, metal-on-metal electrical connection. The usual method, after removing the existing screw, is to use a file, rasp, sandpaper, steel wool, or other tools or materials to abrade the paint or coatings off the chassis, and then reinstall the screw. However, that typically results in unsightly blemishes and an inconsistent surface finish that could oxidize or corrode over time.

There is an easy solution, and you probably even have it in your hardware drawer. The center hole of a metal washer actually makes a fine template for confining the area from which paint or coating is removed from a chassis. The result is a clean, contained, bare metal finish that provides a reliable bonding surface for a screw or ring terminal and maintains a professional appearance.

Note that this process will produce airborne dust particles from the paint or surface coating. Therefore, it is recommended that you wear an appropriate face mask, safety glasses or goggles, and other protective clothing as necessary.

Check that the location of the hole has adequate clearance around it. Find a washer that has an inner diameter that just fits over the head of the existing screw. Ideally, the inside diameter of the washer should be at least 1/32 of an inch, but not more than 1/16 of an inch, larger than the outer diameter of the screw head so that not too much bare metal surface shows. **Table 1** suggests washers to use for various screw sizes. **Figure 36** shows a fender washer that just fits over the screw head. Fender washers typically have significantly larger outer diameters, making them well suited for use as templates, though they might not be readily available in all sizes. If the equipment chassis surface is textured, as shown in the photo, or is easily scratched or damaged, you can protect it with a layer of painter's tape or masking tape applied to the bottom side of the washer.

To start, remove the screw from the chassis.

Insert a Dremel® #443, 1/8-inch steel wire brush (see **Figure 37**) into a Dremel tool. You may also use a Dremel #225 flexible shaft accessory if necessary.

The brush should be new or in very good condition. If it has many missing, twisted, or damaged bristles, it may tend to wander out of the washer and might be ineffective at removing the paint or coating.

Center the washer over the screw hole and hold it in place with your fingertips.

With the Dremel tool set at its lowest speed, hold it (or the flexible shaft) like a pencil as you insert the tip of the brush straight into the center hole of the washer.

Keep these points in mind:

■ Let the tool do the work; excessive speed or pressure could make the tool hard to control.

■ Tilting or angling the brush too much could prematurely deform the bristles.

Using a soft plunging motion, carefully and slowly move the brush around within the center hole of the washer to ensure complete removal of all paint or other coating.

Table 1
Suggested Washer Sizes

Screw Size	Washer Interior Diameter (inches)
#4	7/32
#6	1/4
#8 (sheet metal screws with 1/4-inch hex heads)	11/32
#8	5/16
#10	3/8

Figure 36 — The top side of a flat washer has slightly beveled or rounded edges, while the bottom has sharper edges. Place the bottom side against the chassis. [Mark Kupferschmid, AC9PR, photo]

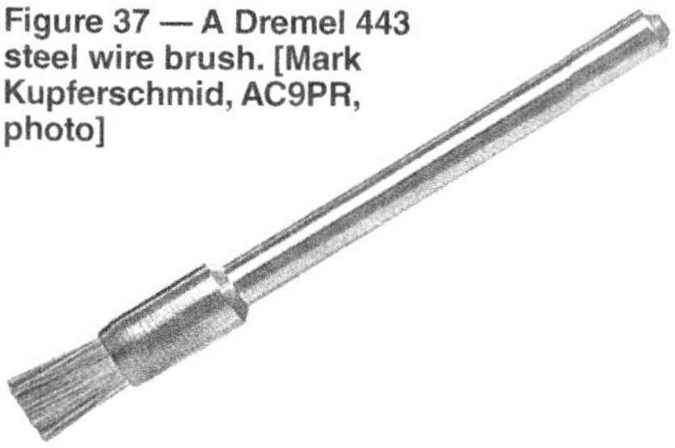

Figure 37 — A Dremel 443 steel wire brush. [Mark Kupferschmid, AC9PR, photo]

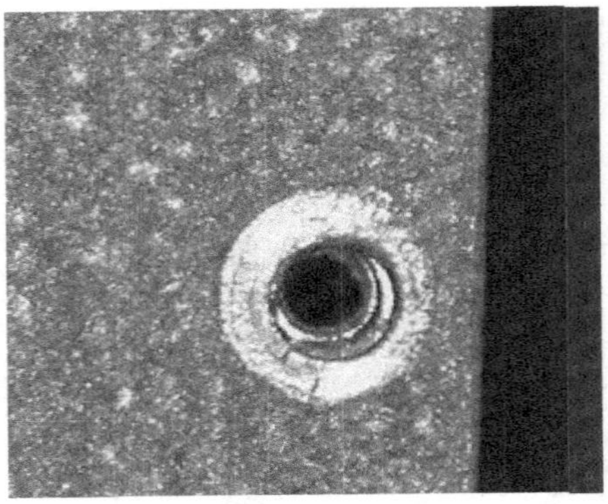

Figure 38 — The paint is totally removed, revealing shiny metal that will ensure a solid electrical connection. [Mark Kupferschmid, AC9PR, photo]

Figure 39 — The screw is reinstalled. [Mark Kupferschmid, AC9PR, photo]

Keeping the washer in place, take the Dremel and brush away and check to see if all the paint or coating is removed yet. If it is all shiny, clean metal (**Figure 38**), you may proceed to the next step; if not, keep working at it with the brush.

Remove the washer and clean the dust away from the area of the screw hole. Install the screw (see **Figure 39**) and repeat the process for any other screw holes used for electrical bonding.

For chassis locations where there is simply not the space to place a large flat washer over a screw hole, one solution may be to fabricate your own templates from soft steel. Cut a length of steel plate and carefully drill a hole of the appropriate size on one end. You can bend the steel plate if needed to make the grip more comfortable. Be sure to remove any burrs and rough edges. — *73, Mark Kupferschmid, AC9PR,* **mkkupferschmid@gmail.com**

Electrolytic Capacitor Restoration

The cost to replace electrolytic "can" capacitors in vintage radios sometimes exceeds the values of the radios themselves. A quad 40 μF, 450 V capacitor in a typical FP style can (twist-lock) can often cost $40 for an original unit, and even then, its condition may be questionable. At best, you'd be gambling on its reliability. While several experimenters have managed to stuff modern replacement capacitors into original FP cans, that can be a messy, time-consuming operation, and the results are often ugly.

An alternative is to fabricate a small look-alike package with modern capacitors glued inside. I bought a bunch of 47 μF radial lead capacitors from numerous vendors on eBay for around 60 cents each, in quantities of 10.

My scheme dispenses with the vintage capacitor can and instead uses a section of square aluminum tubing, or sometimes the can of an old IF transformer. I insert four capacitors into the container, oriented so the negative leads are toward the center, and attach them with E6000 glue (see **Figure 40**). I usually solder turret terminals to the capacitor leads (see **Figure 41**), but you can also form the leads into loops that can be soldered, instead (see **Figure 42**).

I sometimes leave the original FP capacitor in place and attach the new capacitor *under* the chassis with E6000 glue, as it only takes up a couple of square inches of space. This is for aesthetic purposes, because it allows the chassis to appear unchanged, at least when viewed from above. The bond of the glue is quite strong, but the package can be removed by prying with a screwdriver. The scheme also leaves the venting areas of the capacitors open to the atmosphere, should they decide to give up the ghost.

The total cost, not counting the glue, is less than $3. It takes around 30 minutes to fabricate the unit, not including glue setup time. — *73, Robert Miller, KE6F,* **ke6f@arrl.net**

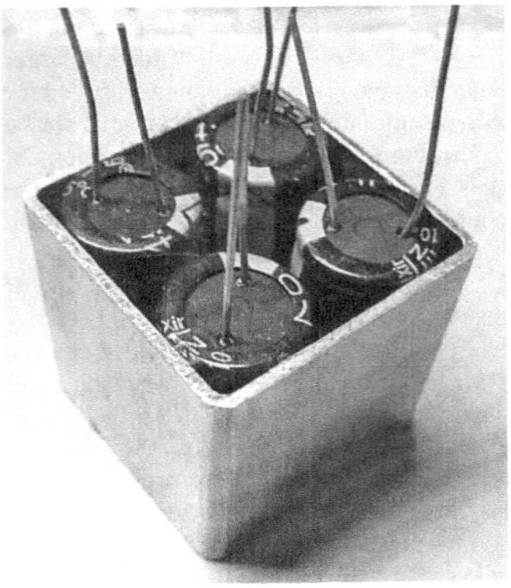

Figure 40 — Four 47 μF radial lead capacitors inserted within the new housing. Glue will hold them in place. I inserted a felt pen in the center (not shown) to hold the capacitors in position while the glue dried. [Robert Miller, KE6F, photo]

Figure 41 — This is a finished package with turret terminals. [Robert Miller, KE6F, photo]

Figure 42 — The finished unit, with the negative leads joined at the center and solder-loop connections for the positive leads. [Robert Miller, KE6F, photo]

Adding Antennas to the Drake MN-2700 Tuner

I have never used the direct wire inputs on any antenna tuner; I just use coax in the shack. Through experimentation, I found that long wires or twinlead into the shack causes RFI on my equipment, so I always put a small tuner or tuning circuit outside and run coax from it to the **ANT** switch inside. This leaves me with two unused **ANT** switch positions on the Drake MN-2700 tuner, which could be converted to two additional SO-239 coax **ANT** terminals. I modified my tuner to add two more switch-selectable coax **ANT** ports (for a total of five) by removing the little fiberglass plate on the back containing the three screw terminals for wires. I then fashioned a small aluminum plate to fill the hole, and I drilled the plate to hold two more SO-239 coax receptacles (see **Figure 43**).

If you look at the **ANT** selector switch inside the MN-2700, the last two selections (clockwise from the **ANT** switch) are for long and balanced wire inputs, but they employ straight wires to the switch. If you unsolder the two wires from the screw terminals on the fiberglass plate, those wires can be extended to the center conductor pins of the two new SO-239 coax receptacles. Then you'll have five selectable coax **ANT** inputs to the tuner, starting from left to right on the back: 1 (dummy load), 2 (**ANT** 1), 3 (**ANT** 2), 4 (first new SO-239 receptacle), and 5 (second new SO-239 receptacle).

Place the holes for the new SO-239s as high as you can on the replacement aluminum plate. The lower SO-239 is close to a lug on the big coil inside the tuner, so the higher up on the plate, the better.

Initially, I was going to buy a $500 automated antenna switch before seeing several hams criticizing the cost and features. Then it occurred to me that there were two unused **ANT** switch positions on the MN-2700, so I converted them to SO-239 ports instead. — *73, Larry Serra, N6NC*, **n6nc@arrl.net**

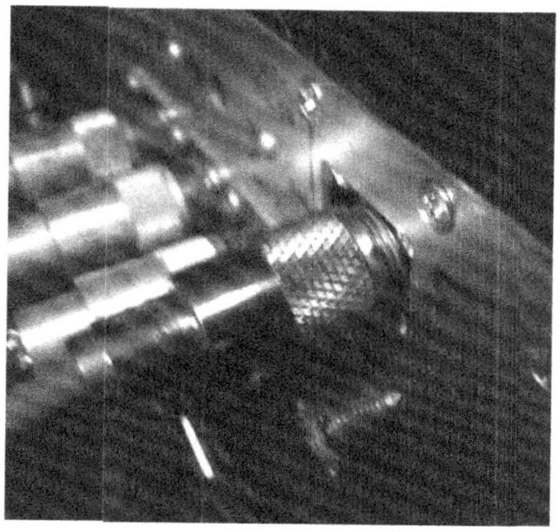

Figure 43 — The modified Drake MN-2700 tuner back panel. [Larry Serra, N6NC, photo]

Hiding Defects on Worn Chrome

The Molotow company (**www.molotow.com**) makes paint markers that simulate chrome plating. Their liquid chrome mirror-effect markers are available in several tip sizes — I prefer to use the 4-millimeter tip. I have found this product to be useful for touching up chrome radio equipment. My microphone had spots on the head where the chrome had flaked off and left the dark base metal exposed (see **Figure 44**). I touched it up with this product, and I think it did an excellent job. I do not think a paint marker could perfectly cover large areas of worn chrome, but it works for chips and scratches. This marker is available from many hobby shops, large retailers, and Amazon. — *73, Tom Sherwood, W8AAZ*, **w8aaz@frontier.com**

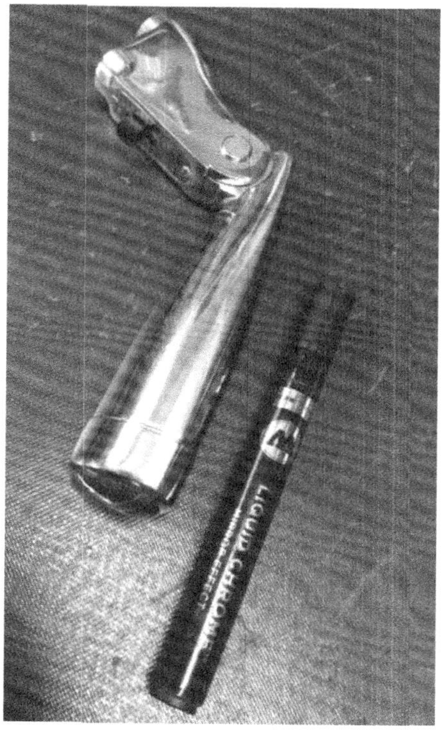

Figure 44 — Molotow liquid chrome mirror-effect paint markers can hide minor damages to radio accessories. [Tom Sherwood, W8AAZ, photo]

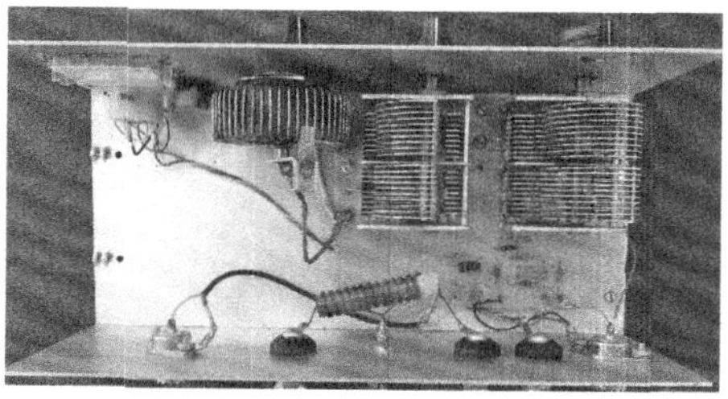

Figure 45— A view inside the TEN-TEC 277 antenna tuner. The toroidal coil and the rotary selector arm are to the left of center. [David Vine, WA1EAW, photo]

Emery Boards to the Rescue

My vintage TEN-TEC 277 antenna tuner uses a novel approach to select the desired inductance. It involves a rotary mechanism that contacts the windings of a toroidal inductor (see **Figure 45**). The contact on the rotary arm had become corroded with age, and oxidation built up on the coil windings. Together, they caused erratic operation on several bands. I tried cleaning the coil with sandpaper, but I couldn't reach the wire surfaces, even when using a flat screwdriver with sandpaper wrapped around the blade.

I considered emery boards used for filing nails — they would be perfect for the job. The boards I acquired were roughly 0.5 inch wide, 4 inches long, and 0.05 inch thick. This gave me plenty of room to clean the coil. To clean the rotating arm, I removed the screw, the nut, and the two plastic insulators located on the left rear side of the tuner. Once finished, I reassembled everything.

I also had trouble with the shunt capacitor. Whenever I turned the knob to the number 7 position and released it, the capacitor would drop to the number 10 position instead. I discovered that the center plate separating the two rotors was the main culprit. It was supposed to provide enough resistance to them, but it no longer did. After cutting an emery board into a small rectangular piece, I jammed it into one side of the center plate to increase the friction. The tuner now works great!

You may not own a TEN-TEC 277, but these techniques can be applied to many older antenna tuners. — *73, Theodore Turk, WB8ADA*, **hammytee@hotmail.com**.

CHAPTER 8

Software and Computers

Take Advantage of the Data Socket

To connect your computer and radio for digital modes, there are several workable options, but some are more convenient than others. When changing from operating SSB to PSK31, do you need to lower your microphone gain and disable the speech compressor? When you later return to sideband, how long does it take you to restore those pileup-busting settings? A good digital interface setup allows you to change modes with minimal adjustments.

On the computer side, you can use the built-in sound card, or add a second sound card dedicated to digital modes. If your shack computer does double duty as a media center, a second sound card means you no longer need to adjust volume and line-in sensitivity whenever you switch from playing music and videos to digital modes and back. In addition, many secondary sound cards connect via a single USB plug, while your main sound card typically requires one cable each for audio input and audio output.

On the radio side, you can connect via the microphone input and headphone/speaker output, or use the six-pin Mini-DIN "data socket" in the back of most current transceivers. The data socket offers fixed-level audio output, allowing you to freely adjust the radio's front panel volume to monitor receive. The data socket audio input is not affected by compression or microphone gain settings, so you can leave them as-is when switching between SSB and digital modes.

The RIGblaster "Advantage" model combines a USB sound card with rig control (CAT). Its manual only describes setups based on the rig microphone and speaker connectors, while I wanted to take advantage of the Mini-DIN in the back of my radio. Upon investigation, I found that hooking up the Advantage to the data socket is actually very easy. All you need is a homebrew cable, a six-pin Mini-DIN connector (Mouser 806-KMDAX-6P), and appropriate jumper settings.

Table 1
Cable Pinout

RJ45, 8P8C	Mini-DIN 6
Pin 8 — Brown	Pin 1 — Data In
Pin 7 — White/brown	Pin 2 — GND
Pin 6 — Green or orange, depending on cable type	Pin 3 — PTT
Pin 4 — Blue	Pin 4 — Not used
	Pin 5 — AF Out
	Pin 6 — Not used

Figure 1 — Connector P1, visible at the upper left, must be configured per Table 2 to match the prepared cable. [Kay Kasemir, AI4IC, photo]

For the cable, start with a CAT 5 computer network cable, cut the connector off one end, and solder the leads to a six-pin Mini-DIN connector using the connections shown in **Table 1**.

Inside the RIGblaster Advantage (see **Figure 1**), use the information in **Table 2** to configure P1, which is the two-row, 13-pin header inside the RIGblaster. Note that we do not use any of the provided "Instant Setup Connectors." Instead, the cable described in Table 1 maps Pins 1, 2, 3, and 5 of the data socket to corresponding pins on the P1 "output" side, and from there we can easily connect them to the microphone (**MIC**), **PTT**, and receive audio (**RCV AUD**) functions of the RIGblaster.

Depending on your radio, you may also want to install a jumper on the two-pin socket located above the left-most audio transformer in the photo. This jumper activates a 20 dB trans-mit volume attenuator. It is required for my Icom IC-746PRO, because the output of

Table 2
RIGblaster Advantage P1 Settings

Output Side Pin	Output Side Function
1 — Pin 1 "Data In"	MIC
2 — Pin 2 "GND"	MIC GND
3 — Pin 3 "PTT"	PTT
5 — Pin 5 "AF Out"	RCV AUD

Additional: Add jumper to the GND TIE column to connect PTT GND with MIC GND.

the RIGblaster is too high otherwise.

Connect the RIGblaster to the computer via its USB cable, and to the radio via the RJ45-to-Mini-DIN cable you just assembled. Optionally, add a CAT cable as described in the RIGblaster manual. Once you configure the audio levels of the RIGblaster sound card to match your computer, you are all set! — *73, Kay Kasemir, AI4IC,* **ai4ic@arrl.net**

Receiving Text Message Alerts from *JTAlert*

The WSJT modes created by Joe Taylor, K1JT, have been popular in the Amateur Radio world for years now. But there are also some slightly less-well-known aspects of those modes, such as the spectacular add-on software *JTAlert*, by Laurie Cowcher, VK3AMA.

JTAlert creates specific audio and visual alerts following the decode of any number of things you are looking for. It's simple to set up, allowing you to create alerts for specific states, DXCC entities, grids, continents, zones, prefixes, or even specific call signs that you're looking to contact. **Figure 2** shows the setup menu for *JTAlert*, with all the potential alerting options. Additionally, the latest versions of *JTAlert* allow for more data to correspond with the ARRL International Grid Chase taking place this year (see **www.arrl.org/international-grid-chase-2018** for more information).

JTAlert allows for easy integration with just about any logging program available. It connects seamlessly to the *N3FJP* logging suite for me and quickly logs contacts I have made. With a few more clicks of the mouse, a *JTAlert* ADIF (Amateur Data Interchange Format) file provides me with information on whether a particular decoded station is a new contact or someone I've worked before on that band or mode. You can also have it create its own ADIF log if it does not support your computer's logging software, or it can connect to any number of online logging platforms (see **Figure 3**).

Even with my modest station, I have had a lot of success on my favorite band (160 meters) over the years. My quest for Worked All States on 160 meters was met with the common hurdles for those of us in the eastern third of the continental US, Hawaii, and Alaska. The hunt was made easier with the use of K1JT's newer digital modes and *JTAlert*. I set up *JTAlert* for 160-meter contacts with Hawaii and Alaska in such a way that, upon my station decoding one of those two entities, it would sound a loud alarm. After several weeks, I had Hawaii confirmed on Logbook of The World (LoTW).

I soon contemplated ways to find Alaska without feeling like I needed to be within hearing range of my shack computer once the band quieted down and lengthened out for the winter. Rather than an audible alert from a speaker, I concocted a simple way to receive an e-mail or text message to my cell phone to alert me regarding a flagged entity being received in my station.

JTAlert allows "user-defined alerts" that trigger another program following an alert being generated. I found a free program called *sendEmail* which, when run from a *Windows* command line, sends an e-mail via an SMTP server. I knew that a text could be sent to almost any mobile carrier with an e-mail address, so I had my solution (see the table, **"Cell Phone Carrier Data,"** for examples). With those pieces in place, this became a simple fix.

To try this text/e-mail alert for yourself, download the TLS-supported version of *sendEmail* at **http://caspian.dotconf.net/menu/Software/SendEmail**. Unpack the program into a folder and move it onto your hard drive. I moved the folder directly to the C: folder and renamed it "Send."

Create a batch file to run the script with your e-mail server's parameters. To make a batch file, just open up the *Notepad* program

Cell Phone Carrier Data

Carrier Network	E-mail Address
AT&T	[yournumber]@txt.att.net
Boost Mobile	[yournumber]@myboostmobile.com
Cricket Wireless	[yournumber]@mms.cricketwireless.net
Sprint	[yournumber]@messaging.sprintpcs.com
T-Mobile	[yournumber]@tmomail.net
U.S. Cellular	[yournumber]@email.uscc.net
Verizon	[yournumber]@vtext.com
Virgin Mobile	[yournumber]@vmobl.com
Republic Wireless	[yournumber]@text.republicwireless.com

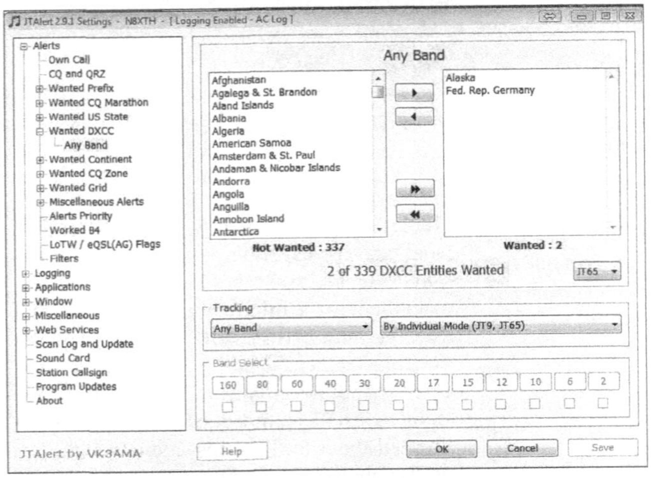

Figure 2 — *JTAlert's* setup menu, featuring options for different alerts. [Deron Boring, N8XTH]

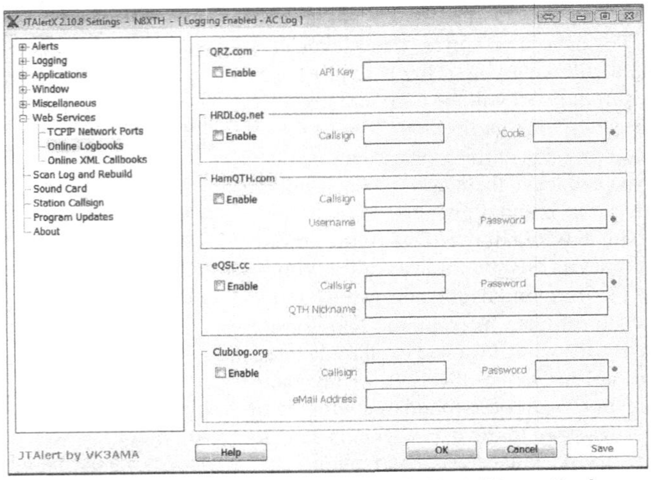

Figure 3 — The online logging menu for JTAlert. [Deron Boring, N8XTH]

```
Blank:

c:\send\sendemail.exe -f [from email address] -t [destination email address] -cc [3308675309@vtext.com]
-u "JTAlert" -m "JTAlert Entity Alert Triggered" -s [smtpserveraddress] -xu [redacted] -xp
[redacted]

Gmail:

c:\send\sendemail.exe -f ****@gmail.com -t *****@gmail.com -cc 3308675309@vtext.com -u "JTAlert"
-m "JTAlert Entity Alert Triggered" -s smtp.gmail.com:587 -xu **** -xp ####### -o tls=yes
```

Figure 4 — The batch file text needed to set up JTAlerts via text message. [Deron Boring, N8XTH]

on your computer and start typing. When you save it, click **FILE**, **SAVE AS**, and name it "sendemail.bat".

The .bat file needs to contain the following user-entered parameters after the directory and .exe file (see **Figure 4**):

- f, the originating e-mail
- t, the destination e-mail
- cc, any carbon-copied destination e-mails you want
- u, message subject
- m, message body
- s, e-mail server address and port
- xs, e-mail username or login
- xp, e-mail password
- o tls=yes (Only if you are using a secure server. If you're using an unprotected server, leave this out of the batch file.)

The e-mail server settings may be a bit tricky if you are using Gmail or Yahoo! Mail. You will need to activate SMTP access through those websites in the e-mail settings. I set up a dummy mailbox on my office's e-mail server to circumvent those issues. If you have an ISP e-mail address (through Roadrunner, AT&T, etc.) that does not require a secured connection for SMTP sending, it may be easier to set it up there.

Save the .bat file in the same directory as the "sendemail.exe" file. I am more than happy to provide you with all the files needed to cover both methods mentioned above.

After you're done with those settings, run the .bat file and watch the program run. You should get an e-mail and/or text message within a few seconds of the script running. If you get an error, note of error and correct the server and/or login parameters in the .bat file until it works.

Once it's working, it's time to set up *JTAlert* to send an e-mail or text message on command. Go to *JTAlert*'s **SETTINGS, ALERTS, MISCELLANEOUS ALERTS, USER-DEFINED ALERTS**, and at the bottom, point *JTAlert* to the .bat file you created. Set up the desired alerts, and give it a try. It should send the alerts out within a few seconds after decoding. Other scripts can be used to send more detailed information such as the triggered entity (state, DXCC entity, etc.), but that was too detailed for my needs. I wanted a simple way to find out when my alert was triggered, which this does. A text message saying "*JTAlert* entity alert triggered" is enough information for me.

Joe Taylor and his team have created an amazing new universe for those of us who love the challenge of propagation in Amateur Radio, especially during the weak part of the solar cycle. *JTAlert* is a very powerful tool you can use to find those difficult contacts and to further your enjoyment of Amateur Radio. — *73, Deron Boring, N8XTH,* **n8xth.oh@gmail.com**

Photos by the author.

Generating Morse Code MP3s from Text Files with *fldigi*

I was interested to read George Allison's, K1IG, article, "Increase Your CW Speed with Wordsworth," in the May 2017 issue of *QST*. George used the outstanding *fldigi* program (created by Dave Freese, W1HKJ, and associates) to translate text files into Morse code, with wide spacing between words to help increase code comprehension at higher speeds.

High-speed CW has been an interest of mine for years. I used *fldigi* to translate text files into Morse code MP3 files that I listen to using my portable MP3 player while taking my daily walk. Here's how.

1) I decided on a text file to translate, then removed punctuation using three spaces for a comma and seven spaces for a period, per a suggestion by Chuck, AAØHW. Then I saved this file.

2) I opened *fldigi*, went to **CONFIGURE**, then **MODEMS**, then **AUDIO** to make sure my computer sound card was in the **PLAYBACK** box.

3) I copied this text file into the blue field of *fldigi* by right-clicking and selecting **INSERT FILE**. That way, the text file to be converted to Morse was in the blue field. (Make sure to toggle off the **T/R** box at the lower right-hand corner of *fldigi* or the program will start playing the file in Morse.)

4) I set the tone, code speed, and shaping to my liking. Then I went to the **FILE** command in the upper left-hand corner of *fldigi* and selected **AUDIO**, followed by **TX GENERATE**. This allowed me to select the destination file for the .wav file soon to be generated by the program.

5) I toggled the **T/R** command to transmit and *fldigi* started sending the text file in the blue box in Morse. I watched while the program worked through the text file, because *fldigi* does not stop automatically after the text is completely translated. I stopped it by toggling **T/R** to receive mode.

6) The .wav file was then available for transfer to my MP3 player. If you need an .mp3 file for your audio player, you can use one of the many .wav to .mp3 converters available online.

My goal is to copy word sounds as described by George in the article mentioned earlier. Then I can keep up with some of the high-speed CW operators I run into on the bands. — *73, Joe Gerry, W3GW,* **jliving2001@yahoo.com**

Automating *fldigi*'s Macro Functions

I have tried plenty of Morse code practice software and had little success with the spaces between words. The programs seemed to skip all spaces after the first, which negated the use of spaces for timing. I have used *fldigi* for a long time, but never thought to check the timing of spaces. George's, K1IG, article in the March 2017 issue inspired me to set up and test a prototype automating the use of *fldigi* using the macro function to select the files and speeds that have been set up for practice.

I first worked out a standard macro set using the macro editor in *fldigi*, then made a directory structure to hold the edited files. One *fldigi* macro set can contain 48 macros, so I was able to handle two sets of practice files made up of 24 speed selections. The speed selections are part of the file name to keep things less confusing. The text files are stored in their own directory after editing.

To keep things easy, I created a directory structure in my C: drive called "CW Practice Files." Each subdirectory is given a descriptive name for the files they contain. There are 12 speed subdirectories in each of the 12 speed directories (see **Figure 5**). These speed directories each contain two edited files that have been spaced to the requirements of the spacing chart in George's article.

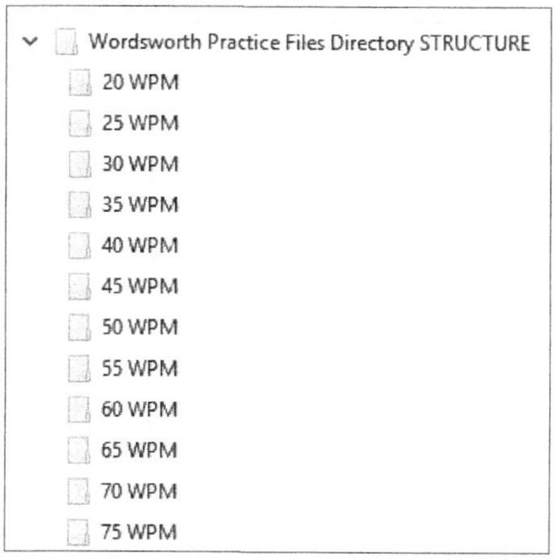

Figure 5 — The author's organized "CW Practice Files" directory. [Bill Hunt, WA8OKR, photo]

Here's a name and path example: **C:\CW Practice Files\100 most words\20 WPM \100MostUsedWords 20WPM 2sec.txt**

This method allows for easy editing of the macro to get the correct path information into the macro.

When the macros are constructed, one general master is made and then pasted into the macro buttons for editing. The macro buttons are given names that are descriptive of the contents, so there are fewer errors when editing.

The macro files are stored in a macro directory within the *fldigi* data structure.

For example, **C:\Users\Wa8okr2\fldigi.files\macros_WW100 Numbers #1 and Rand Calls.mdf**

I have found that the free text editor *Textpad* has features that allow the arranging and editing of the practice files. The two most useful features of the text editor are the ability to show the spaces with dots and the end of the line as a very special character. Some other text editors don't allow the visual display of spaces and end-of-line characters.

As George's article shows, the single space can be replaced with the correct number using the search and replace function in the text editor. It is essential to use a text editor only, because the formatting and other features in word processors interfere with the processing of the characters in the file in use. — *73, Bill Hunt, WA8OKR,* **wa8okr@yahoo.com**

An Easy Way to Document Computer Settings

Before I started to update *fldigi* on my PC, I decided to record all the current settings so I would have a baseline 0-for the install, and a record just in case I wanted to change something back.

Rather than writing everything down, I decided to take screenshots of the settings by using the **PRINT SCREEN** function on my PC, as follows:

1) I opened a new document using my word processor. I use *Libre Office*, but this also works for *Word*, and probably most other word processors. I then saved the blank document and reduced it.

2) I started the program that I wanted to document (in this case, *fldigi*) and went to the first configuration screen.

3) When I had what I wanted to save on the monitor, I pressed the **PRINT SCREEN** key on my keyboard.

4) I then expanded the document, clicked on it to get the cursor, and pasted in the screen image.

5) Because I only wanted a small part of the image, I clicked on it to select it, then cropped it down to the part I wanted.

6) After pasting, I reduced the document to get back to *fldigi* and the next configuration screen.

7) I then moved to the next configuration screen and repeated steps 3 – 6. I was able to save six configuration screens on one page.

I now have a record of the settings without a lot of work. — *73, Joe Birsa, N3TTE,* **jjbirsa@yahoo.com**

Using a Smart Watch to Find Grayline DX

I found one function of the Apple Watch very useful for DXers. In the watch's menus, you can choose to display several time zones on the main screen (for example: local time, UTC, etc.). In the world clock menu, the day and night areas can be displayed. The grayline zone on the map can now be used as a handy indicator of when a DX contact through grayline may be possible. Also, the sunrise and sunset times are shown, in addition to the visual indication on the map (see **Figure 6**).

When turning the crown, the user can select between different cities (which must be pre-defined on the iPhone the watch is paired with). I intended to use that feature for phone conferences with Europe and Asia to find good time slots. In addition, it seems to be a hidden feature for DXers. Other smartwatches can likely do the same, so look out for these features on your own smartwatch. — *73, Matthias Zapatka, AJ4BB/DL6ZM/DU3ZM,* **zapamatt@yahoo.com**

Fixing Com Ports After an Update

If you're a Microsoft Windows 10 user, chances are that you've experienced errors after an update reconfigured your com ports. Maybe you started your contest logger and an error message popped up, saying that the com port no longer existed, or the logging program would no longer connect to your rig. These can be a pain to get straightened out again, but I have investigated the information available on the system com ports and discovered how to figure out where each device belongs.

Figure 6 — The grayline zone on a smartwatch's world map can be used as a quick indicator of when to try making DX contacts. [Matthias Zapatka, AJ4BB, photo]

I found that each serial device in use on my system uses a unique identifier in Windows Device Manager. By taking a few minutes to track down and record that information in a spreadsheet, I can easily reassign com ports back to the settings stored in my control software (*N1MM+ Logger*, *WSJT-X*, *Ham Radio Deluxe,* and so on).

To begin, open the Device Manager by typing "Device" into the taskbar search box and selecting **DEVICE MANAGER** from the results. In the Device Manager, scroll down and expand the **PORTS (COM & LPT)** section to show all the active serial devices. In my system, I have 12 ports in use, including six just for my Timewave Navigator. Looking further, I can see that "Navigator WinKey & FSK" use ports COM6 and COM7 (both report the same name). By right-clicking on COM6 and selecting **PROPERTIES**, I can see by the "Location" that this is the port for the WinKey (see **Figure 7**). If *Windows Update* changes my com ports, I can find out which com port says "Location: on Navigator WinKey" and change the port number back to what I want it to be.

The real difficulty begins with all the serial ports labeled "USB Serial Port (COMxx)." It seems hard to tell those apart. I know that my Elecraft KAT500 uses COM11, and my KPA500 uses COM9. However, the Location in **PROPERTIES** only indicates "on USB Serial Converter." That doesn't help, so we need to dig a bit deeper.

With the com port **PROPERTIES** window open, select the **DETAILS** tab. In the **PROPERTIES** dropdown box, select **DEVICE INSTANCE PATH**. The value listed is unique to the serial device on that port. My KPA500 has a value of FTDIBUS\VID_0403+PID_6001+A70316XVA\0000. My KAT500 has a value of FTDIBUS\VID_0403+PID_6001+A105HX85A\0000.

No matter what com port an update may reassign the KPA500 to, that value is constant and identifies it as the KPA500, and the same goes for the KAT500.

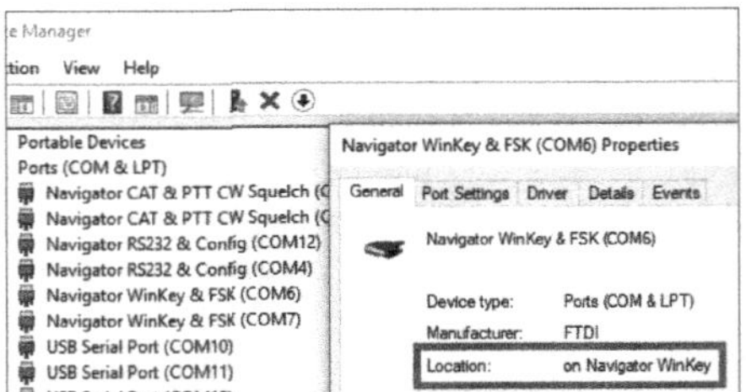

Figure 7 — The "Location" information for the com ports. [Dennis Moore, NJ6G, photo]

	A	B	C	D	E
1	Com Ports	Name	Location	Device Instance Path	Device
6	COM6	Navigator WinKey & FSK	Navigator WinKey	FTDIBUS\VID_0403+PID_B811+00000001A\0000	Navigator
7	COM7	Navigator WinKey & FSK	Navigator FSK	FTDIBUS\VID_0403+PID_B811+00000001B\0000	Navigator
8	COM3	Silicon Labs CP210x USB to UART Bridge	Port_#0002.Hub_#0013	USB\VID_10C4&PID_EA60\05500000FF72	Kenwood TS-590
9	COM11	USB Serial Port	USB Serial Converter	FTDIBUS\VID_0403+PID_6001+A105HX85A\0000	KAT500
10	COM9	USB Serial Port	USB Serial Converter	FTDIBUS\VID_0403+PID_6001+A70316XVA\0000	KPA500

Figure 8 — A spreadsheet keeping track of which com ports connect to your devices makes things easier to restore after updates. [Dennis Moore, NJ6G, photo]

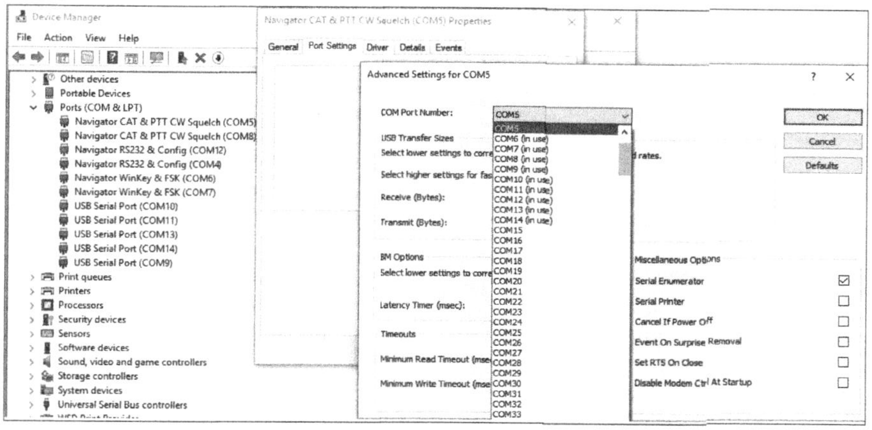

Figure 9 — Choosing a new com port number. [Dennis Moore, NJ6G, photo]

Record that information in a spreadsheet for future reference (see **Figure 8**). Note that if you have two identical devices, like two KPA500s, the Device Instance Path may not distinguish between the two. While you're in the **PROPERTIES** tab for the com port, also record the information listed in the **PORT SETTINGS** tab (Speed, Data Bits, Parity, etc.). Repeat the steps for all the other serial devices. With this data re-corded, you are ready for when an update changes your port numbers.

When an update changes your com ports, check the Device Instance Path for each of the new assignments to find out what device it is, and determine which com port that device should be on. If, for example, I check the Device Instance Path on COM5 and see that the value is for my KPA500, I know I need to change that port back to COM9 from COM5 (see **Figure 9**).

To change the com port number back to what it was before the update, open the **PROPERTIES** window for the port you want to reconfigure (right-click on the port and select **PROPERTIES**). Select the **PORT SETTINGS** tab and click the **ADVANCED** button. The COM Port Number dropdown lists the available com ports in your system.

If you're lucky, the com port you want to use is open and you can just select that port. If it is already in use, you'll need to free it up by reassigning that in-use port to one that isn't in use, then come back and assign this port to the now-open port. Repeat the process until all your com ports are back to where they should be. When you're done, reboot your system. After rebooting, go back into Device Manager and check to make sure your com ports stayed where you assigned them. — *73, Dennis Moore, NJ6G,* **dennis@mail-4life.net**

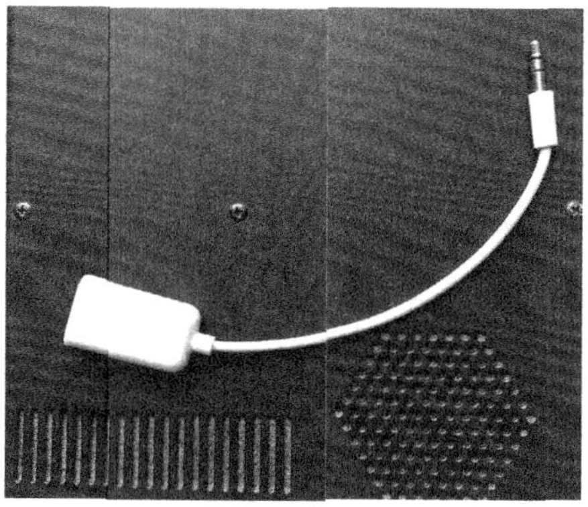

Figure 10 — An ordinary headphone audio splitter can be used to share a single CI-V control port on an Icom transceiver with two software applications. [Max Hochschild, AC1MX, photo]

An Inexpensive CI-V Splitter

As a fan of computer control for rigs, I need at least two CI-V control ports for the logging software I use with my Icom IC-706 transceiver. I reasoned that if the only two connections to the rig I need are serial and ground, then an audio cable splitter should suffice. Splitters also function as signal combiners. The two former input ports are isolated from each other, so neither logging program will be affected by commands the other sends.

I bought an audio splitter (see **Figure 10**) from a local store (a headphone splitter would work) and plugged my two CI-V control interfaces into the splitter's two output ports. I then connected the splitter to the rig's CI-V plug and powered on the station PC. Both programs immediately began to track the radio's frequency, mode, and band as if there were only one cable connected. I could find no changes in performance from the original setup. This should work with any splitter, whether mono or stereo, and any rig that uses a 1⁄8-inch plug for CI-V. — *73, Max Hochschild, AC1MX,* **ac1mx@outlook.com**

Forcing Windows to "Forget" a COM Port

When you install a device that requires a COM port in Windows for the first time, Windows assigns that device to COM 1. The next device you install will be assigned to COM port 2 and additional devices will be assigned consecutive numbers. However, Windows never forgets the ports numbers it has assigned unless you initiate a special procedure and force Windows to uninstall them. Simply uninstalling a port in the "normal" Device Manager will not totally uninstall it.

Assume you have devices on COM ports 1, 2, 3, and 4, and then you decide to remove one of those devices — the one on COM port 3. You'd go to the Device Manager and uninstall COM port 3. Now when you plug in a new device, you're puzzled by the fact that Windows assigned it to COM port 5. If you return to the Device Manager, find the device on COM port 5, then attempt to change it back to COM port 3, Windows will tell you that COM 3 is still in use.

Again, Windows never forgets a COM port unless you force it to do so. While you can manually reassign COM 5 to COM 3, this may not cure the problem because Windows will still be sending data to the original COM 3.

Here's the fix for Windows 10, the most commonly used version of Windows.

Start by logging into Windows as an Administrator. Next, find the **Command Prompt**. Click the **START** icon in the lower-left corner, then scroll through the programs until you see the **Windows System** folder. Click on this folder and, among all the items displayed, you'll see **Command Prompt** (see **Figure 11**).

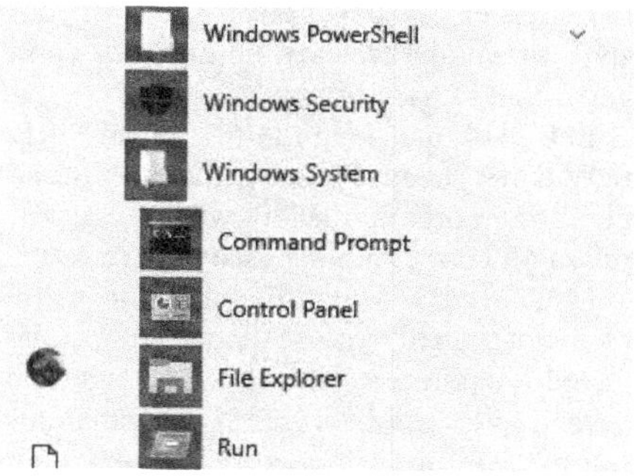

Figure 11 — In Windows 10, the Command Prompt is in the Windows System folder. [Alan Hoffmaster, WA3EKL, photo]

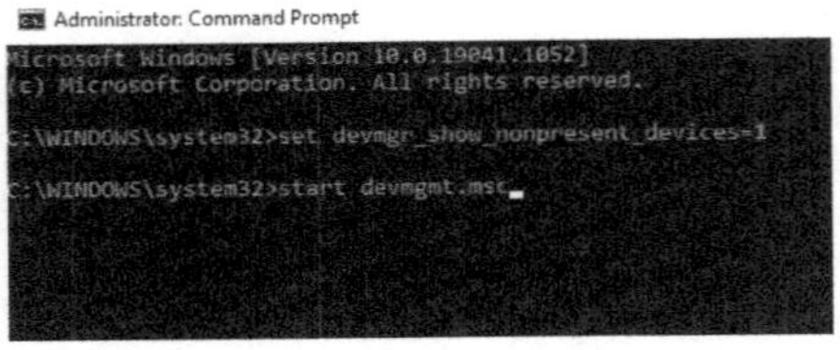

Figure 12 — The Command Prompt screen in Windows 10. [Alan Hoffmaster, WA3EKL, photo]

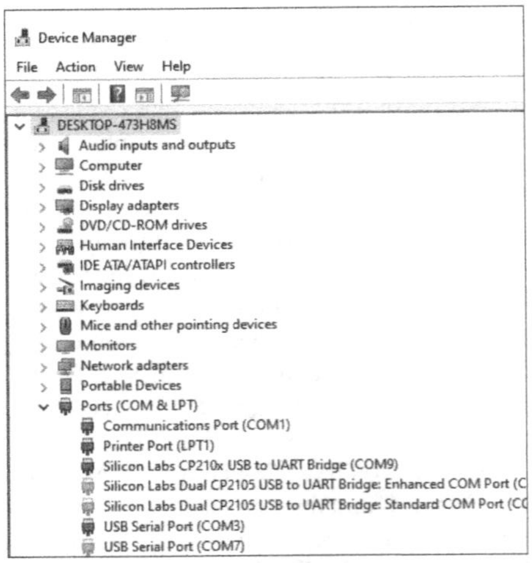

Figure 13 — In the Device Manager, expand the Ports list and notice which COM ports are grayed out. [Alan Hoffmaster, WA3EKL, photo]

Right-click the **Command Prompt** icon on the desktop, followed by **More**, and a window will pop up. Left-click **Run as Administrator**. A black command window will appear.

At the blinking cursor, type the following (all in lowercase letters):

set devmgr_show_nonpresent_devices=1

Then push the **ENTER** key on your keyboard. If you typed the line correctly, you will see the blinking cursor once again. Type:

start devmgmt.msc and push the **ENTER** key again (see **Figure 12**).

A slightly different version of the Device Manager will appear. Left-click the **View** tab below the words "Device Manager" at the top of the new window and another window will pop up.

Left-click **Show hidden devices**. All your devices should now show up like they do in the normal Device Manager.

Left click the > sign to the left of **Ports (COM & LPT)** to expand the list. Now you can see all the ports that Windows has ever installed on your computer (see **Figure 13**).

Most likely, you will have the same number assigned to more than one port. The grayed-out ports are the original ports, or the currently non-used ports, and you can uninstall them in this window by right-clicking the port and then left-clicking **Uninstall**.

Be careful that you don't uninstall a port you need, or you'll have to reinstall that device again. For example, I use *LP-Bridge* to allow *N1MM+* and *WSJT-X* to talk to my Elecraft K3 transceiver at the same time. *N1MM+* is on virtual COM port 5 and *WSJT-X* is on virtual COM port 7 within *LP-Bridge*. If those programs aren't running when I do the above procedure, ports 5 and 7 appear to be grayed out. Don't make my mistake and uninstall a port just because it's grayed out.

Write down all the ports you're using before you start uninstalling them. When you're finished uninstalling, close everything, restart your computer, and you may discover that your COM port issues have disappeared. — *73, Alan Hoffmaster, WA3EKL,* **alanbh@cablespeed.com**

Mapping Your Contacts

Whenever I tell someone that I am a ham radio operator, the first thing they ask is how far I can "talk." I usually rattle off a few places, but it would be more impressive to show them a map of my ARRL Logbook of The World (LoTW) confirmed contacts.

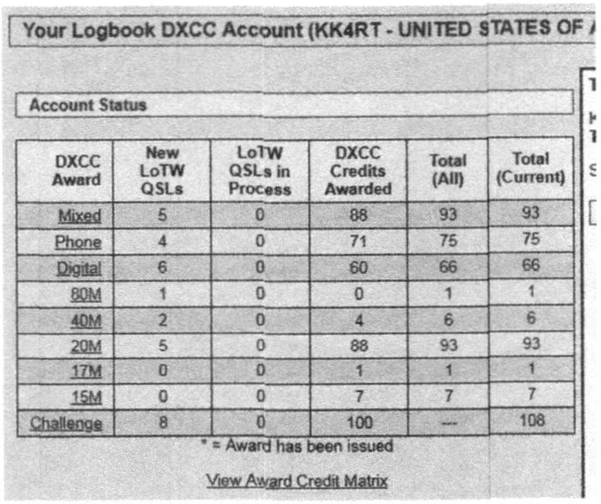

Your Logbook DXCC Account (KK4RT - UNITED STATES OF

Account Status

DXCC Award	New LoTW QSLs	LoTW QSLs in Process	DXCC Credits Awarded	Total (All)	Total (Current)
Mixed	5	0	88	93	93
Phone	4	0	71	75	75
Digital	6	0	60	66	66
80M	1	0	0	1	1
40M	2	0	4	6	6
20M	5	0	88	93	93
17M	0	0	1	1	1
15M	0	0	7	7	7
Challenge	8	0	100	---	108

* = Award has been issued

View Award Credit Matrix

Figure 14 — Start with your Logbook of The World account, click on your DXCC Award, and then click on View Award Credit Matrix below the table of confirmed contacts. You can see the link just below the table. [Scott Phillips, KK4RT, photo]

I knew I could easily customize a Google Map, all I needed was a list of countries or states/provinces to upload.

I started with my LoTW account, clicked on my DXCC Award, and then clicked on **View Award Credit Matrix** below the table of confirmed contacts, as shown in **Figure 14**. (You could also use a list from your favorite logging program.)

Next, I clicked on the **No Scroll** button at the upper right (see **Figure 15**) to cause the table to fully expand and allow me to highlight everything I wanted to copy. With copying accomplished, I pasted the result into Microsoft *Excel*, where I could sort it. (You could use also Google Docs.)

Google Maps accepts many types of files for import. I sorted mine so that I was left with a list of my confirmed contacts for all bands. This is the list of just country names, with the column heading **Entity**. You can add separate layers to the map to show different bands, modes, unconfirmed, and so on.

After going to **www.google.com/maps**, I clicked on the menu icon represented by the three horizontal lines on the left. Finally, I clicked on **Your Places**, followed by **Maps**, and then **Create Map**.

An untitled map appeared, and I gave it a name I would remember. Using the **Import** function under **Untitled layer**, I uploaded the country list that I had just created from LoTW. The Google wizard is easy to use and guides you through the process.

I clicked on the three vertical dots in the upper right area of the window shown in **Figure 16** to see the options for setting the default view, printing, or embedding the map on a website. I could also click on **Share** to allow others to view the map from a link. **Figure 17** shows my results.

There are countless layers you can add to your map, so follow my steps and have fun. *— 73, Scott Phillips, KK4RT;* **kk4rt@arrl.net**

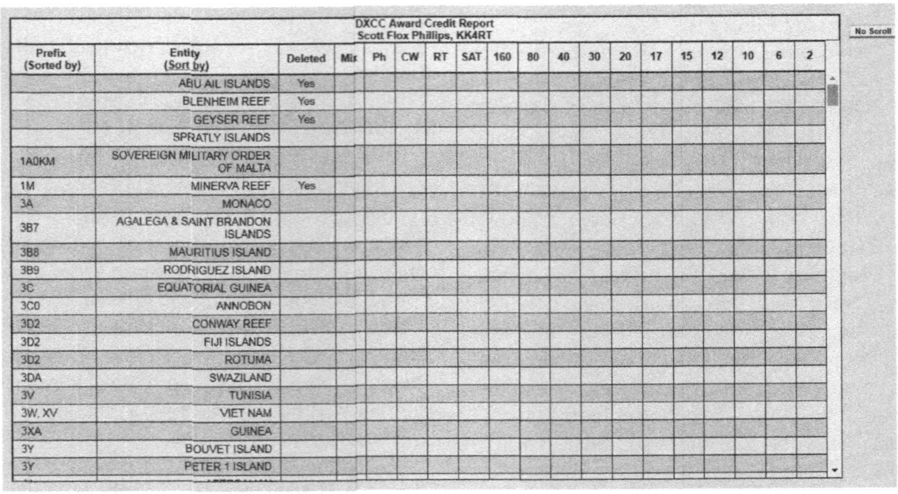

DXCC Award Credit Report
Scott Flox Phillips, KK4RT

No Scroll

Prefix (Sorted by)	Entity (Sort by)	Deleted	Mix	Ph	CW	RT	SAT	160	80	40	30	20	17	15	12	10	6	2
	ABU AIL ISLANDS	Yes																
	BLENHEIM REEF	Yes																
	GEYSER REEF	Yes																
	SPRATLY ISLANDS																	
1A0KM	SOVEREIGN MILITARY ORDER OF MALTA																	
1M	MINERVA REEF	Yes																
3A	MONACO																	
3B7	AGALEGA & SAINT BRANDON ISLANDS																	
3B8	MAURITIUS ISLAND																	
3B9	RODRIGUEZ ISLAND																	
3C	EQUATORIAL GUINEA																	
3C0	ANNOBON																	
3D2	CONWAY REEF																	
3D2	FIJI ISLANDS																	
3D2	ROTUMA																	
3DA	SWAZILAND																	
3V	TUNISIA																	
3W, XV	VIET NAM																	
3XA	GUINEA																	
3Y	BOUVET ISLAND																	
3Y	PETER 1 ISLAND																	

Figure 15 — Click the No Scroll button at the upper right corner and the table will fully expand. [Scott Phillips, KK4RT, photo]

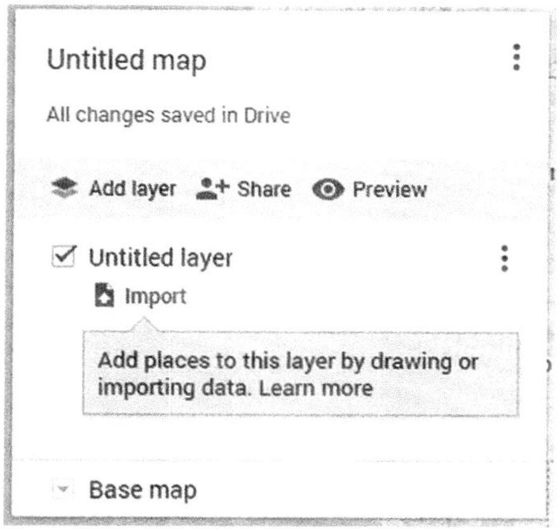

Figure 16 — Click on the three vertical dots in the upper right area of this window for options. [Scott Phillips, KK4RT, photo]

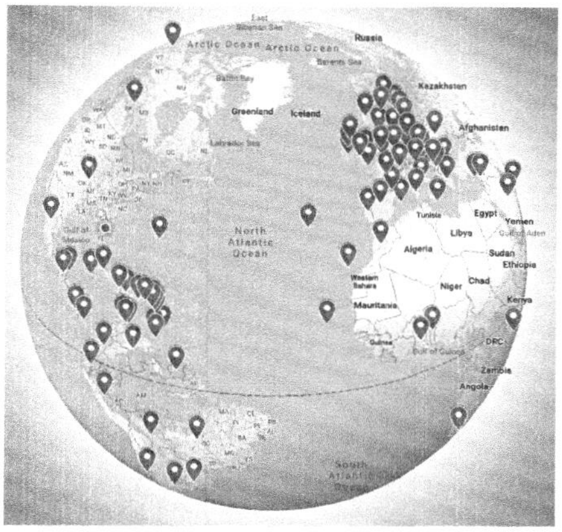

Figure 17 — The result of all the clicking and loading is an attractive map you can share. [Scott Phillips, KK4RT, photo]

Morse Ringtones

Most cell phones can be configured to play a canned ringtone for different contacts, which can be handy for identifying the caller without looking at the screen. But I began to have difficulty remembering which ringtone was for whom, especially after upgrading phones.

I noticed that my new Android 11-equipped phone lets me create my own ringtones, so I made .wav files with Morse code messages instead of music for each of my contacts. I chose the speed and tone most comfortable for me, and I kept the messages brief. This hack works great, and it is a terrific conversation starter about ham radio. — *73, Todd Thuss, K4XG,* **k4xg@arrl.net**

CHAPTER 9

Troubleshooting/Test Gear

Load Bank

Most everything in my shack, except the computer, runs on multiple 12 V dc external power supplies. When one of these supplies began to behave erratically while transmitting, I brought it to the workbench for testing. This power supply is a straightforward series-pass design with overvoltage protection and foldback current limiting. The overvoltage protection circuit was working. The current limiting seemed to be the problem, but how to test it? Under no-load conditions, the internal voltages were all within reasonable limits. Resistance checks were also within limits. Obviously, I needed to test it under load.

I didn't have power resistors capable of handling the full load. All I could find were some wire-wound resistors in the range of 100 Ω and up. One possibility was to use the rig itself to load up the supply, but I deemed that too risky.

I found an answer at **mpja.com** in the form of 2 Ω, 120 W resistors for about $6 each. I calculated that with three of these resistors and some series/parallel combinations, I could test the supply at several points. However, this would leave me with a mess of wire and loose, fragile resistors. I decided to build a load bank and keep it on hand for future use.

I mounted the resistors on a piece of 1 × 12-inch board and added a 30 A terminal strip (see **Figure 1**). By attaching the resistors to one side of the strip, I could use small jumpers on the output side and make up several combinations. Now, I had a load bank with a 120 W minimum to 360 W maximum rating. [Good safety practice would be to add a fuse in series with one of the input lines. One of the terminal strip shorting bars could be removed and replaced with a leaded fuse. — *Ed.*] Using the terminal strip, I could set up combinations of 6, 4, 3, 2, 1, and 0.66 Ω. The load bank could be used for other than 12 V power sources, such as an emergency battery bank, though I'd stay below 24 V and watch the power rating of various combinations. — *73, Lionel Booth, N5LB,* **n5lb@arrl.net**

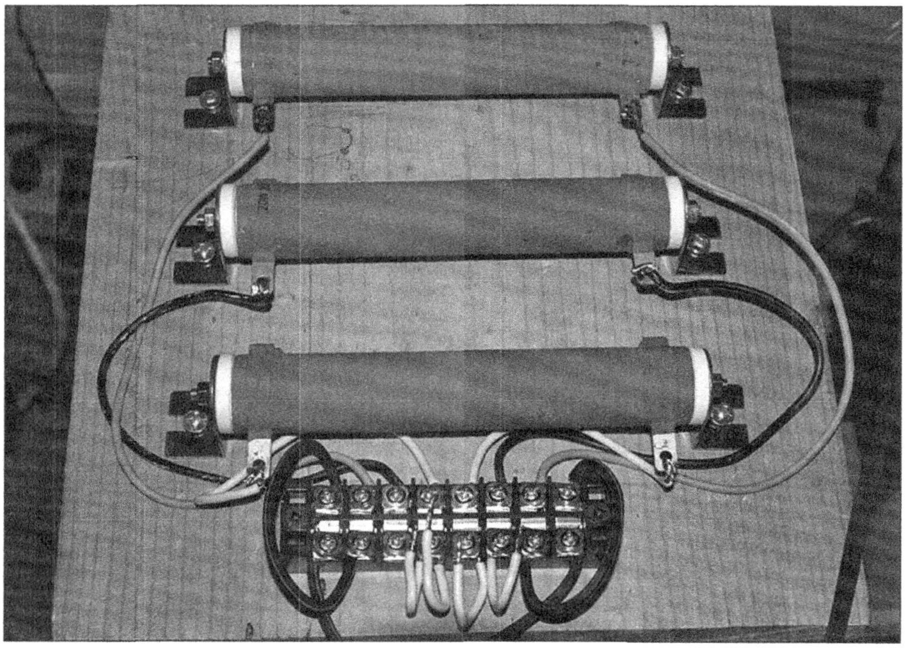

Figure 1 — This simple and adjustable resistor bank can be a big asset when testing power supplies or battery banks. [Lionel Booth, N5LB, photo]

Phase and Zero Beat Indicator

This very simple, inexpensive circuit provides a display of the phase angle between two frequency sources that are zero beat (±45°). It will also show whether the frequency of the test signal is higher or lower than the reference and the frequency difference, up to about 4 Hz — your visual counting limit. This is accomplished by comparing a signal to a sequentially delayed reference source using light-emitting diodes (LEDs) to display the results.

Figure 2 shows the concept in schematic form. The circuit uses three 90° networks in series totaling 270°. Four LEDs are physically connected to display the phase angle or, if the phase or frequency is changing, the LEDs will display a visually rotating pattern.

When the LED pattern rotates clockwise, the test signal is higher than the reference. A counterclockwise rotation indicates that the test signal is lower than the reference. In either case, the faster the LED rotation, the greater the frequency difference between the test signal and the reference.

The frequency difference, or drift rate, is related to the time that it takes for the LED pattern to step completely around the four LEDs. If the reference was at exactly 10.000 MHz, one rotation (360°) per second would indicate a test signal accuracy of 0.1 PPM; that is, a time clock would gain or lose 1 second in 4 months. One rotation in 16.7 minutes would indicate an accuracy of 0.0001 PPM.

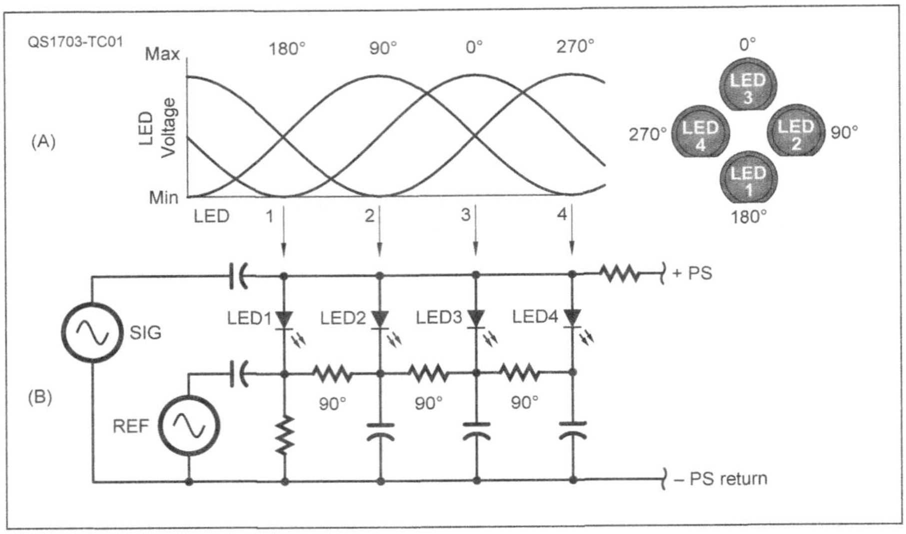

Figure 2 — This shows the overall concept. "A" shows the phase relationship between the test signal and the reference signal. "B" shows the circuit design concept.

Operating Theory

Referring to **Figures 2A** and **2B**, the test signal (SIG) is applied to all the LEDs in parallel. The reference (REF) travels down the series of delays with one complete cycle along the length of the delay networks, as the REF signal's (not shown) negative peak moves down the delay line (to the right). The individual LEDs will light only *when* the SIG is positive and *where* the REF is negative. There is only one point along the delay lines where the voltage is at maximum across the LEDs. That point is displayed as the phase difference between the two frequencies. [Note that, in the following discussion, the phase detection variation of up to ±45° should be taken into account when comparing the oscilloscope traces in Figures 3, 4, and 5, with Figure 2A. — *Ed.*]

When the signal and reference are zero beat and *in phase* (see Figure 2A, 0° and Figure 3), LED3 (0°) will be bright with the maximum potential across it due to the reference's negative peak being delayed 180° by the two 90° networks. LEDs 2 (90°) and 4 (270°) will be dimly lit. LED1 (180°) will have a low potential across it and will be dark. Conversely, when the signal and reference are at zero beat and *out of phase* by 180°, LED1 will be bright while LED3 will be dark.

When the phase of the signal *lags* the reference by 90° (see Figure 2A, 270° and Figure 4), LED4 will have a maximum potential across it and be bright, and LED2 will have a low potential across it and be dark. LEDs 1 and 3 will be dimly lit. Conversely, if the signal *leads* the reference by 90° (see Figure 2A, 90° and Figure 5), LED2 will be bright and LED4 will be dark.

Construction and Operation

The schematic of **Figure 6** uses only one 90° network. The center-tapped transformer provides the alternate polarities of the test signal so that the 0° and 180° points may be displayed without a delay network. Adding one 90° network and two LEDs that are connected across the alternate polarity signal will display the 90° leading or 270° lagging points in a manner similar to that explained above.

In Figure 6, T1 is hand-wound on an Amidon FT-50-61 ferrite core. The primary is 10 turns of #26 AWG and the secondary is 10 turns center-tapped.

To use the indicator, set the power supply voltage for maximum LED brightness while keeping one LED dark. Use red LEDs with the lowest operating voltage and all of the same type for smooth operation. Each input should be 3 – 5 V_{RMS}.

A short video showing the unit's operation is available on the ARRL website. Also on the website is an SMD version of the circuit with a PC board layout and construction photos. This additional information can be found at www.arrl.org/qst-in-depth. — *73, John Clark,* **K2AOP, k2aop@arrl.net**

Dollar Dummy Load

I would like to share this simple idea for an RF dummy load. I experimented with the unique and inexpensive RFP1109, a beryllium oxide high-temperature resistor that I found on eBay. It is a 50 Ω, non-inductive resistor in a small package that looks similar to an RF final amplifier transistor used in most transceivers (see **Figure 8**). It can handle 100 W

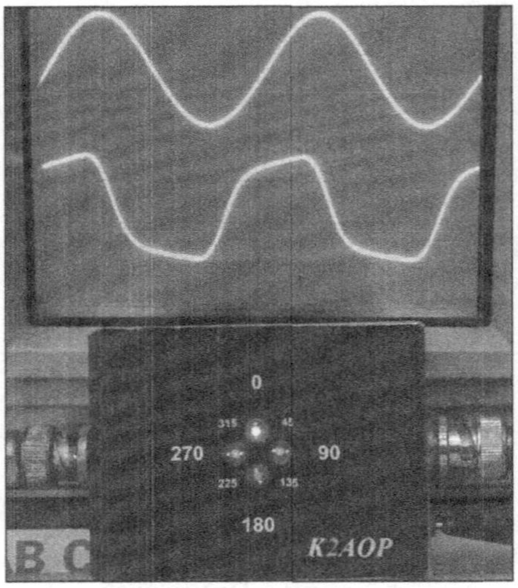

Figure 3 — This oscilloscope view shows the test signal (top) and the reference signal in phase at 0°, with the tester's display below. [John Clark, K2AOP, photo]

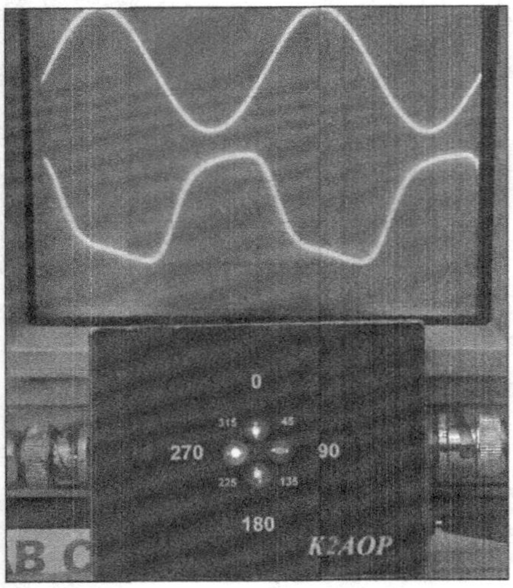

Figure 4 — This oscilloscope view shows the test signal (top) lagging the reference signal by 90°. [John Clark, K2AOP, photo]

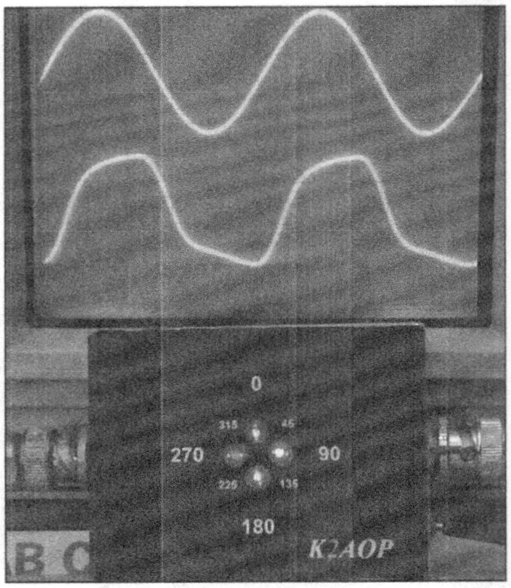

Figure 5 — This oscilloscope view shows the test signal (top) leading the reference signal by 90°. [John Clark, K2AOP, photo]

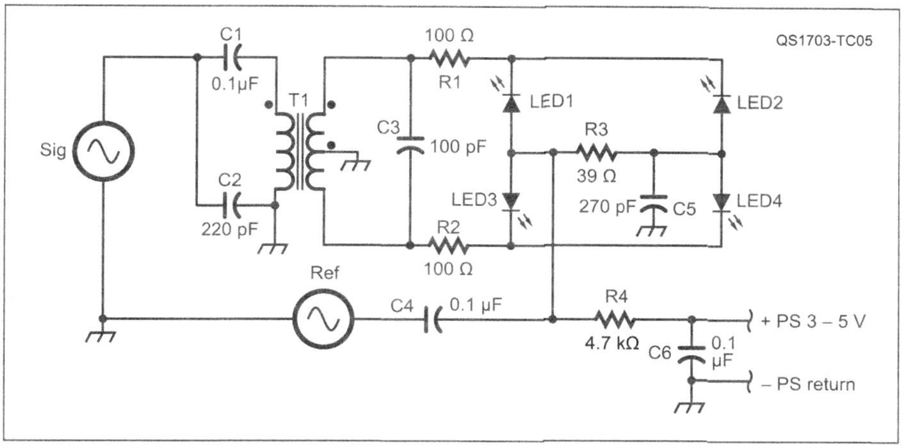

Figure 6 — This is a working circuit designed for leaded components.

C1, C4, C6 — Capacitors, 0.1 μF
C2 — Capacitor, 220 pF
C3 — Capacitor, 100 pF
C5 — Capacitor, 270 pF
LED1, LED2, LED3, LED4 — Digi-Key C503B-RCS-CW0Z0AA1-ND
R1, R2 — Resistor, 100 Ω, 1/10 W, 5%
R3 — Resistor, 39 Ω, 1/10 W, 5%
R4 — Resistor, 4.7 kΩ, 1/10 W, 5%
T1 — See text

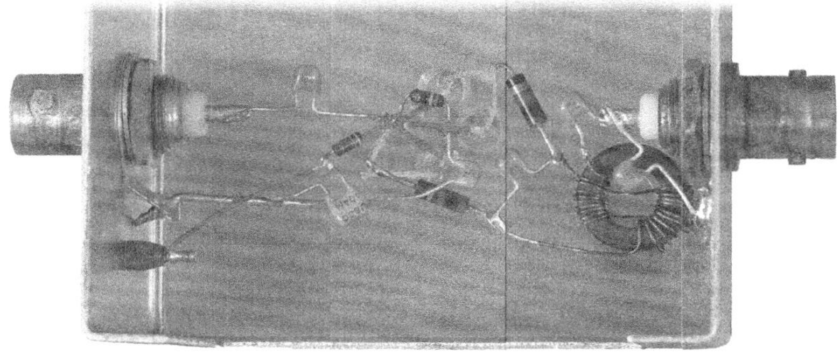

Figure 7 — The circuit of Figure 6 implemented with leaded components. The red shading in the center indicates the LED positive leads and the green shading indicates the LED negative leads. [John Clark, K2AOP]

and is specified for use from dc to 3 GHz [see **www.mouser.com/ds/2/21/G100N50W4%20 Datasheet_Rev2-371340.pdf** for a similar device's data sheet, which is rated at 100 W. However, attention must be paid to the power derating curve, which plots the power that the resistor can handle as a function of its solder interface temperature: at 100°C, it's rated at 100% power; at 125°C, it looks to be rated at 50% power, and at 150°C, the show is over (0% rating). — *Ed.*].

The resistor is too small to handle the high heat levels generated during RF transmitting, so I mounted it on an old CPU heatsink with a fan for cooling. I tested the dummy load on the HF, VHF, and UHF bands with exceptional results. SWR is flat — near 1:1 throughout. I did not even use a low-loss cable, such as RG-8 or RG-58, but RG-174 for the purpose of flexibility and twisting. Be careful when soldering the device to a heat sink because it is very delicate and easy to break. — *73, Dandy Lachica, K6ZRH,* **k6zrh@arrl.net**

Test Infrared LEDs with Your Digital Camera

How do you know if your infrared LED, such as that on an LED-based remote control, is functioning? Your cell phone camera and almost any digital camera is sensitive to infrared emissions. **Figure 9A** shows the business end of a remote control when no buttons are pressed, or after the batteries have run down. In **Figure 9B,** a button is pressed, and you can clearly see normal operation of two infrared LEDs lit up that are not visible to the naked eye. You can test your infrared project LEDs the same way. — *73, Kai Siwiak, KE4PT,* **ke4pt@arrl.org**

Figure 8 — The RFP1109 is a 50 Ω, non-inductive resistor made from beryllium oxide, which allows it to operate at high temperatures. The device pictured is rated to 100 W. [Dandy Lachica, K6ZRH, photo]

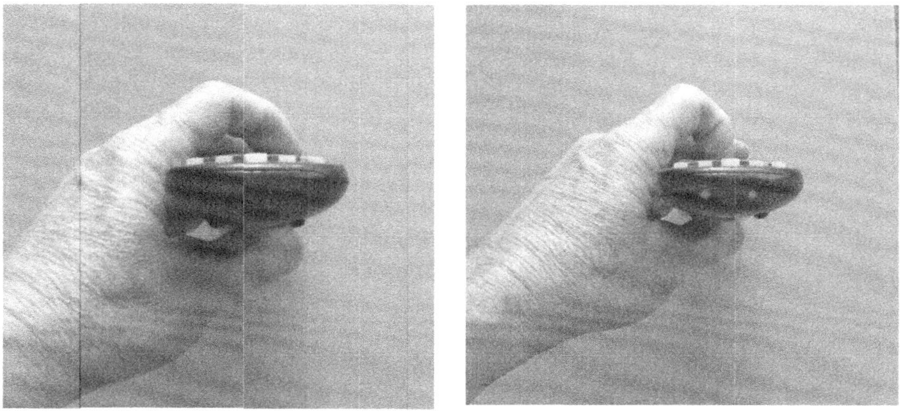

(A) (B)

Figure 9 — (A) shows a remote control when no buttons have been pressed. (B) shows two infrared LEDs lit up once a button has been pressed. [Kai Siwiak, KE4PT, photo]

APRS Image Reception

Trying to listen to your APRS signal on 144.390 MHz when the air is flooded with signals is like trying to use a magnet to find an aluminum needle in a haystack. That's the quandary I found myself in when checking out an intermittent problem with my APRS setup, a standalone assembly mounted in my trunk. Trying to use my voice transceiver in the front of the car yielded a myriad of signals — which one was mine?

Typically, you would use an attenuator to reduce the level of outside signals, but because my radio is installed in the dash, with all the coax neatly hidden away, connecting an attenuator would have been difficult. Using a little imagination, I realized I could monitor my signal via listening to the image frequency. My Kenwood TM-742 radio has low side injection of the 10.7 MHz local oscillator, meaning the local oscillator is 10.7 MHz below the frequency on the dial. [*Note: The IF frequency of a handheld or mobile transceiver can be anywhere from roughly 10 – 60 MHz, or higher. Many newer transceivers don't publish the first IF frequency in the specifications, and the ARRL Lab has noted a few where the published IF was incorrect. Also, some receivers are direct conversion with no IF stage at all. Be sure to use a radio that can receive the image frequency for testing.— Ed.*]

The image frequency, for a low side injection local oscillator, is twice the IF frequency plus the frequency of interest. In the case of APRS at 144.390 MHz, we get (10.7 MHz × 2) + 144.39 MHz = 165.79 MHz.

Usually, you want to have good image rejection to prevent the radio hearing signals that you are not tuned to. The image frequency works for this purpose because modern receivers employ image rejection filters that attenuate the receiver's image frequency from 76 to over 130 dBs. Using the image frequency essentially renders the receiver extremely insensitive to image frequency signals, hence blocking out all but the strongest signals on the frequency. Tuning to the image frequency made receiving my own signal easy while attenuating almost all the other signals on frequency.

Using the image frequency, you will probably hear only one APRS signal (yours) and then be able to hear how your signal sounds and how your system is performing, including pegging of turns. I have the APRS image frequency, 165.79 MHz, saved in a memory for ease of use. — *73, Allen Wolff, KC7O,* **kc7o@arrl.net**

Diagnosing Problems with Sound-Card-to-Radio Connections

When I was in college, I worked at a stereo equipment repair shop. One of our most useful tools was a "signal tracer," which was simply an audio amplifier with a probe. With this tool, I could determine where in the audio signal path the signal disappeared and isolate the problem.

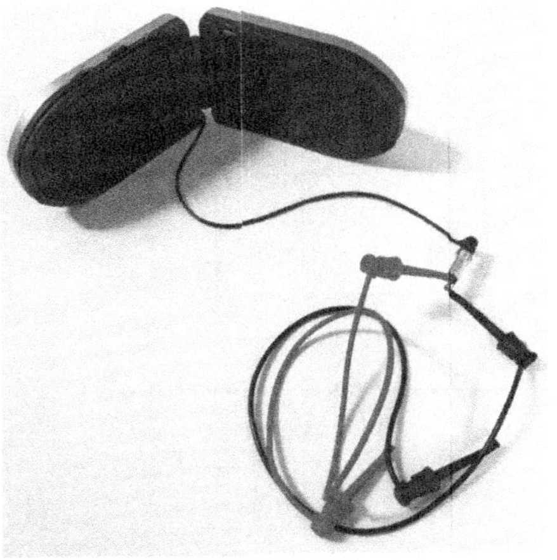

Figure 10 — The battery-powered audio amplifier, with mini grabbers for troubleshooting. [Bill Pellerin, KE5XV, photo]

More recently, Rick Hall, K5GZR, and I were helping our mutual friend, Mike Davidson, N5MT, diagnose and fix a problem with his radio and an external sound card device. It occurred to me that a signal tracer would be the right tool to get to the bottom of the problem.

I didn't own one, but I did have a battery-powered audio amplifier, meant to be used with a portable music device. The audio amplifier has a ⅛-inch phone plug as its input. Different amplifiers may have different connectors, but the ⅛-inch phone plug is common. If you get a ⅛-inch phone socket, you will be able to connect probes or wires to the tabs on the socket (either permanently or temporarily). **Figure 10** shows a pair of mini grabbers connected to the socket — black to ground and red to either of the stereo (in this case) channels.

From then on, diagnosing the problem consists of seeing where signals are present and where they are not present. If you pick an audio mode in your software (RTTY, PSK31, or any of the JT tune modes) and put that software in transmit (or tune) mode, there should be audio on the connector that connects to the radio (usually a DIN plug of some sort). If there's no audio, the problem could be in the configuration of the interface or in the setup of the software and sound card settings in the computer. If you hear the signal on your computer's speaker, it's because the software is configured to send the audio to the internal sound card and not the external one.

Likewise, you can check for audio out of the radio. Almost always in modern radios, the **DATA** port on the back of the radio provides a fixed level of audio output. Your homebrew signal tracer can amplify that signal if present and you should hear it.

If you are not using an external sound card, you can still use this approach to isolate the problem to the computer, the software setup, the connection from the computer to the radio, or the radio settings. Be sure the setting on the radio is such that it receives the audio from the back-panel **DATA** connector and not from the microphone. — *73, Bill Pellerin KE5XV,* **ke5xv@arrl.net**

An MFJ-259C Modification

The MFJ-259C antenna analyzer has been very useful to me as a new ham. In the March 2017 "Product Review" column, Phil Salas, AD5X, wrote about the excessive sensitivity of the tuning knob on MFJ analyzers (particularly regarding the 269C, but the front configuration of the 259C seems to be almost identical to the 269C). Being mechanically inclined, it frustrated me that the 259C tuning knob overreacts when adjustments are attempted. I solved the problem as follows (see **Figure 11**):

1) I cut a slightly less than 2-inch-diameter circle of sealed cell foam. Its outer diameter just clears the adjacent band selector knobs.

2) I then cut out the center of the foam circle so that it snugly and firmly fits over the MFJ-259C tuning knob.

3) Finally, I pressed the foam disk onto the tuning knob. This provides the equivalent of a reduction gear when turning the MFJ-259C's overly sensitive tuning knob, and it should work equally well on the 269C.

The sealed cell foam disk extends the effective knob diameter from ½ to 2 inches, thus providing a much longer linear distance through which one turns the knob to cover the same degree equivalent. That lengthened arc greatly slows the central shaft's turning rate, making it easier to tune.

This inexpensive, quick fix makes frequency tuning more gradual, and less jumpy on these analyzers. — *73, Craig Szwed, KC1ELZ,* **kc1elz@arrl.net**

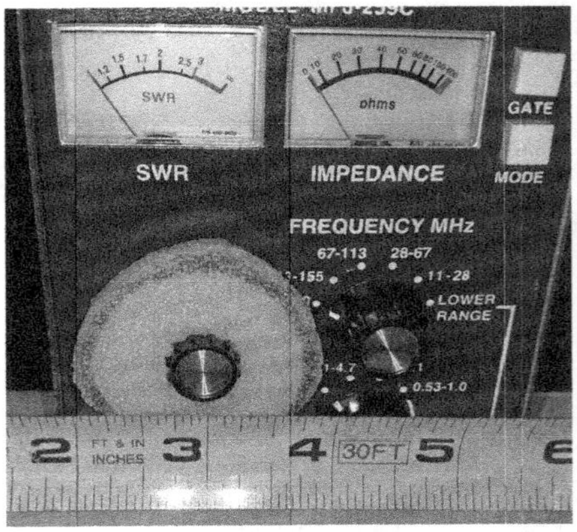

Figure 11 — A disc of foam can expand the effective diameter of the tuning knob, making frequency adjustments more controllable. [Craig Szwed, KC1ELZ, photo]

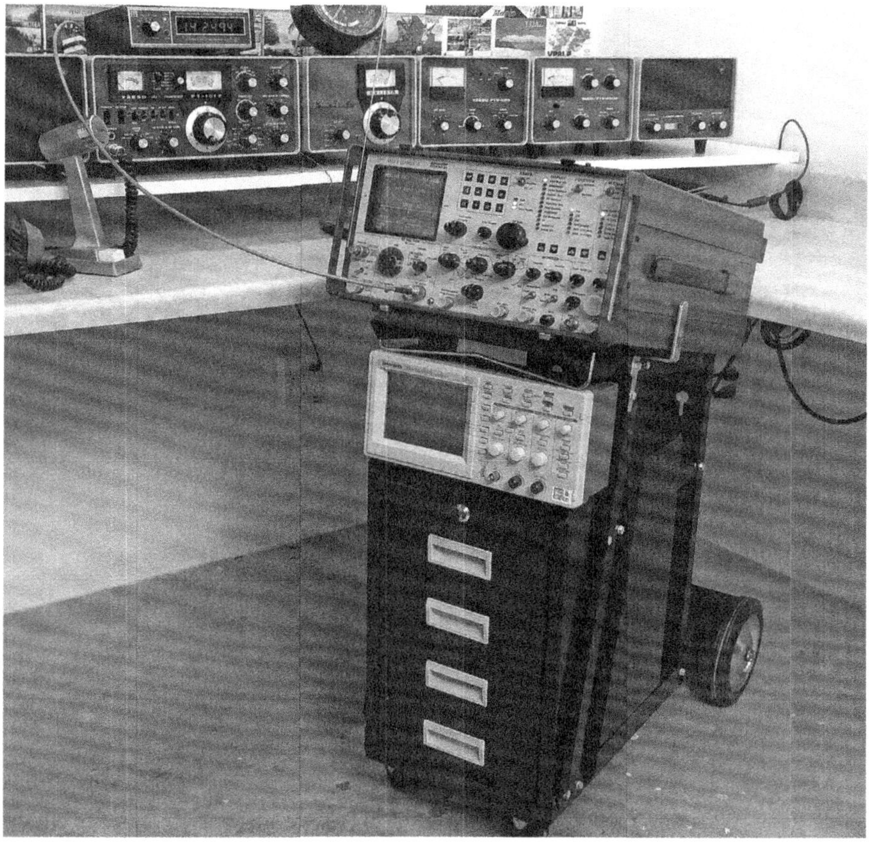

Figure 12 — Al's, NW2M, ham station hosting a service monitor and oscilloscope. [Al Rabassa, NW2M, photo]

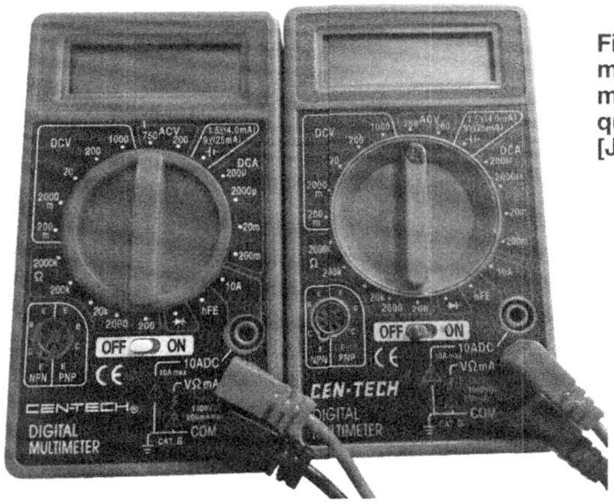

Figure 13 — Marking a multimeter with a permanent marker can make it easier to quickly check its settings. [James Cooper, NE6O, photo]

A Unique Storage Solution

A welding cart can provide a unique storage solution for any large or bulky pieces of test equipment. The unit I use has four lockable steel drawers, heavy-duty wheels, and a rear shelf. It's great for an isolation transformer or battery USP that older pieces of equipment may need. The drawers can hold all the probes, cables, connectors, adapters, and manuals that I will ever need. The cart shown in **Figure 12** cost $40 and provided a very tidy solution to a bulky problem. — *73, Al Rabassa, NW2M,* **nw2m@arrl.net**

Clearly Marked Meters

In my experience, I've found that the Cen-Tech digital multimeter is accurate enough for most tasks I perform, and its low cost allows the use of multiple meters on the same circuit. However, the selection dial is one color (red) and shows the function selected only as a slope in the plastic. Because of this, it is not clear at a glance which of the two opposite functions is selected.

While checking out a power transformer, I used the meter to both ohm out the windings and to measure the ac secondary voltages. The 200 Ω range and the 750 V ac range are opposite each other on the selector. I accidently connected a meter set to 200 Ω while measuring the transformer secondary voltage. At a glance, the selector switch appeared to be on the correct setting. When I powered up the transformer, however, the meter blew up. Luckily, nothing was damaged other than the meter. Still, it was a scary moment!

To avoid this problem in the future, I marked the selector with a permanent marker (see **Figure 13**). Now it's easier to see the actual setting of the meter. I like these meters and will continue to use them, but will be more careful about how they are set. — *73, James Cooper, NE6O,* **james_cooper@cox.net**

An RF Tap Fabrication

I thought I'd share how to fabricate a simple tap, so if you ever had to measure the frequency of your transmitter you could do it without having to interrupt your RF connections. All you need is a panel-mount BNC connector, a 220 Ω ½ W resistor, a piece of hobby-type copper tubing (about 7 inches long), and a little bit of hardware (see **Figure 14**).

Pick off a very small bit of your trans-mit power to provide it to your frequency counter (just a signal value large enough to trigger the counter's input, but not to overload it). I found this technique very useful when I first started participating in ARRL's Frequency Measuring Tests, and when verifying the relationship of my rig's frequency-shift keying mark and space tones when operating RTTY.

This technique should work with just about any dummy load, as long as there is sufficient space in the cabinet to mount the copper tubing and resistor. Allow enough space between the tubing and the load resistor to preclude any flashovers when you're operating with high power.

Figure 14 — The modified front panel of the author's dummy load, showing a BNC connector that can be connected directly to a frequency counter. [Dino Papas, KLØS, photo]

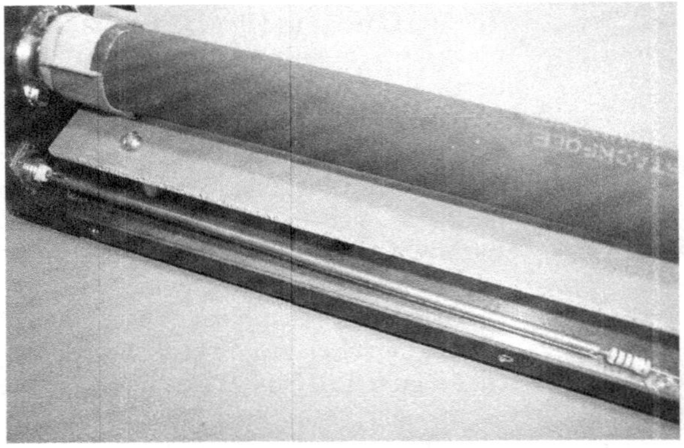

Figure 15 — An internal view of the author's pick-off tubing. [Dino Papas, KLØS, photo]

Figure 17 — The repaired probe. [Michael Janis, KE3OQ, photo]

The tap could not be any simpler. However, I've seen this same technique used with a single loop of wire connected directly to the case with its other end providing the pick-off point to the BNC connector's center conductor. If you want to try that first, it could save you a trip to the store.

You should be able to get the small-diameter copper tubing at your local hobby shop. The circuit is sheer simplicity, with the copper tubing connected to the case via a 220 Ω resistor (this is not critical).

As for the connections, simply solder the resistor to the inner surface of the tubing. Slit the tubing slightly at the BNC connector end, slip it over the center conductor, and solder it. Drill a hole in the case and install a nut, bolt, and ring-type terminal to provide the ground end connection (see **Figure 15**). — *73, Dino Papas, KLØS,* **kl0s@cox.net**

Multimeter Battery Replacement

I own several analog and digital multimeters, but I always use my Simpson 260. However, I recently noticed a slight instability in the higher resistance ranges, so I checked all the leads, contacts, and the 9 V battery. The battery seemed fine, but I replaced it anyway and that solved the problem. I did at times notice the D cell had some corrosion, so I would always clean it up and clean its contacts. The D cell could be too big for the meter, so it outlasts its usefulness and corrodes.

I replaced the D cell and put in a single AA battery holder with two-sided tape in its place and tacked the leads to the D-cell holder (see **Figure 16**). I think the AA battery will last long enough to indicate it's failing long before it starts to leak and corrode. This should give me at least a few years of maintenance-free service.

A decent alkaline AA battery is good for 2,500 mAh. We could call it 1,200 mAh to be on the safe side. That provides 12 hours of shorted leads and other resistance measuring, which should be more than enough for any ham. — *73, George Keller, W2DM,* **georgenann@aol.com**

Harbor Freight Probe Repair

I've had issues with some Harbor Freight probes, stemming from the fact that the wire is not soldered or otherwise securely attached to the metal tip. With use, the wire often comes disengaged from the probe, making the probe useless.

To correct this issue, place the tip of the probe against something solid like a vise and push inward, being careful not to hurt yourself. The tip and wire will push out the back of the probe and you'll notice not much bare wire on the probe cable. You will also see a small flat (about ⅛ inch) in the base of the tip where the wire was probably wound.

Next, strip the wire to reveal about 1 inch of bare copper. The wire is very thin, so be careful stripping it. Tightly wind the strands so that the stiffer wire will enter the probe and the bare wire sticks out the front of the probe.

Clamp the probe tip in a vise, with the non-point end sticking out. Take a file and extend that ⅛-inch flat out to the end. Don't take off too much; just match the cut depth that's there.

Push the wire into the plastic probe, with the bare copper sticking all the way out. Carefully place the metal tip with the flat against the wire. This will allow the wire and tip to re-enter the probe when pushed inward (be sure to push the tip against something solid). Push it in to match the distance desired. Finally, wind the excess copper lead around the base of the tip and secure with a drop of clear nail polish.

You now have a refurbished probe (see **Figure 17**). I have also noticed that the resistance of the probes is much less after repair. — 73, Michael Janis, KE3OQ, **mjanis@ptd.net**

Mic Cleaning Reminder

I recently spilled hot sauce on my microphone, so I thought it would be a good idea to take some antibacterial cleaner to the mic. While cleaning the mic with a paper towel and antibacterial cleaner, I was shocked to discover how much filth came off.

I wrapped a tiny, flat-head screwdriver in a paper towel that had been doused in the cleaner so I could dig into the fine corners of the microphone. Yikes! The dirt was really settled into every nook and cranny. After that, I rubbed down the mic with isopropyl alcohol, which loosened even more junk. I again used my screwdriver wrapped in the paper towel to dig into all the buttons, the switches on the side, and the PTT switch. When no more gunk came off, I did a final wipe-down and the mic looked almost new.

It occurred to me that we speak into our microphones endlessly and never consider

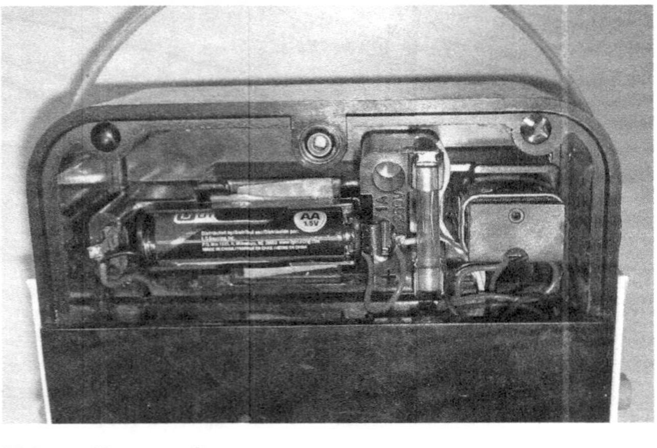

Figure 16 — Replacing a multimeter's D battery with an AA battery should prevent corrosion. [George Keller, W2DM, photo]

Figure 18 — A jumper wire soldered from the can to the lid got rid of the dummy load's RF. [Al Forbes, KJ4YEV, photo]

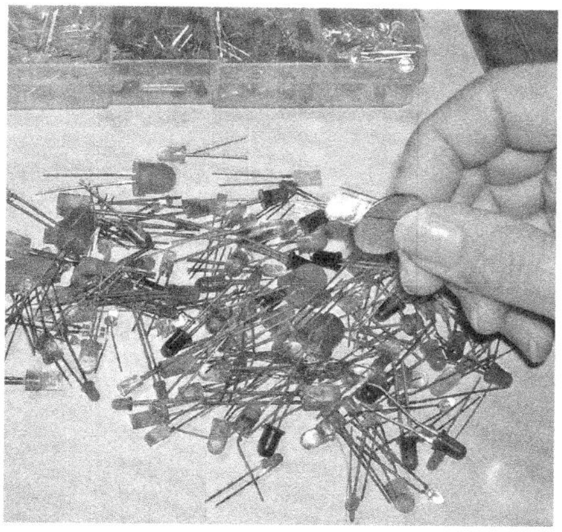

Figure 19 — Sliding a 3.6 V battery between the leads of an LED provides an easy method to test whether the LEDs are working. [William Gerhold, K2WH, photo]

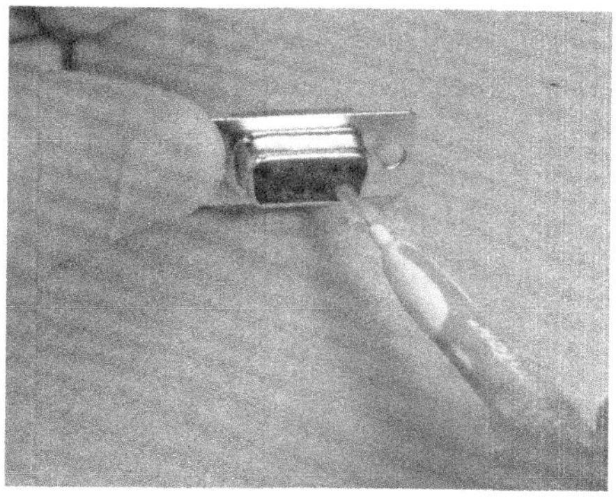

Figure 20 — Sockets on small connectors, like the DB9 shown here, can be cleared of dirt and contaminants with an interdental brush. [Henry Boze, N4HB, photo]

what bacteria may be building up. Remember to clean your microphone occasionally. — *73, Rev. Dr. Richard R. Stewart, KB1NOX,* **stwricha@aol.com**

Fixing Dummy Load RF

I built a dummy load similar to the "Build an Inexpensive 150 W Dummy Load with Wattmeter" project by Jack Purdum, W8TEE, and Al Peter, AC8GY, featured in the November 2018 issue of *QST* (although mine did not have the wattmeter). I wanted to share something I encountered.

The first time I tried to use my dummy load, multiple devices in my home had RF interference that did not occur with an antenna. I discovered that the dummy load was radiating RF because the can was not shielding the resistors. There is a grey coating on the inside of the lid and inside the can (I assume it is there to prevent oxidation). Because of this coating, there was no continuity between the grounded lid and the can.

I resolved the issue by soldering a jumper wire from the lid to the can (see **Figure 18**). Fortunately, the can was made of a material that takes solder easily and this fixed my RF problem. — *73, Al Forbes, KJ4YEV,* **alphaal3@bellsouth.net**

Testing LEDs Fast

Over the years, I've amassed about 1,000 LEDs from many sources, in all different colors and sizes. In the past, I tested them with a 12 V dc power supply, dropping resistor, and a few clip leads, which was a hassle.

I discovered a simpler, more convenient way to test them. All it takes is one quarter-sized, very thin, 3.6 V battery. I just slide the battery between the LED leads and the LED will either light up or it won't, indicating a bad LED or the need to flip the LED to the other polarity (see **Figure 19**). The speed and ease of this method has helped me test many LEDs without the need for a power supply, resistors, or clip leads. — *William Gerhold, K2WH,* **k2wh@optonline.net**

Mic Audio Solution

After years of trouble-free operation, one day my Kenwood MC-60A desk microphone wouldn't put out audio on my Kenwood TS-790E transceiver. The PTT switching worked perfectly, but there was no sound. It happened often when starting a transmission.

I experienced this problem on all combinations, between different transceivers and mics. A few other Kenwood users reported the same issues. Most of them were also heavy users and used the same sets. I did some more testing but could not find anything consistent.

Finally, I bought a new curled mic cable and immediately noticed the difference. I had no trouble at all. When I examined the new cable, I noticed the tight fitting of the cable plug.

This inspired me to work on my old curled cable plugs to improve the fitting. When I tried them out, I had no issues. Apparently, the spring-force of the female contacts in the cable plugs of the mic cable lose strength when used frequently. This leads to loss of audio during transmitting.

Bending the female contacts back to improve contact force helped only temporarily, so if you're experiencing this issue, a new plug is the real solution. — *Frank Laanen, PE1EWR,* **pe1ewr@zeelandnet.nl**

Clean that Connector!

DB9s and similar connectors can become intermittent due to the buildup of dirt and oxidization on the socket surfaces. An inexpensive and effective way to clean the sockets without having to remove the connector is to use a standard interdental cleaner with a tapered flexible wire brush (see **Figure 20**). These interdental brushes can be found at most drug stores.

To clean the contacts, add a small amount of electrical tuner cleaner or other suitable electronic cleaning solvent to the brush and clean out the plugs as required. The brush fibers should be periodically cleaned with a cloth or paper towel to remove any contamination. This process works well and has saved many female-type plugs, particularly those that are used in damp environments. — *Henry Boze, N4HB,* **henry.boze@gmail.com**

Yaesu MH-36B6JS Microphone Repair

I have a Yaesu FT-100 mobile radio. Recently, I had an issue with the coiled cable breaking at the strain-relief, so I bought and installed a replacement.

Everything worked fine until the push-to-talk (PTT) started acting up. It seemed as though I had to click the PTT several times (with the radio off, of course) in order to make the switch work. Then it would work intermittently, not engage at all, or disengage toward the end of a transmission. If I wiggled the PTT button, pushed it to one side while transmitting, or simply mashed it, it would work for a bit, but then become intermittent again.

Rather than buying a replacement microphone, I decided to take a homebrew approach. I found a great deal on microswitches on Amazon — a bag of 20 for just $7. These switches have side terminals where the original in the microphone had end terminals. This was not an issue because the terminals sit way above any surface-mount component.

The two bottom terminals are NO (normally open) terminals that close when the switch is activated. The top terminal is an NC (normally closed) terminal that opens when the switch is pressed. I trimmed the NC terminal off the new switch and soldered two new wires to the remaining terminals. I took the mounting screw out of the original switch and de-soldered the connecting wires from the printed circuit board.

Next, I soldered the wires from the new switch to the board. I then positioned the new switch over the mounting hole and screwed it in place. The original switch had a locator pin that kept it from rotating on the circuit board. The new switch does not, but it doesn't try to rotate when activating the PTT. Also, the end of the switch rests against one of the channel up/down switches, which gives it more support.

After re-installing the mic button, I re-assembled the mic and tested it during our weekly RACES/SKYWARN net. It works flawlessly. The microphone is back to "original" condition with a new cable and a new PTT switch, all for much less cost than expected. — *Doug Birky, KB8M,* **kb8m@arrl.net**

Repairing Headphones

Recently, I found that one side of my headphones had gone silent. I was able to take apart the two headphone speakers and, by testing continuity between the speakers, determined that the issue was a discontinuity in the shield of the very thin cable that runs up through the headband from one side to the other. I first assumed that this would be the end of the headphones, because finding a replacement for the tiny cable would be difficult and would require disassembly of the entire headband.

Then I had the idea of trying to use a needle to probe the coax. I got a $1\frac{1}{2}$-inch needle from our sewing supplies and securely grasped it with the mini-test clip (see **Figure 21**) of my multimeter. I set my multimeter to resistance measurement with the buzzer enabled for continuity. With one end of the meter clipped to the shield termination on one speaker, I walked my way down the cable, gently pushing the pin and test clip combination into the wire. It didn't take much pressure to get a beep. Sure enough, after I continued along, the beep stopped and I had found the spot where the cable was broken. It was at a swivel joint of the headset earpiece. I removed a section of insulation, soldered a small wire to re-unite the separated cable, and then covered everything with a small amount of heat-shrink tubing. It took a while to get everything back together again, but my headphones sound as good as ever.

I think this would have worked if it was the inner wire rather than the coax shield that was broken, but it would take more pressure and finesse with the needle. — *73, Dick Grote, K6PBF,* **k6pbf@arrl.net**

Figure 22 — The MFJ-1702 coaxial switch has a common port switched between two separate ports. [Clint Millett, VE3CMQ, photo]

Figure 23 — A rubber O-ring washer across the connections increases the holding tension. [Clint Millett, VE3CMQ, photo]

MFJ-1702 Coaxial Switch Modification

I recently discovered a problem with my MFJ Enterprises MFJ-1702 coaxial switch. The connection to the incoming coax connection was intermittent, resulting in poor receive and likely poor transmitting as well. After opening it up, I discovered that the port 1 and 2 contacts were bent and not making a decent connection with the common port. The common port's contact had become twisted and the switched contacts were barely connecting. The small set screw holding the common port in place had also gotten loose, allowing the connector to twist.

To fix these issues, I started by removing the common port. This involved loosening the set screw and removing the connector. With the common connector out of the way, I gained access to the port 1 and 2 contacts (see **Figure 22**).

I placed a rubber O-ring washer with a $\frac{7}{16}$-inch inside diameter (a $\frac{3}{8}$-inch ring should work as well) across these connections and then reinstalled the common port, making sure that the contact was square with the common port pin. I also checked that the set screw was firmly in place so this connector couldn't move.

I checked the operation of the switch and found it to be much better than before. The contacts made a firm and complete connection because the O-ring created additional pressure for the contacts without overstressing the contact leaves (see **Figure 23**).

This modification only took a few minutes, cost almost nothing, and would likely work on similar switches. Note, however, that performing this modification may violate the manufacturer's warranty, so don't attempt the modification unless you're sure the warranty has expired. — *73, Clint Millett, VE3CMQ,* **ve3cmq@secrs.com**

An Easy Handheld Dummy Load and Output Tester

Five-watt handheld transceivers are the most popular variety by far. Whether digital or analog, they are mainstays of VHF and UHF communication, especially in public service applications.

Before you deploy on a public service activation, it helps to be able to test the radio and make sure it is generating full output. The dummy load and indicator shown in **Figure 24** will do the job easily and inexpensively. The circuit (see **Figure 25**) provides a safe load for the transceiver and the lamp glows brightly to indicate full 5 W output. The reason for using resistors in series and parallel rather than a single resistor in series with the lamp is to protect the radio if the lamp filament opens. If this occurs, the radio will still "see" a sufficient load. — *73, William Rynone, WB2EIQ,* **rynone.eng@gmail.com**.

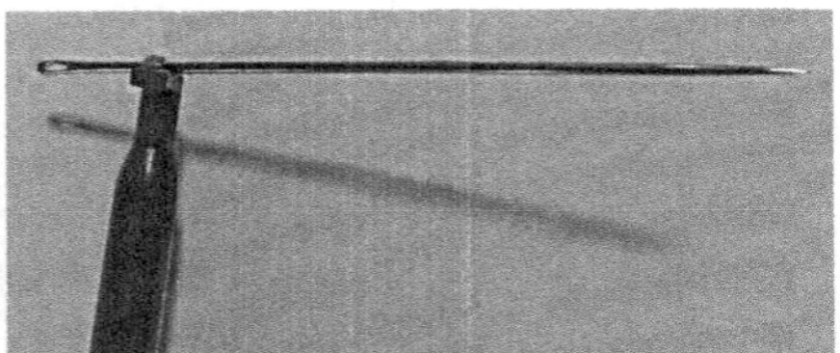

Figure 21 — A sewing needle can be put to work as a continuity probe in tight spaces. [Dick Grote, K6PBF, photo]

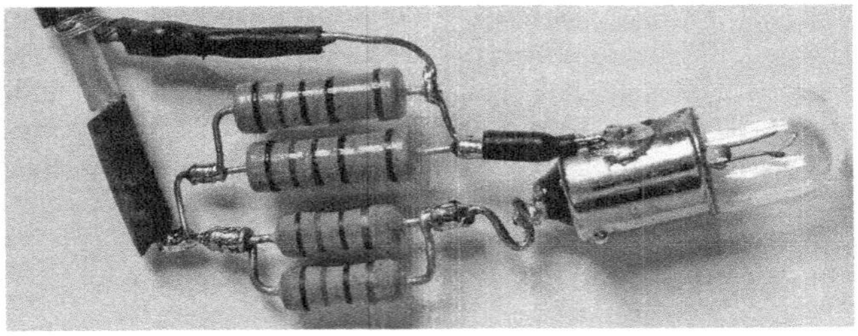

Figure 24 — This easy-to-build dummy load provides a quick output indication for 5 W handheld transceivers. [William Rynone, WB2EIQ, photo]

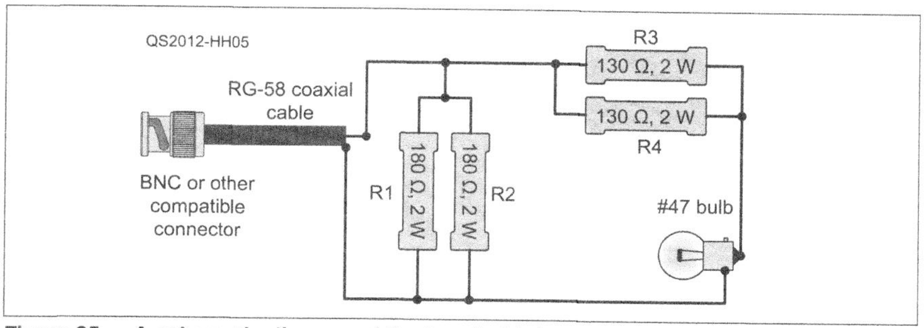

Figure 25 — A schematic diagram of the handheld dummy load. Parts are available from Mouser Electronics at www.mouser.com.

R1 — 180 Ω 2 W resistor (Mouser 594-5083NW180R0JA100)
R2 — 180 Ω 2 W resistor (Mouser 594-5083NW180R0JA100)
R3 —130 Ω 2 W resistor (Mouser 594-5083NW130R0JA100)
R4 — 130 Ω 2 W resistor (Mouser 594-5083NW130R0JA100)
Lamp — Type 47 (Mouser 606-CM47)

DIY Meter Stand

I have several digital multimeters that I'm constantly propping up, only to have them tumble off again. I decided to make a small stand for the meter, which solved my problem.

I wanted to make the stand foldable for storage. The viewing angle is set at 60 degrees. The stand consists of the following components (note that the base and support were made from scrap wood): base (½ × 3 × 5 inches); metal cabinet hinge (2½ inches); metal strap hinge (3 inches), and the meter support (½ × 3 × 6 inches). I nailed three small side pieces made from a paint-stirring stick to the support with ½-inch carpet nails.

I connected the support plate to the base using the metal cabinet hinge (see **Figure 26**). I used a round file to make a small notch to cradle the hinge pin, so the hinge lays flat on the base.

I used a small 3-inch strap hinge to hold the meter support plate at a 60-degree angle. I cut one end off and put a small bend in the other end to align with the angle. This hinge is installed so that the hinge strap is at 90 degrees to the base plate when the meter support is at 60 degrees.

I installed a small stop screw in the back of the support plate to set the angle. Note that the head of the screw may have to be filed down to a diameter that will easily enter the hole in the hinge strap. The finished stand is shown in **Figure 27**.

Because most multimeters are lightweight, I secured the meter to the support plate to avoid pulling it off the stand using a rubber band. Hook-and-loop fasteners or glue-on magnets may be other options. — *73, Michael Janis, KE3OQ,* **mjanis@ptd.net**

Figure 26 — A rear view of the meter stand showing the support hinge. [Michael Janis, KE3OQ, photo]

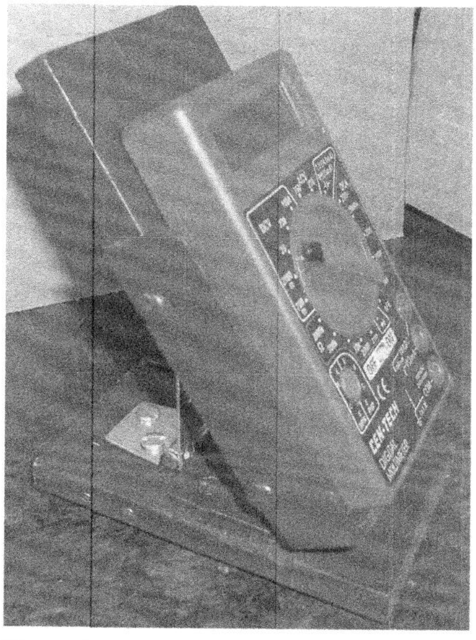

Figure 27 — The finished stand supports a lightweight meter and folds up for storage. [Michael Janis, KE3OQ, photo]

Measuring Current Draw in Battery-Powered Gear

I came up with this solution to prevent frustration when trying to hold the probes of my multimeter while testing current drawn by battery-operated equipment. Take a piece of double-sided printed circuit board and solder two banana jacks, one to each side, as shown in **Figure 28**. To measure the current draw, detach one battery terminal from the device, insert the board, and then move the battery back into position to apply pressure and make solid contact with both sides of the board. In the case of series batteries, spread one of the batteries apart from its neighbor and insert the board between them. Now connect your multimeter probes to the banana jacks and measure the current flow. — *73, Bob Tait, VK3XP,* **vk3xp@internode.on.net**

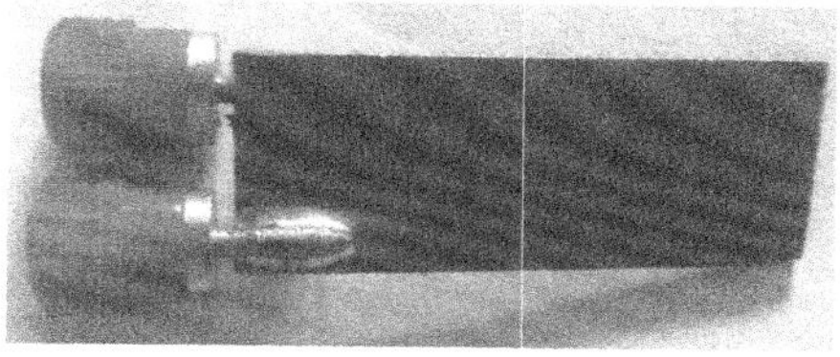

Figure 28 — Make a tool to measure battery current draw with a 1 × 2 inch double-sided printed circuit board and two banana backs. [Bob Tait, VK3XP, photo]

A Low-Voltage AC Dummy Load

An adjustable dc dummy load is a useful tool for testing power supplies. Such a circuit draws a specified current from a supply and measures the resulting output voltage, to describe how the latter varies with the former. Less frequently seen is an adjustable ac load, but such a unit can identify the relationship between current and output voltage for transformers. This is important information when choosing a transformer to, say, light up the filaments of a tube-based project, or when determining the properties of a hamfest find.

Power MOSFETs are often used in dc dummy loads. The voltage applied to a MOSFET's gate determines its resistance. Here is an ac load that contains two MOSFETs back-to-back. The same gate voltage is applied to both, so that an ac voltage across them sees a consistent resistance. The gate voltage may be varied continuously from a moderate negative voltage to an equal positive value to adjust that resistance. The circuit diagram is shown in **Figure 29.**

To measure the current drawn from a transformer and the resulting voltage, there are three possibilities, complicated by the fact that low-voltage ac meters are less common than dc devices. One option is analog meters, which are available from a variety of sources, but which may not offer as much precision as we want.

Another option is digital meters, which are more precise — or at least repeatable — but less common for low ac voltages. The third option measures the current drawn and the resulting voltage with external meters, such as digital multimeters with ac ranges.

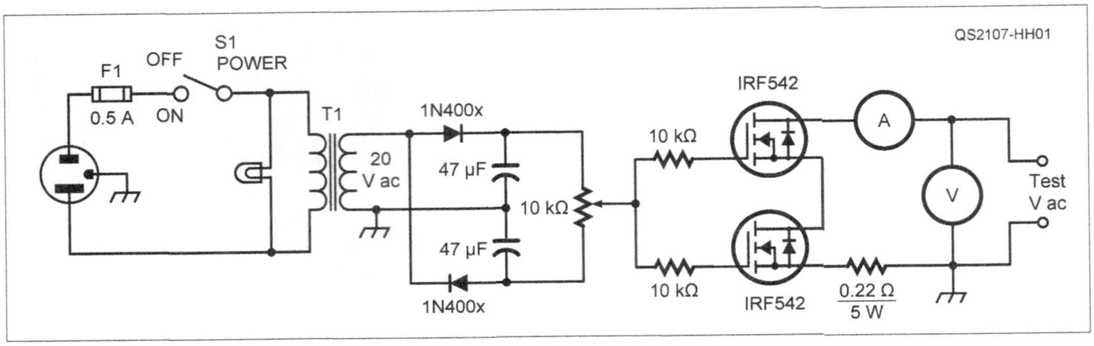

Figure 29 — A schematic diagram of the low-voltage ac load.

Figure 30 — The assembled unit. [Bryant Julstrom, KCØZNG, photo]

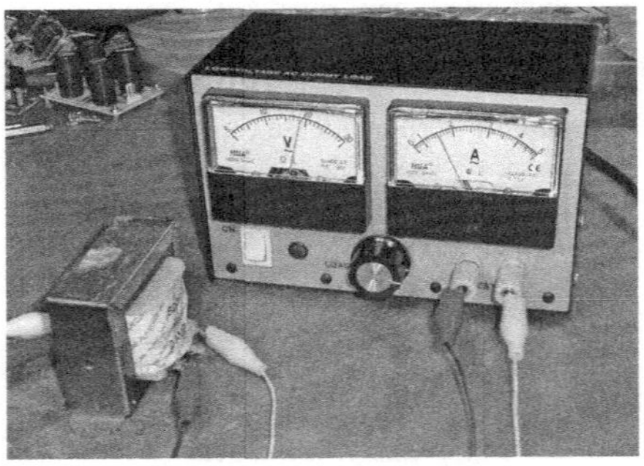

Figure 31 — Testing a transformer with the load. [Bryant Julstrom, KCØZNG, photo]

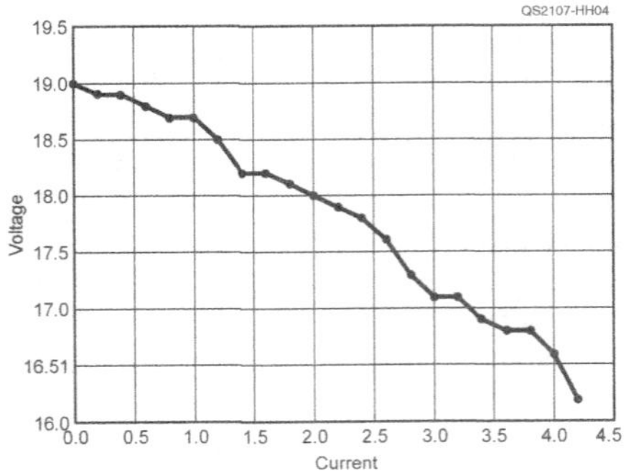

Figure 32 — The transformer's output voltage as a function of the current drawn from it.

The circuit in **Figure 29** is built around a pair of inexpensive analog meters (0 – 30 V ac and 0 – 5 A) and parts that I had on hand. A simple power supply uses a small 20 V transformer (T1) and almost any rectifiers to provide the adjustable gate voltage to a pair of IRF542 MOSFETs. Most of the circuitry occupies a piece of a RadioShack protoboard.

The unit's enclosure is made from several pieces of sheet aluminum. Its size was dictated by the meters, which occupy most of the front panel, along with the power switch, indicator, load potentiometer, and two binding posts to connect to a transformer being tested. The load control is a multiturn potentiometer, to make precise setting of the load easier. A heat sink occupies much of the rear panel, to which the MOSFETs are attached with mica insulators and shoulder washers. The heat sink should be large. When the unit draws, say, 4 A from a transformer at 12 V, it must dissipate 48 W. **Figure 30** shows the assembled unit from the back.

To use the dummy load, connect the secondary of the transformer being tested to the load and power up both the load and the transformer, being careful of the primary voltage. With the load's control, set the current to be drawn from the transformer, and read the resulting voltage. **Figure 31** shows a transformer being tested. **Figure 32** presents a graph of a transformer's output voltage as a function of the current drawn from it. As expected, that voltage diminishes as the current increases.

Some MOSFETs can handle large voltages, so an ac dummy load using such transistors should be able, with appropriate changes in metering, to analyze high-voltage transformers as well. In that case, be careful of the high voltage being tested. — *73, Bryant Julstrom, KCØZNG*, **kc0zng@arrl.net**

A Line Splitter to Measure AC Current

To measure ac current with a clamp-on ammeter, many amateurs resort to purchasing a so-called "line splitter" accessory. However, you can save some money by simply making your own in just a few minutes.

I fashioned a line splitter with nothing more than a short length (about 12 inches) of #14 AWG extension cord, an ac plug, and a socket. When fabricating this type of device, be sure to use a wire gauge sufficient to carry the current to be measured. As shown in

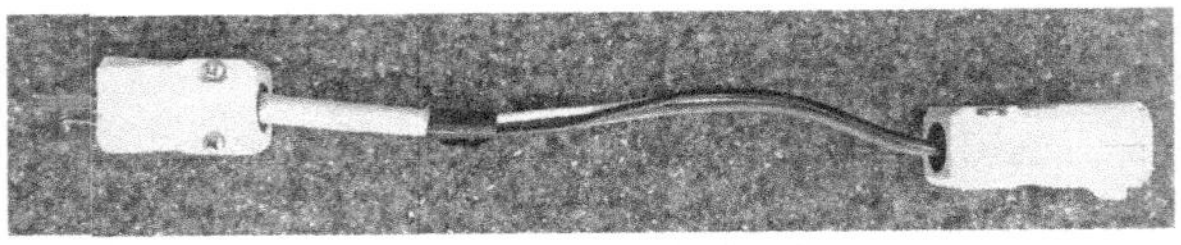

Figure 33 — The outer jacket of the extension cord is stripped away to reveal the ground (green), hot (black), and neutral (white) wires. [Terry White, VE5TLW, photo]

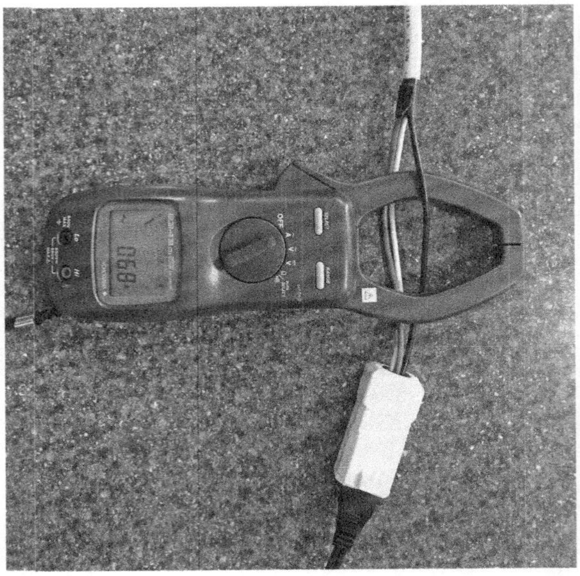

Figure 34 — To measure ac current, clamp the ammeter around either of the current-carrying wires (black or white). [Terry White, VE5TLW, photo]

Figure 33, the outer jacket of the extension cord is stripped away to reveal the three individual wires inside: ground (green), hot (black), and neutral (white). Install the plug and socket in the usual manner.

To measure ac current, clamp the ammeter to either of the current-carrying wires, black or white (see **Figure 34**). The 6.8 A reading on the ammeter was the current draw of an electric toaster. — *73, Terry White, VE5TLW,* **twhite@sasktel.net**.

Test Equipment Terminal Adapters

Here is a simple idea: adapting the terminals of a piece of test equipment to accept banana jacks. I recently added a new piece of test equipment to my bench, a Rigol DL3021A Precision DC Electronic Load. A dc load is a particularly useful device, as it lets you test batteries and power supplies under conditions that you select — constant current, voltage, resistance, or power parameters.

The new Rigol load only has screw-on input terminals for connecting to a Device Under Test, or DUT (see **Figure 35**); these posts will accept either spade or ring-style wire terminals, but unfortunately there is no provision for banana plug-style connectors, unlike my BK Precision electronic load. I wanted the flexibility of using banana plugs, so I built a pair of adapters that would securely attach to the Rigol load's terminals.

I wasn't sure if there was a commercial solution to this problem so, like any good ham, I set off to the junk box to see what I could find. I ended up with two panel-mount banana jacks and some white plastic spacers.

I found that the end of a ring terminal could be inserted into the banana plug and it fit well with a bit of percussive persuasion. The banana plug was a bit long when fully seated, so the white plastic washer filled in the gap nicely. I then soldered the terminal to the body of the plug for a good electrical and mechanical connection.

Repeating the process for the black plug completed a solid pair of adapters I can use with the load, as you can see in **Figure 36**. Red and black permanent markers were used to complete the color scheme and give the adapters a professional look. — *73, Dino Papas, KLØS;* **kl0s@arrl.net**

Single-Evening Voltage Reference

How often have you been misled by an out-of-calibration VOM? I propose an inexpensive, DIY way to check your dc voltage meters and ensure accurate measurements. This voltage reference (see **Figure 37**) is even more critical for calibrating older mechanical meters because such meters have the propensity to "stick."

An Analog Devices AD584K voltage reference IC can fit the bill. It is a low-cost, highly accurate device. Three of the dc voltage outputs are provided at 2.5, 5.0, and 10.0 V, and are within ±3.5, ±6, and ±10 mV, respectively.

Referring to the schematic in **Figure 38**, pushbuttons are used to select the appropriate programming pin of the AD584K. The selected voltage is presented at a front-panel connector (in this case, a BNC). An NTE968 voltage regulator provides a stable 15 V dc to the AD584K.

The power may be supplied by two 9 V batteries wired in series. I used a wall wart for power. Since the calibrator is not used every day, I didn't want to go through the exercise of replacing batteries or cleaning up leaky ones, so I chose the ac supply instead. The power supply should provide 17.5 – 18 V dc. The calibrator only requires 4 – 5 mA of current, so a small, low-current supply is all that is required. If using batteries, an on/off power switch is recommended.

This is a nice single-evening pro-ject. The metal work is more time-consuming than the electronic wiring. The AD584K (or an alternative, such as the AD584JNZ) is available at most major distributors and offered in a DIP configuration. — *73, Bruce Lent, K6HRU,* **k6hru@arrl.net**

Low-Power 50 Ω Dummy Load

We built this in order to check our coax and troubleshoot multiband antennas (see **Figures 39** and **40**). When a multiband vertical or tribander beam suddenly shows a high SWR, you know something bad has happened, but it's not always clear if the problem is the coax or the beam.

Some hams use a 50 Ω dummy load to check out their HF beams and coax. They climb the tower, attach the coax to the

Figure 35 — The new Rigol load only has screw-on input terminals for connecting to a Device Under Test. I needed adapters that would allow me to use banana plugs instead. [Dino Papas, KLØS, photo]

Figure 37 — This voltage reference is easy to build and provides three dc voltage outputs at 2.5, 5.0, and 10.0 V. [Bruce Lent, K6HRU, photo]

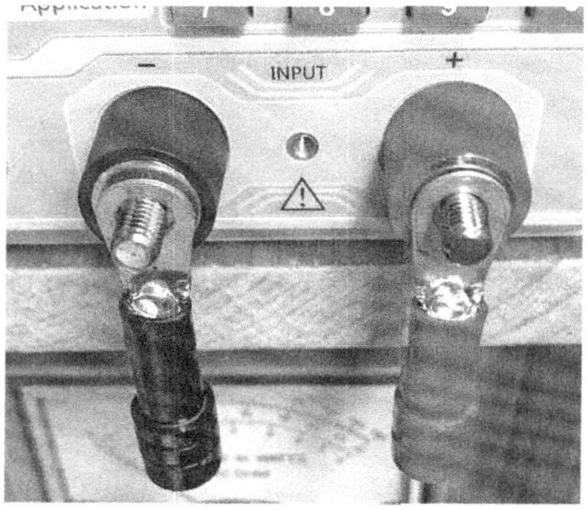

Figure 36 — My efforts resulted in two solid adapters. [Dino Papas, KLØS, photo]

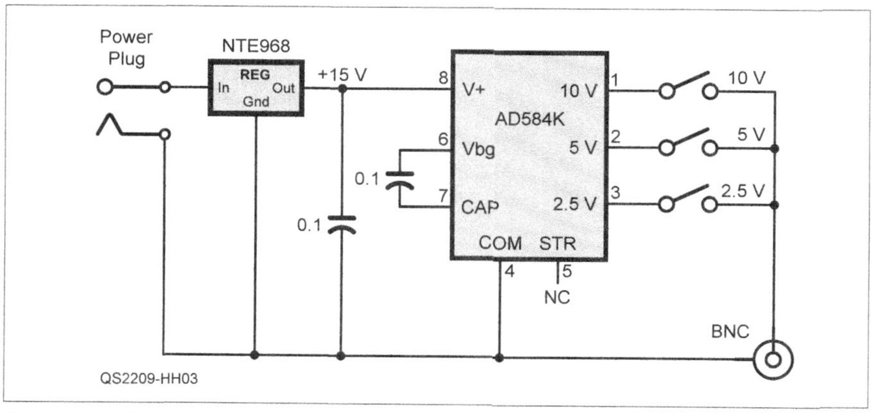

Figure 38 — Schematic diagram of the voltage reference.

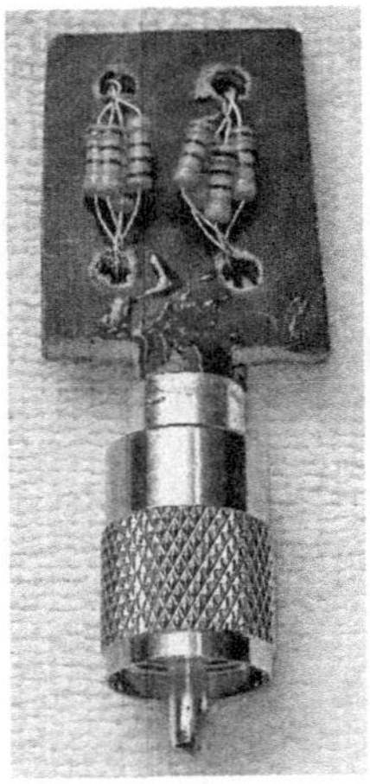

Figure 39 — The front of the 50 Ω dummy load. [Bob Doherty, K1VV, photo]

Figure 40 — The back of the 50 Ω dummy load. [Bob Doherty, K1VV, photo]

dummy load, and put an antenna analyzer on the other end of the coax. If it shows 50 Ω and an SWR of 1:1, the beam is the problem. If it shows a high SWR, then something is wrong with the coax. It could be a short, a broken center conductor, or the braid. The beam can then be checked using a short section of coax with the analyzer to make sure it is not both.

Many hams do not have a dummy load in their shack. This low-power dummy load cost me $4 to make. It consists of two banks of 4-each parallel-connected 100 Ω ½ W carbon (do not use wire wound) resistors. Two of these in series present a total 50 Ω non-inductive load. The resistors are arranged on an old 1 × 2-inch piece of Bakelite, and they are wired and soldered to a PL259. We put some two-part epoxy into the end of the PL259 on the Bakelite to hold it in place. This little dummy load can handle 2 W of RF. — *73, Bob Doherty, K1VV*, **k1vv@comcast.net**

Appendix

RF Connectors and Transmission Lines

There are many different types of transmission lines and RF connectors for coaxial cable, but the three most common for amateur use are the UHF, Type N and BNC families. The type of connector used for a specific job depends on the size of the cable, the frequency of operation and the power levels involved. **Table 1** shows the characteristics of many popular transmission lines,

1 UHF Connectors

The so-called UHF connector (the series name is not related to frequency) is found on most HF and some VHF equipment. It is the only connector many hams will ever see on coaxial cable. PL-259 is another name for the UHF male, and the female is also known as the SO-239. These connectors are rated for full legal amateur power at HF. They are poor for UHF work because they do not present a constant impedance, so the UHF label is a misnomer. PL-259 connectors are designed to fit RG-8 and RG-11 size cable (0.405-inch OD). Adapters are available for use with smaller RG-58, RG-59 and RG-8X size cable. UHF connectors are not weatherproof. Most PL-259s are nickel plated, but silver-plated connectors are much easier to solder and only slightly more expensive.

Figure 1 shows how to install the solder type of PL-259 on RG-8 cable. Proper preparation of the cable end is the key to success. Follow these simple steps. Measure back about ¾-inch from the cable end and slightly score the outer jacket around its circumference. With a sharp knife, cut through the outer jacket, through the braid and through the dielectric — almost to the center conductor. Be careful not to score the center conductor. Cutting through all outer layers at once keeps the braid from separating. (Using a coax stripping tool with preset blade depth makes this and subsequent trimming steps much easier.)

Pull the severed outer jacket, braid and dielectric off the end of the cable as one piece. Inspect the area around the cut, looking for any strands of braid hanging loose and snip them off. There won't be any if your knife was sharp enough. Next, score the outer jacket about 5⁄16-inch back from the first cut. Cut through the jacket lightly; do not score the braid. This step takes practice. If you score the braid, start again. Remove the outer jacket.

Tin the exposed braid and center conductor, but apply the solder sparingly and avoid melting the dielectric. Slide the coupling ring onto the cable. Screw the connector body onto the

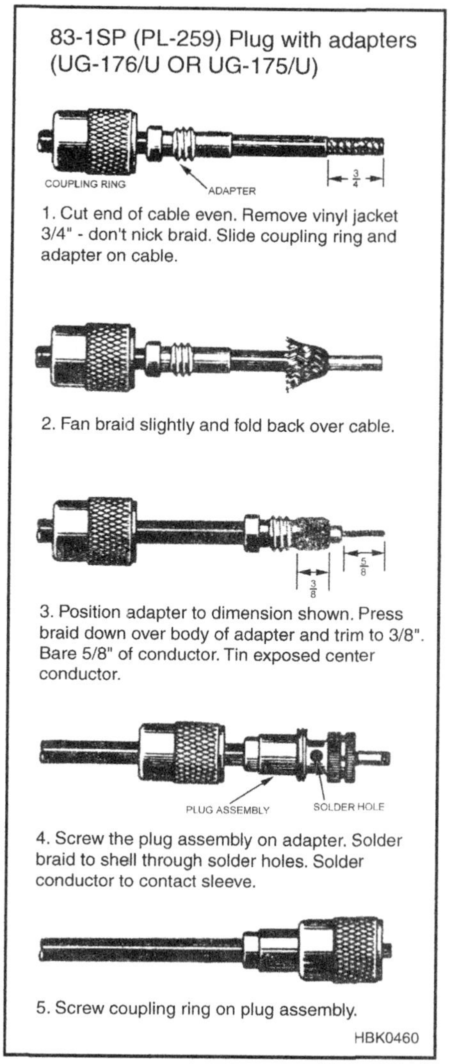

Figure 2 — Installing PL-259 plugs on RG-58 or RG-59 cable requires the use of UG-175 or UG-176 adapters, respectively. The adapter screws into the plug body using the threads of the connector that grip the jacket on larger cables. (*Courtesy Amphenol Electronic Components*)

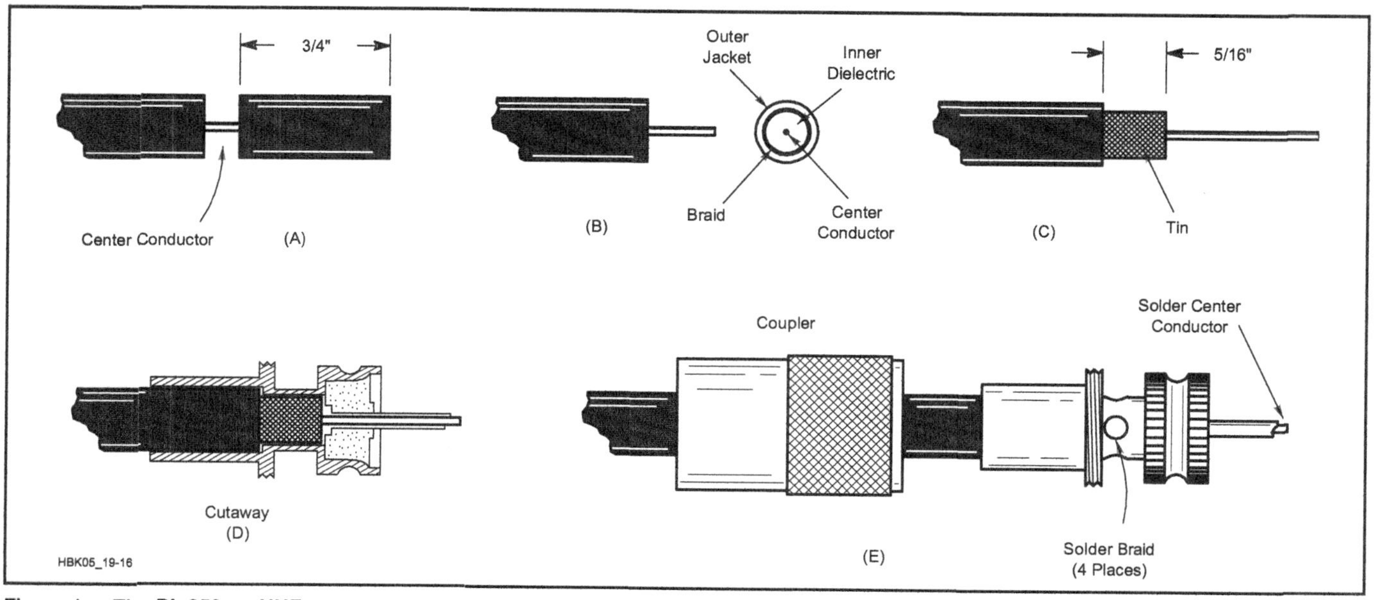

Figure 1 — The PL-259, or UHF, connector is almost universal for amateur HF work and is popular for equipment operating in the VHF range. Steps A through E are described in detail in the text.

cable. If you prepared the cable to the right dimensions, the center conductor will protrude through the center pin, the braid will show through the solder holes, and the body will actually thread onto the outer cable jacket. A very small amount of lubricant on the cable jacket will help the threading process.

Solder the braid through the solder holes. Solder through all four holes; poor connection to the braid is the most common form of PL-259 failure. A good connection between connector and braid is just as important as that between the center conductor and connector. Use a large soldering iron for this job. With practice, you'll learn how much heat to use. If you use too little heat, the solder will bead up, not really flowing onto the connector body. If you use too much heat, the dielectric will melt, letting the braid and center conductor touch. After soldering, to prevent distorting the dielectric do not move the connector or cable until both are cool. A damp rag or sponge can be used to cool the connector quickly.

Solder the center conductor to the center pin. The solder should flow on the inside, not the outside, of the center pin. If you wait until the connector body cools off from soldering the braid, you'll have less trouble with the dielectric melting. Trim the center conductor to be even with the end of the center pin. Use a small file to round the end, removing any solder that built up on the outer surface of the center pin. Use a sharp knife, very fine sandpaper or steel wool to remove any solder flux from the outer surface of the center pin. Screw the coupling ring onto the body, and you're finished.

Figure 2 shows how to install a PL-259 connector on RG-58 or RG-59 cable. An adapter is used for the smaller cable with standard RG-8 size PL-259s. (UG-175 for RG-58 and UG-176 for RG-59.) Prepare the cable as shown. Once the braid is prepared, screw the adapter into the PL-259 shell and finish the job as you would a PL-259 on RG-8 cable.

Figure 3 shows the instructions and dimensions for crimp-on UHF connectors that fit all common sizes of coaxial cable. While amateurs have been reluctant to adopt crimp-on connectors, the availability of good quality connectors and inexpensive crimping tools make crimp technology a good choice, even for connectors used outside. Soldering the center conductor to the connector tip is optional.

UHF connectors are not waterproof and must be waterproofed whether soldered or crimped as shown in the section of the **Safety** chapter on Antenna and Tower Safety.

2 BNC, N and F Connectors

The BNC connectors illustrated in **Figure 4** are popular for low power levels at VHF and UHF. They accept RG-58 and RG-59 cable, and are available for cable mounting in both male and female versions. Several different styles are available, so be sure to use the dimensions for the type you have. Follow the installation instructions carefully. If you prepare the cable to the wrong dimensions, the center pin will not seat properly with connectors of the opposite gender. Sharp scissors are a big help for trimming the braid evenly. Crimp-on BNC connectors are also available, with a large

UHF Connectors

Braid Crimp - Solder Center Contact

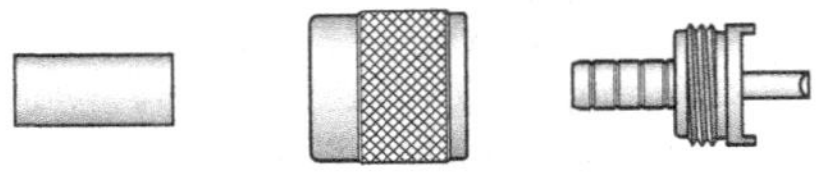

Amphenol	Cable RG-/U	Cable Attachment		Hex Crimp Data			Stripping Dims, inches (mm)		
		Outer	Inner	Cavity for Outer Ferrule	Die Set Tool 227-994	CTL Series Tool No.	a	b	c
83-58SP	58, 141	Crimp	Solder	0.213(5.4)	227-1221-11	CTL-1	1.14 (29.0)	0.780 (19.9)	0.250 (6.4)
83-58SP-1002	400	Crimp	Solder	0.213(5.4)	227-1221-11	CTL-1	1.14 (29.0)	0.780 (19.9)	0.250 (6.4)
83-59DCP-RFX	59	Crimp	Solder	0255(6.5)	227-1221-13	CTL-1	1.22 (30.9)	0.890* (22.6)	0.543 (13.8)
83-58SCP-RFX	58	Crimp	Solder	0.213(5.4)	227-1221-11	CTL-1	1.22 (30.9)	0.890* (22.6)	0.543 (13.8)
83-59SP	59	Crimp	Solder	0.255(6.5)	227-1221-13	CTL-1	1.22 (30.9)	0.890* (22.6)	0.543 (13.8)
83-8SP-RFX	8	Crimp	Solder	0.429(10.9)	227-1221-25	CTL-3	1.22 (30.9)	0.890* (22.6)	0.543 (13.8)

See www.AmphenolRF.com for assembly instructions for all other connectore types. These dimensions only apply to Amphenol connectors and may not be correct for other manufacturers.
* Manufacturer's assembly dimensions incorrectly show 0.574 inches.

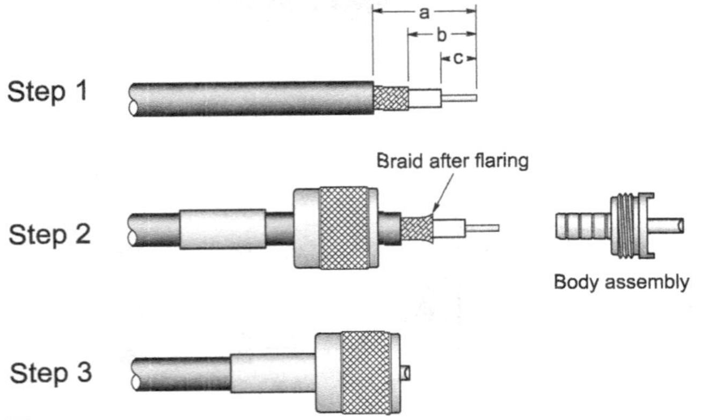

Step 1 Cut end of cable even. Strip cable to dimensions shown in table. All cuts are to be sharp and square. Do not nick braid, dielectric or center conductor. Tin center conductor avoiding excessive heat.

Step 2 Slide coupling nut and ferrule over cable jacket. Flair braid slightly as shown. Install cable into body assembly, so inner ferrule portion slides under braid, until braid butts shoulder. Slide outer ferrule over braid until it butts shoulder. Crimp ferrule with tool and die set indicated in table.

Step 3 Soft solder center conductor to contact. Avoid heating contact excessively to prevent damaging insulator. Slide/screw coupling nut over body.

HBK0475

Figure 3 — Crimp-on UHF connectors are available for all sizes of popular coaxial cable and save considerable time over soldered connectors. The performance and reliability of these connectors is equivalent to soldered connectors, if crimped properly. (*Courtesy Amphenol Electronic Components*)

Table 1

Nominal Characteristics of Commonly Used Transmission Lines

RG or Type	Part Number	Nom. Z_0 Ω	VF %	Cap. pF/ft	Cent. Cond. AWG	Diel. Type	Shield Type	Jacket Matl	OD inches	Max V (RMS)	Matched Loss (dB/100') 1 MHz	10	100	1000
RG-6	Belden 1694A	75	82	16.2	#18 Solid BC	FPE	FC	P1	0.275	300	0.3	.7	1.8	5.9
RG-6	Belden 8215	75	66	20.5	#21 Solid CCS	PE	D	PE	0.332	2700	0.4	0.8	2.7	9.8
RG-8	Belden 7810A	50	86	23.0	#10 Solid BC	FPE	FC	PE	0.405	300	0.1	0.4	1.2	4.0
RG-8	TMS LMR400	50	85	23.9	#10 Solid CCA	FPE	FC	PE	0.405	600	0.1	0.4	1.3	4.1
RG-8	Belden 9913	50	84	24.6	#10 Solid BC	ASPE	FC	P1	0.405	300	0.1	0.4	1.3	4.5
RG-8	CXP1318FX	50	84	24.0	#10 Flex BC	FPE	FC	P2N	0.405	600	0.1	0.4	1.3	4.5
RG-8	Belden 9913F	50	83	24.6	#11 Flex BC	FPE	FC	P1	0.405	300	0.2	0.6	1.5	4.8
RG-8	Belden 9914	50	82	24.8	#10 Solid BC	FPE	FC	P1	0.405	300	0.2	0.5	1.5	4.8
RG-8	TMS LMR400UF	50	85	23.9	#10 Flex BC	FPE	FC	PE	0.405	600	0.1	0.4	1.4	4.9
RG-8	DRF-BF	50	84	24.5	#9.5 Flex BC	FPE	FC	PE	0.405	600	0.1	0.5	1.6	5.2
RG-8	WM CQ106	50	84	24.5	#9.5 Flex BC	FPE	FC	P2N	0.405	600	0.2	0.6	1.8	5.3
RG-8	CXP008	50	78	26.0	#13 Flex BC	FPE	S	P1	0.405	600	0.1	0.5	1.8	7.1
RG-8	Belden 8237	52	66	29.5	#13 Flex BC	PE	S	P1	0.405	3700	0.2	0.6	1.9	7.4
RG-8X	Belden 7808A	50	86	23.5	#15 Solid BC	FPE	FC	PE	0.240	300	0.2	0.7	2.3	7.4
RG-8X	TMS LMR240	50	84	24.2	#15 Solid BC	FPE	FC	PE	0.242	300	0.2	0.8	2.5	8.0
RG-8X	WM CQ118	50	82	25.0	#16 Flex BC	FPE	FC	P2N	0.242	300	0.3	0.9	2.8	8.4
RG-8X	TMS LMR240UF	50	84	24.2	#15 Flex BC	FPE	FC	PE	0.242	300	0.2	0.8	2.8	9.6
RG-8X	Belden 9258	50	82	24.8	#16 Flex BC	FPE	S	P1	0.242	300	0.3	0.9	3.2	11.2
RG-8X	CXP08XB	50	80	25.3	#16 Flex BC	FPE	S	P1	0.242	300	0.3	1.0	3.1	14.0
RG-9	Belden 8242	51	66	30.0	#13 Flex SPC	PE	SCBC	P2N	0.420	5000	0.2	0.6	2.1	8.2
RG-11	Belden 8213	75	84	16.1	#14 Solid BC	FPE	S	PE	0.405	300	0.1	0.4	1.3	5.2
RG-11	Belden 8238	75	66	20.5	#18 Flex TC	PE	S	P1	0.405	300	0.2	0.7	2.0	7.1
RG-58	Belden 7807A	50	85	23.7	#18 Solid BC	FPE	FC	PE	0.195	300	0.3	1.0	3.0	9.7
RG-58	TMS LMR200	50	83	24.5	#17 Solid BC	FPE	FC	PE	0.195	300	0.3	1.0	3.2	10.5
RG-58	WM CQ124	52	66	28.5	#20 Solid BC	PE	S	PE	0.195	1400	0.4	1.3	4.3	14.3
RG-58	Belden 8240	52	66	29.9	#20 Solid BC	PE	S	P1	0.193	1400	0.3	1.1	3.8	14.5
RG-58A	Belden 8219	53	73	26.5	#20 Flex TC	FPE	S	P1	0.195	300	0.4	1.3	4.5	18.1
RG-58C	Belden 8262	50	66	30.8	#20 Flex TC	PE	S	P2N	0.195	1400	0.4	1.4	4.9	21.5
RG-58A	Belden 8259	50	66	30.8	#20 Flex TC	PE	S	P1	0.192	1400	0.5	1.5	5.4	22.8
RG-59	Belden 1426A	75	83	16.3	#20 Solid BC	FPE	S	P1	0.242	300	0.3	0.9	2.6	8.5
RG-59	CXP 0815	75	82	16.2	#20 Solid BC	FPE	S	P1	0.232	300	0.5	0.9	2.2	9.1
RG-59	Belden 8212	75	78	17.3	#20 Solid CCS	FPE	S	P1	0.242	300	0.2	1.0	3.0	10.9
RG-59	Belden 8241	75	66	20.4	#23 Solid CCS	PE	S	P1	0.242	1700	0.6	1.1	3.4	12.0
RG-62A	Belden 9269	93	84	13.5	#22 Solid CCS	ASPE	S	P1	0.240	750	0.3	0.9	2.7	8.7
RG-62B	Belden 8255	93	84	13.5	#24 Flex CCS	ASPE	S	P2N	0.242	750	0.3	0.9	2.9	11.0
RG-63B	Belden 9857	125	84	9.7	#22 Solid CCS	ASPE	S	P2N	0.405	750	0.2	0.5	1.5	5.8
RG-83	WM165	35	66	44.0	#10 Solid BC	PE	S	P2	0.405	2000	0.23	0.8	2.8	9.6
RG-142	CXP 183242	50	69.5	29.4	#19 Solid SCCS	TFE	D	FEP	0.195	1900	0.3	1.1	3.8	12.8
RG-142B	Belden 83242	50	69.5	29.0	#19 Solid SCCS	TFE	D	TFE	0.195	1400	0.3	1.1	3.9	13.5
RG-174	Belden 7805R	50	73.5	26.2	#25 Solid BC	FPE	FC	P1	0.110	300	0.6	2.0	6.5	21.3
RG-174	Belden 8216	50	66	30.8	#26 Flex CCS	PE	S	P1	0.110	1100	0.8	2.5	8.6	33.7
RG-213	Belden 8267	50	66	30.8	#13 Flex BC	PE	S	P2N	0.405	3700	0.2	0.6	2.1	8.0
RG-213	CXP213	50	66	30.8	#13 Flex BC	PE	S	P2N	0.405	600	0.2	0.6	2.0	8.2
RG-214	Belden 8268	50	66	30.8	#13 Flex SPC	PE	D	P2N	0.425	3700	0.2	0.7	2.2	8.0
RG-216	Belden 9850	75	66	20.5	#18 Flex TC	PE	D	P2N	0.425	3700	0.2	0.7	2.0	7.1
RG-217	WM CQ217F	50	66	30.8	#10 Flex BC	PE	D	PE	0.545	7000	0.1	0.4	1.4	5.2
RG-217	M17/78-RG217	50	66	30.8	#10 Solid BC	PE	D	P2N	0.545	7000	0.1	0.4	1.4	5.2
RG-218	M17/79-RG218	50	66	29.5	#4.5 Solid BC	PE	S	P2N	0.870	11000	0.1	0.2	0.8	3.4
RG-223	Belden 9273	50	66	30.8	#19 Solid SPC	PE	D	P2N	0.212	1400	0.4	1.2	4.1	14.5
RG-303	Belden 84303	50	69.5	29.0	#18 Solid SCCS	TFE	S	TFE	0.170	1400	0.3	1.1	3.9	13.5
RG-316	CXP TJ1316	50	69.5	29.4	#26 Flex BC	TFE	S	FEP	0.098	1200	1.2	2.7	8.0	26.1
RG-316	Belden 84316	50	69.5	29.0	#26 Flex SCCS	TFE	S	FEP	0.096	900	0.8	2.5	8.3	26.0
RG-393	M17/127-RG393	50	69.5	29.4	#12 Flex SPC	TFE	D	FEP	0.390	5000	0.2	0.5	1.7	6.1
RG-400	M17/128-RG400	50	69.5	29.4	#20 Flex SPC	TFE	D	FEP	0.195	1400	0.4	1.3	4.3	15.0
LMR500	TMS LMR500UF	50	85	23.9	#7 Flex BC	FPE	FC	PE	0.500	2500	0.1	0.4	1.2	4.0
LMR500	TMS LMR500	50	85	23.9	#7 Solid CCA	FPE	FC	PE	0.500	2500	0.1	0.3	0.9	3.3
LMR600	TMS LMR600	50	86	23.4	#5.5 Solid CCA	FPE	FC	PE	0.590	4000	0.1	0.2	0.8	2.7
LMR600	TMS LMR600UF	50	86	23.4	#5.5 Flex BC	FPE	FC	PE	0.590	4000	0.1	0.2	0.8	2.7
LMR1200	TMS LMR1200	50	88	23.1	#0 Copper Tube	FPE	FC	PE	1.200	4500	0.04	0.1	0.4	1.3
Hardline														
1/2"	CATV Hardline	50	81	25.0	#5.5 BC	FPE	SM	none	0.500	2500	0.05	0.2	0.8	3.2
1/2"	CATV Hardline	75	81	16.7	#11.5 BC	FPE	SM	none	0.500	2500	0.1	0.2	0.8	3.2
7/8"	CATV Hardline	50	81	25.0	#1 BC	FPE	SM	none	0.875	4000	0.03	0.1	0.6	2.9
7/8"	CATV Hardline	75	81	16.7	#5.5 BC	FPE	SM	none	0.875	4000	0.03	0.1	0.6	2.9
LDF4-50A	Heliax – ½"	50	88	25.9	#5 Solid BC	FPE	CC	PE	0.630	1400	0.02	0.2	0.6	2.4
LDF5-50A	Heliax – ⅞"	50	88	25.9	0.355" BC	FPE	CC	PE	1.090	2100	0.03	0.10	0.4	1.3
LDF6-50A	Heliax – 1¼"	50	88	25.9	0.516" BC	FPE	CC	PE	1.550	3200	0.02	0.08	0.3	1.1
Parallel Lines														
TV Twinlead (Belden 9085)		300	80	4.5	#22 Flex CCS	PE	none	P1	0.400	**	0.1	0.3	1.4	5.9
Twinlead (Belden 8225)		300	80	4.4	#20 Flex BC	PE	none	P1	0.400	8000	0.1	0.2	1.1	4.8
Generic Window Line		450	91	2.5	#18 Solid CCS	PE	none	P1	1.000	10000	0.02	0.08	0.3	1.1
WM CQ 554		440	91	2.7	#14 Flex CCS	PE	none	P1	1.000	10000	0.04	0.01	0.6	3.0
WM CQ 552		440	91	2.5	#16 Flex CCS	PE	none	P1	1.000	10000	0.05	0.2	0.6	2.6
WM CQ 553		450	91	2.5	#18 Flex CCS	PE	none	P1	1.000	10000	0.06	0.2	0.7	2.9
WM CQ 551		450	91	2.5	#18 Solid CCS	PE	none	P1	1.000	10000	0.05	0.02	0.6	2.8
Open-Wire Line		600	0.95-99***	1.7	#12 BC	none	none	none	**	12000	0.02	0.06	0.2	—

(continued on next page)

Approximate Power Handling Capability (1:1 SWR, 40°C Ambient):

	1.8 MHz	*7*	*14*	*30*	*50*	*150*	*220*	*450*	*1 GHz*
RG-58 Style	1350	700	500	350	250	150	120	100	50
RG-59 Style	2300	1100	800	550	400	250	200	130	90
RG-8X Style	1830	840	560	360	270	145	115	80	50
RG-8/213 Style	5900	3000	2000	1500	1000	600	500	350	250
RG-217 Style	20000	9200	6100	3900	2900	1500	1200	800	500
LDF4-50A	38000	18000	13000	8200	6200	3400	2800	1900	1200
LDF5-50A	67000	32000	22000	14000	11000	5900	4800	3200	2100
LMR500	18000	9200	6500	4400	3400	1900	1600	1100	700
LMR1200	52000	26000	19000	13000	10000	5500	4500	3000	2000

Legend:

**	Not Available or varies	N	Non-Contaminating
***	Varies with spacer material and spacing	P1	PVC, Class 1
ASPE	Air Spaced Polyethylene	P2	PVC, Class 2
BC	Bare Copper	PE	Polyethylene
CC	Corrugated Copper	S	Single Braided Shield
CCA	Copper Cover Aluminum	SC	Silver Coated Braid
CCS	Copper Covered Steel	SCCS	Silver Plated Copper Coated Steel
CXP	Cable X-Perts, Inc.	SM	Smooth Aluminum
D	Double Copper Braids	SPC	Silver Plated Copper
DRF	Davis RF	TC	Tinned Copper
FC	Foil + Tinned Copper Braid	TFE	Teflon®
FEP	Teflon ® Type IX	TMS	Times Microwave Systems
Flex	Flexible Stranded Wire	UF	Ultra Flex
FPE	Foamed Polyethylene	WM	Wireman
Heliax	Andrew Corp Heliax		

Figure 4 (below) — BNC connectors are common on VHF and UHF equipment at low power levels. (*Courtesy Amphenol Electronic Components*)

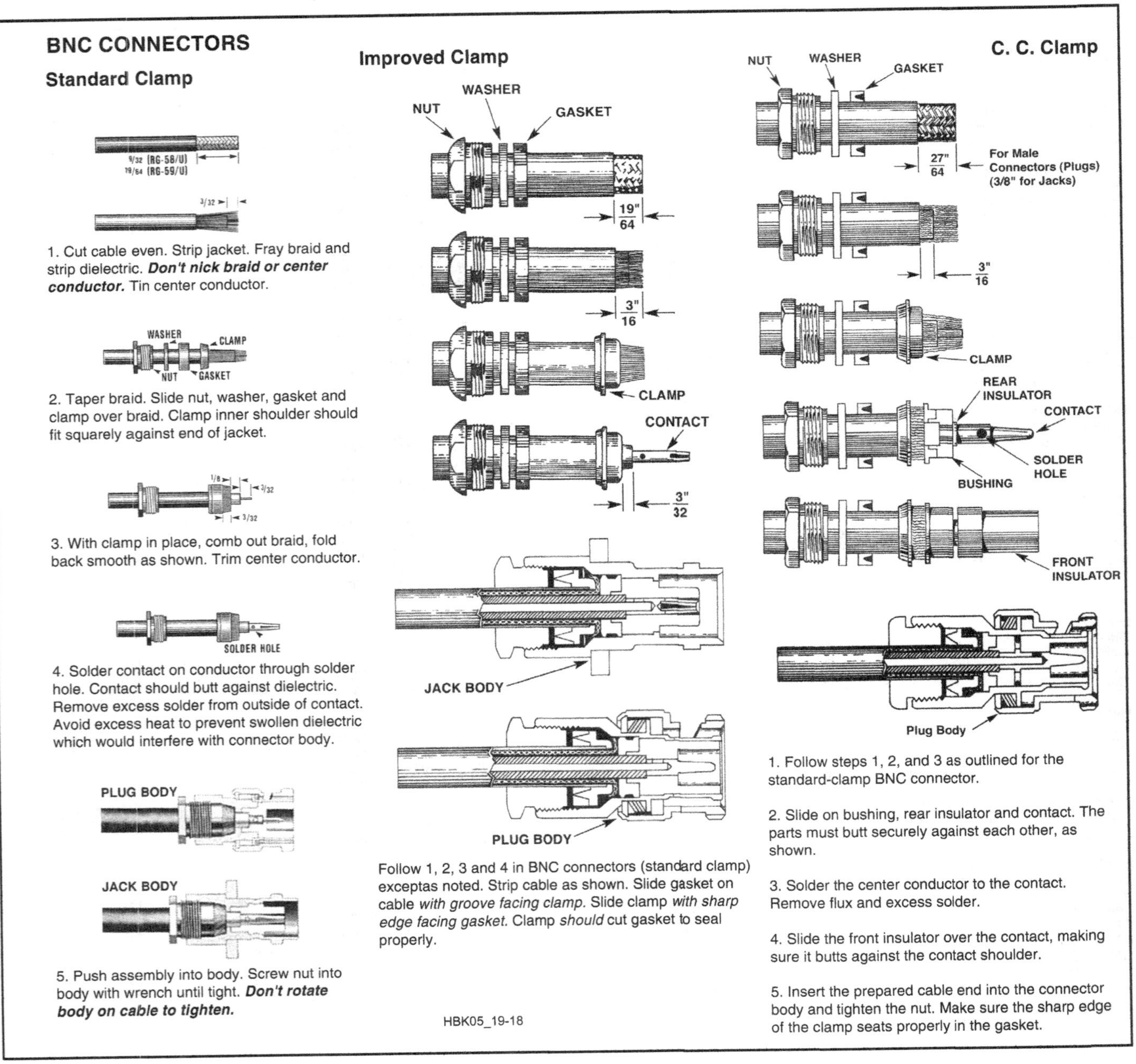

Table 2
Coaxial Cable Connectors

UHF Connectors

Military No.	Style	Cable RG- or Description
PL-259	Str (m)	8, 9, 11, 13, 63, 87, 149, 213, 214, 216, 225
UG-111	Str (m)	59, 62, 71, 140, 210
SO-239	Pnl (f)	Std, mica/phenolic insulation
UG-266	Blkhd (f)	Rear mount, pressurized, copolymer of styrene ins.
Adapters		
PL-258	Str (f/f)	Polystyrene ins.
UG-224,363	Blkhd (f/f)	Polystyrene ins.
UG-646	Ang (f/m)	Polystyrene ins.
M-359A	Ang (m/f)	Polystyrene ins.
M-358	T (f/m/f)	Polystyrene ins.
Reducers		
UG-175		55, 58, 141, 142 (except 55A)
UG-176		59, 62, 71, 140, 210

Family Characteristics:

All are nonweatherproof and have a nonconstant impedance. Frequency range: 0-500 MHz. Maximum voltage rating: 500 V (peak).

N Connectors

Military No.	Style	Cable RG-	Notes
UG-21	Str (m)	8, 9, 213, 214	50 Ω
UG-94A	Str (m)	11, 13, 149, 216	70 Ω
UG-536	Str (m)	58, 141, 142	50 Ω
UG-603	Str (m)	59, 62, 71, 140, 210	50 Ω
UG-23, B-E	Str (f)	8, 9, 87, 213, 214, 225	50 Ω
UG-602	Str (f)	59, 62, 71, 140, 210	—
UG-228B, D, E	Pnl (f)	8, 9, 87, 213, 214, 225	—
UG-1052	Pnl (f)	58, 141, 142	50 Ω
UG-593	Pnl (f)	59, 62, 71, 140, 210	50 Ω
UG-160A, B, D	Blkhd (f)	8, 9, 87, 213, 214, 225	50 Ω
UG-556	Blkhd (f)	58, 141, 142	50 Ω
UG-58, A	Pnl (f)		50 Ω
UG-997A	Ang (f)		50 Ω

Panel mount (f) with clearance above panel

M39012/04-	Blkhd (f)	Front mount hermetically sealed
UG-680	Blkhd (f)	Front mount pressurized

N Adapters

Military No.	Style	Notes
UG-29,A,B	Str (f/f)	50 Ω, TFE ins.
UG-57A.B	Str (m/m)	50 Ω, TFE ins.
UG-27A,B	Ang (f/m)	Mitre body
UG-212A	Ang (f/m)	Mitre body
UG-107A	T (f/m/f)	—
UG-28A	T (f/f/f)	—
UG-107B	T (f/m/f)	—

Family Characteristics:

N connectors with gaskets are weatherproof. RF leakage: –90 dB min @ 3 GHz. Temperature limits: TFE: –67° to 390°F (–55° to 199°C). Insertion loss 0.15 dB max @ 10 GHz. Copolymer of styrene: –67° to 185°F (–55° to 85°C). Frequency range: 0-11 GHz. Maximum voltage rating: 1500 V P-P. Dielectric withstanding voltage 2500 V RMS. SWR (MIL-C-39012 cable connectors) 1.3 max 0-11 GHz.

BNC Connectors

Military No.	*Style*	*Cable RG-*	*Notes*
UG-88C	Str (m)	55, 58, 141, 142, 223, 400	

Military No.	*Style*	*Cable RG-*	*Notes*
UG-959	Str (m)	8, 9	
UG-260,A	Str (m)	59, 62, 71, 140, 210	Rexolite ins.
UG-262	Pnl (f)	59, 62, 71, 140, 210	Rexolite ins.
UG-262A	Pnl (f)	59, 62, 71, 140, 210	nwx, Rexolite ins.
UG-291	Pnl (f)	55, 58, 141, 142, 223, 400	
UG-291A	Pnl (f)	55, 58, 141, 142, 223, 400	nwx
UG-624	Blkhd (f)	59, 62, 71, 140, 210	Front mount Rexolite ins.
UG-1094A	Blkhd		Standard
UG-625B	Receptacle		
UG-625			

BNC Adapters

Military No.	Style	Notes
UG-491,A	Str (m/m)	
UG-491B	Str (m/m)	Berylium, outer contact
UG-914	Str (f/f)	
UG-306	Ang (f/m)	
UG-306A,B	Ang (f/m)	Berylium outer contact
UG-414,A	Pnl (f/f)	# 3-56 tapped flange holes
UG-306	Ang (f/m)	
UG-306A,B	Ang (f/m)	Berylium outer contact
UG-274	T (f/m/f)	
UG-274A,B	T (f/m/f)	Berylium outer contact

Family Characteristics:

Z = 50 Ω. Frequency range: 0-4 GHz w/low reflection; usable to 11 GHz. Voltage rating: 500 V P-P. Dielectric withstanding voltage 500 V RMS. SWR: 1.3 max 0-4 GHz. RF leakage –55 dB min @ 3 GHz. Insertion loss: 0.2 dB max @ 3 GHz. Temperature limits: TFE: –67° to 390°F (–55° to 199°C); Rexolite insulators: –67° to 185°F (–55° to 85°C). "Nwx" = not weatherproof.

HN Connectors

Military No.	Style	Cable RG-	Notes
UG-59A	Str (m)	8, 9, 213, 214	
UG-1214	Str (f)	8, 9, 87, 213, 214, 225	Captivated contact
UG-60A	Str (f)	8, 9, 213, 214	Copolymer of styrene ins.
UG-1215	Pnl (f)	8, 9, 87, 213, 214, 225	Captivated contact
UG-560	Pnl (f)		
UG-496	Pnl (f)		
UG-212C	Ang (f/m)		Berylium outer contact

Family Characteristics:

Connector Styles: Str = straight; Pnl = panel; Ang = Angle; Blkhd = bulkhead. Z = 50 Ω. Frequency range = 0-4 GHz. Maximum voltage rating = 1500 V P-P. Dielectric withstanding voltage = 5000 V RMS SWR = 1.3. All HN series are weatherproof. Temperature limits: TFE: –67° to 390°F (–55° to 199°C); copolymer of styrene: –67° to 185°F (–55° to 85°C).

Cross-Family Adapters

Families	Description	Military No.
HN to BNC	HN-m/BNC-f	UG-309
N to BNC	N-m/BNC-f	UG-201,A
	N-f/BNC-m	UG-349,A
	N-m/BNC-m	UG-1034
N to UHF	N-m/UHF-f	UG-146
	N-f/UHF-m	UG-83,B
	N-m/UHF-m	UG-318
UHF to BNC	UHF-m/BNC-f	UG-273
	UHF-f/BNC-m	UG-255

Type N assembly instructions

HBK05_19-19

CLAMP TYPES

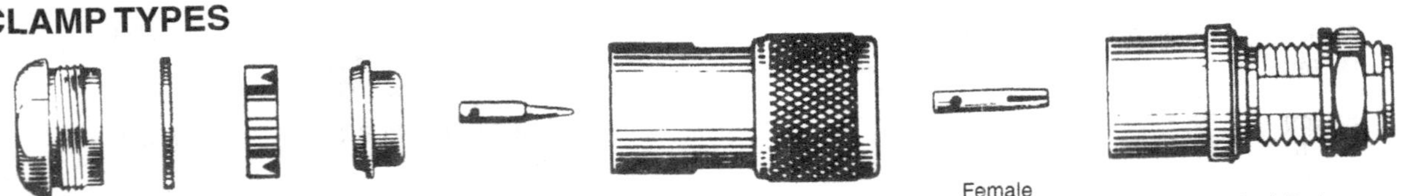

Nut Washer Gasket Clamp Male Contact Plug Body Female Contact Jack Body

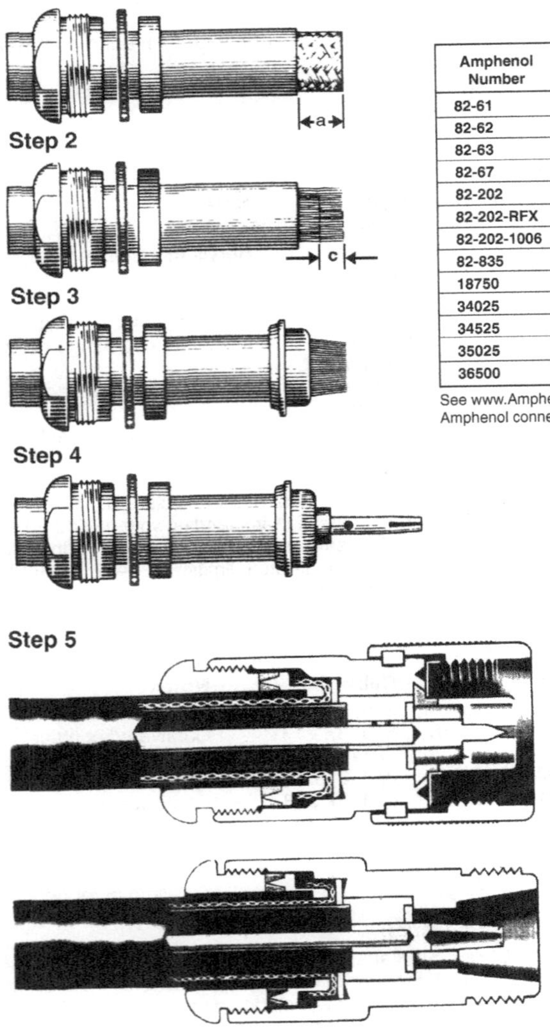

Amphenol Number	Connector Type	Cable RG-/U	Strip Dims., inches (mm)	
			a	c
82-61	N Plug	8, 9, 144, 165, 213, 214, 216, 225	0.359(9.1)	0.234(6.0)
82-62	N Panel Jack		0.312(7.9)	0.187(4.7)
82-63	N Jack	8, 9, 87A, 144, 165, 213, 214, 216, 225	0.281(7.1)	0.156(4.0)
82-67	N Bulkhead Jack			
82-202	N Plug	8, 9, 144, 165, 213, 214, 216, 225	0.359(9.1)	0.234(6.0)
82-202-RFX	N Plug	8, 213, 214	0.315(8.0)	0.177(4.5)
82-202-1006	N Plug	Belden 9913	0.359(9.1)	0.234(6.0)
82-835	N Angle Plug	8, 9, 87A, 144, 165, 213, 214, 216, 225	0.281(7.1)	0.156(4.0)
18750	N Angle Plug	58, 141, 142	0.484(12.3)	0.234(5.9)
34025	N Plug		0.390(9.9)	0.203(5.2)
34525	N Plug	59, 62, 71, 140, 210	0.410(10.4)	0.230(5.8)
35025	N Jack	58, 141, 142	0.375(9.5)	0.187(4.7)
36500	N Jack	59, 62, 71, 140, 210	0.484(12.3)	0.200(5.1)

See www.AmphenolRF.com for assembly instructions for all other connector types. These dimensions only apply to Amphenol connectors and may not be correct for other manufacturers.

Step 1 Place nut, washer, and gasket, with "V" groove toward clamp, over cable and cut off jacket to dimension a.

Step 2 Comb out braid and fold out. Cut off cable dielectric to dim. c as shown.

Step 3 Pull braid wires forward and taper toward center conductor. Place clamp over braid and push back against cable jacket.

Step 4 Fold back braid wires as shown, trim braid to proper length and form over clamp as shown. Solder contact to center conductor.

Step 5 Insert cable and parts into connector body. Make sure sharp edge of clamp seats properly in gasket. Tighten nut.

Figure 5 — Type N connectors are a must for high-power VHF and UHF operation. (*Courtesy Amphenol Electronic Components*)

number of variations, including a twist-on version. A guide to installing these connectors is available with the downloadable supplemental content.

The Type N connector, illustrated in **Figure 5**, is a must for high-power VHF and UHF operation. N connectors are available in male and female versions for cable mounting and are designed for RG-8 size cable. Unlike UHF connectors, they are designed to maintain a constant impedance at cable joints. Like BNC connectors, it is important to prepare the cable to the right dimensions. The center pin must be positioned correctly to mate with the center pin of connectors of the opposite gender. Use the right dimensions for the connector style you have. Crimp-on N connectors are also available, again with a large number of variations. A guide to installing these connectors is available with the downloadable supplemental content.

Type F connectors, used primarily on cable TV connections, are also popular for receive-only antennas and can be used with RG-59

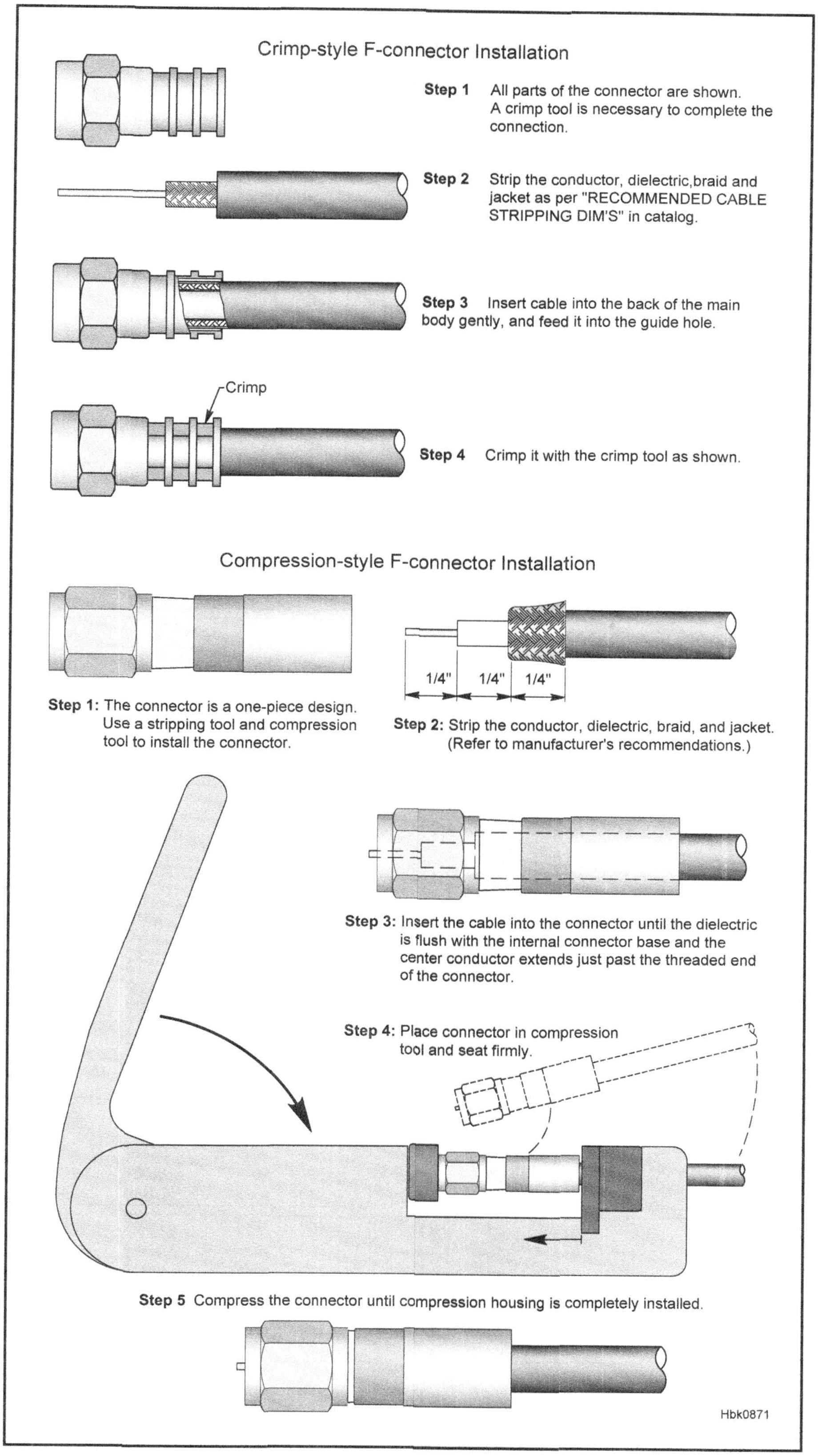

Figure 6 — Type F connectors, commonly used for cable TV connections, can be used for receive-only antennas with inexpensive RG-59 and RG-6 cable. The older crimp-style connectors have largely been replaced by the more reliable and weather-resistant compression-style connectors. Follow the manufacturer's instruction for the cable and connector selected. [Courtesy Amphenol Electronic Components and Technetix, Inc.]

or the increasingly popular RG-6 cable available at low cost. Crimp-on connectors are the only option for these connectors and **Figure 6** shows a general guide for installing them. The exact dimensions vary between connector styles and manufacturers — information on crimping is generally provided with the connectors. There are two styles of crimp; ferrule and compression. The ferrule crimp method is similar to that for UHF, BNC, and N connectors in which a metal ring is compressed around the exposed coax shield. The compression crimp forces a bushing into the back of the connector, clamping the shield against the connector body. In all cases, the exposed center conductor of the cable — a solid wire — must end flush with the end of the connector. A center conductor that is too short may not make a good connection.

3 Connector Identifier and Range Chart

The following pages of figures provide dimensions and side views to help identify the different types of connectors used for RF through microwave frequencies. Dimensions are provided in both imperial and metric units as appropriate. For mm-wave and microwave connectors, calipers or a micrometer may be required to provide an accurate measurement capable of distinguishing between similar connectors. These specifications are intended for connector identification only and should not be the sole dimensions used when laying out a circuit board or drilling a mounting hole.

These figures and chart were provided by Pasternack (**pasternack.com**), a major distributor of coaxial connectors, cable, tools, and other RF materials and supplies.

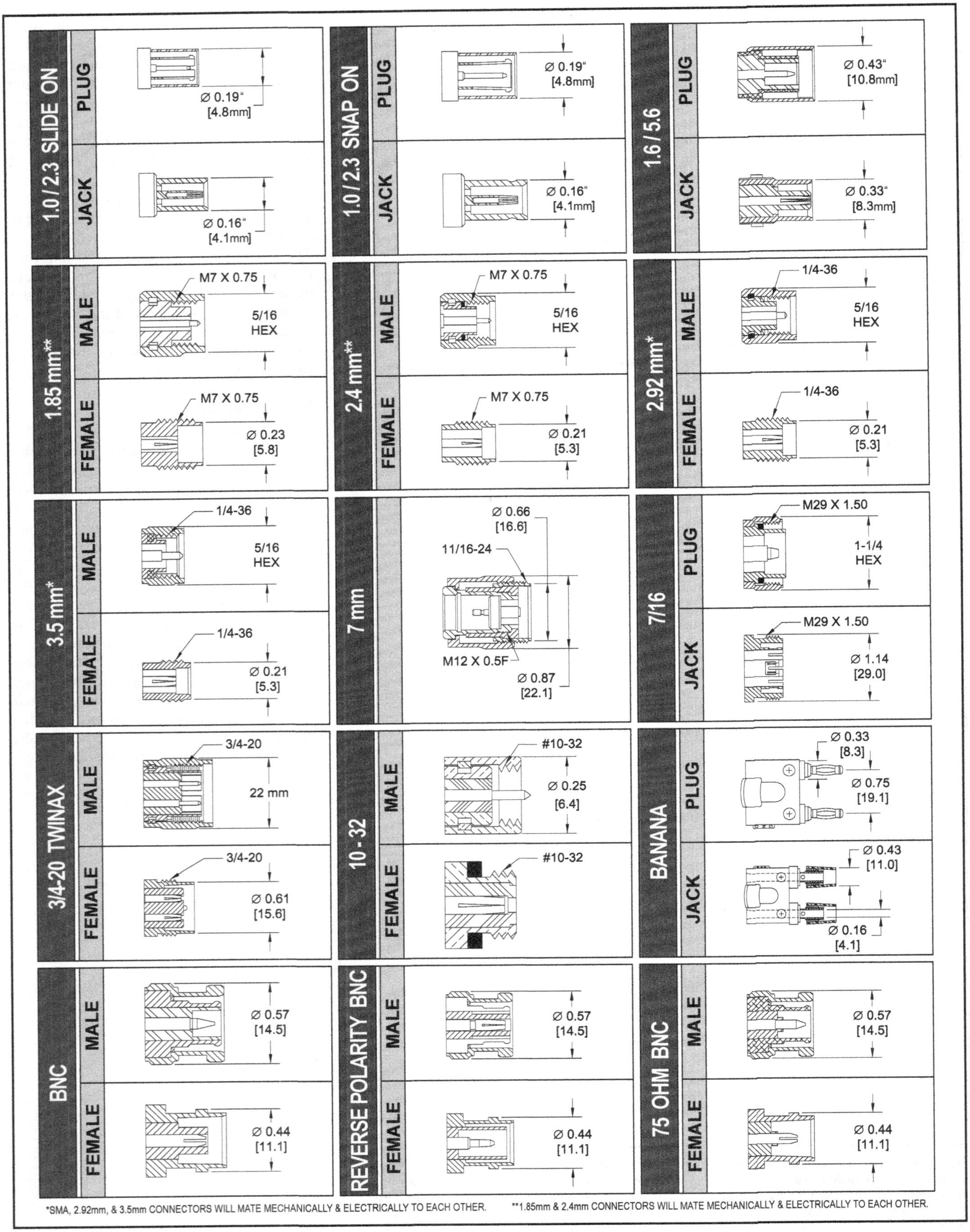

Figure 7A — Connector side views, set 1 of 4. [Courtesy of Pasternack]

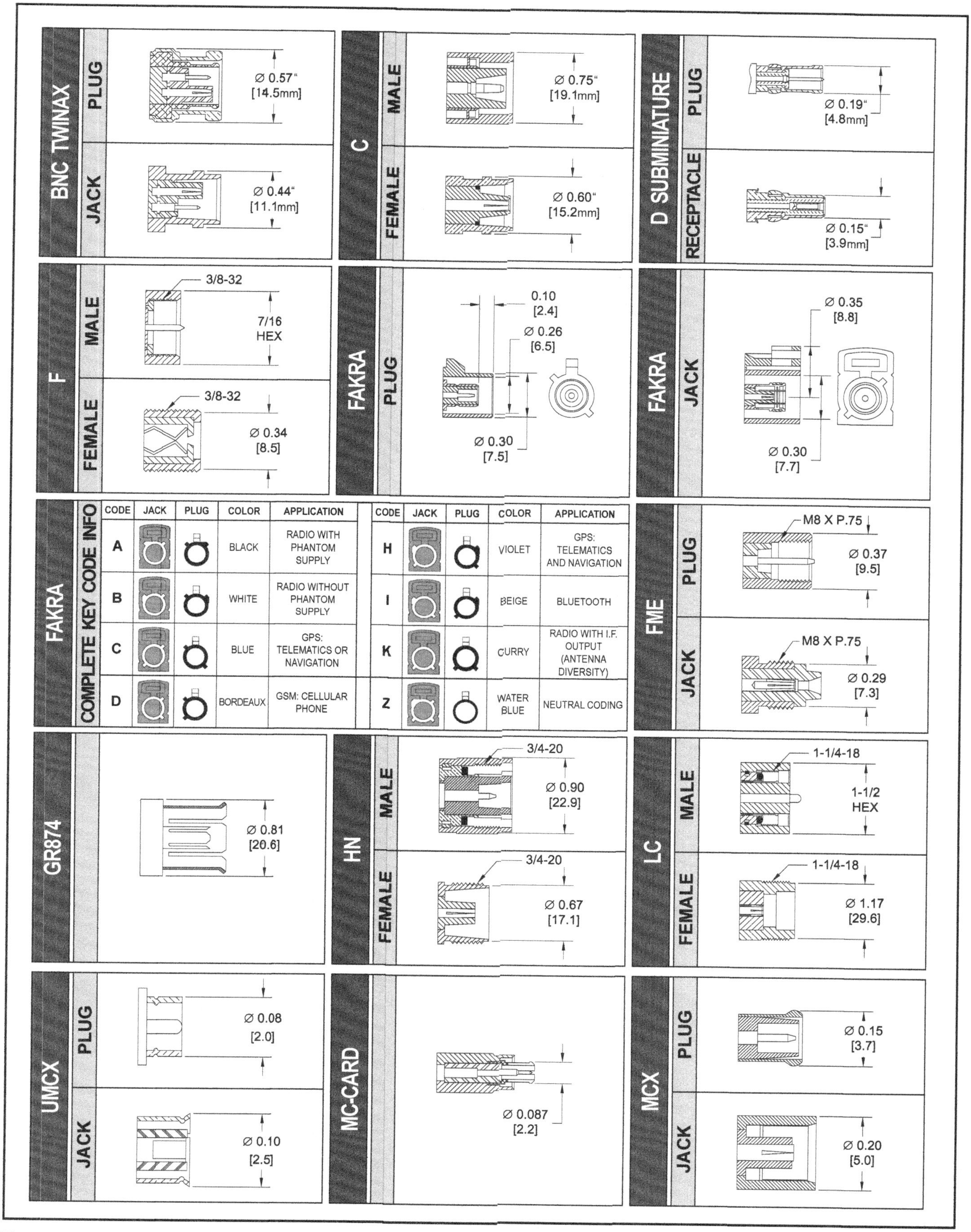

FAKRA — COMPLETE KEY CODE INFO

CODE	JACK	PLUG	COLOR	APPLICATION
A			BLACK	RADIO WITH PHANTOM SUPPLY
B			WHITE	RADIO WITHOUT PHANTOM SUPPLY
C			BLUE	GPS: TELEMATICS OR NAVIGATION
D			BORDEAUX	GSM: CELLULAR PHONE
H			VIOLET	GPS: TELEMATICS AND NAVIGATION
I			BEIGE	BLUETOOTH
K			CURRY	RADIO WITH I.F. OUTPUT (ANTENNA DIVERSITY)
Z			WATER BLUE	NEUTRAL CODING

Figure 7B — Connector side views, set 2 of 4. [Courtesy of Pasternack]

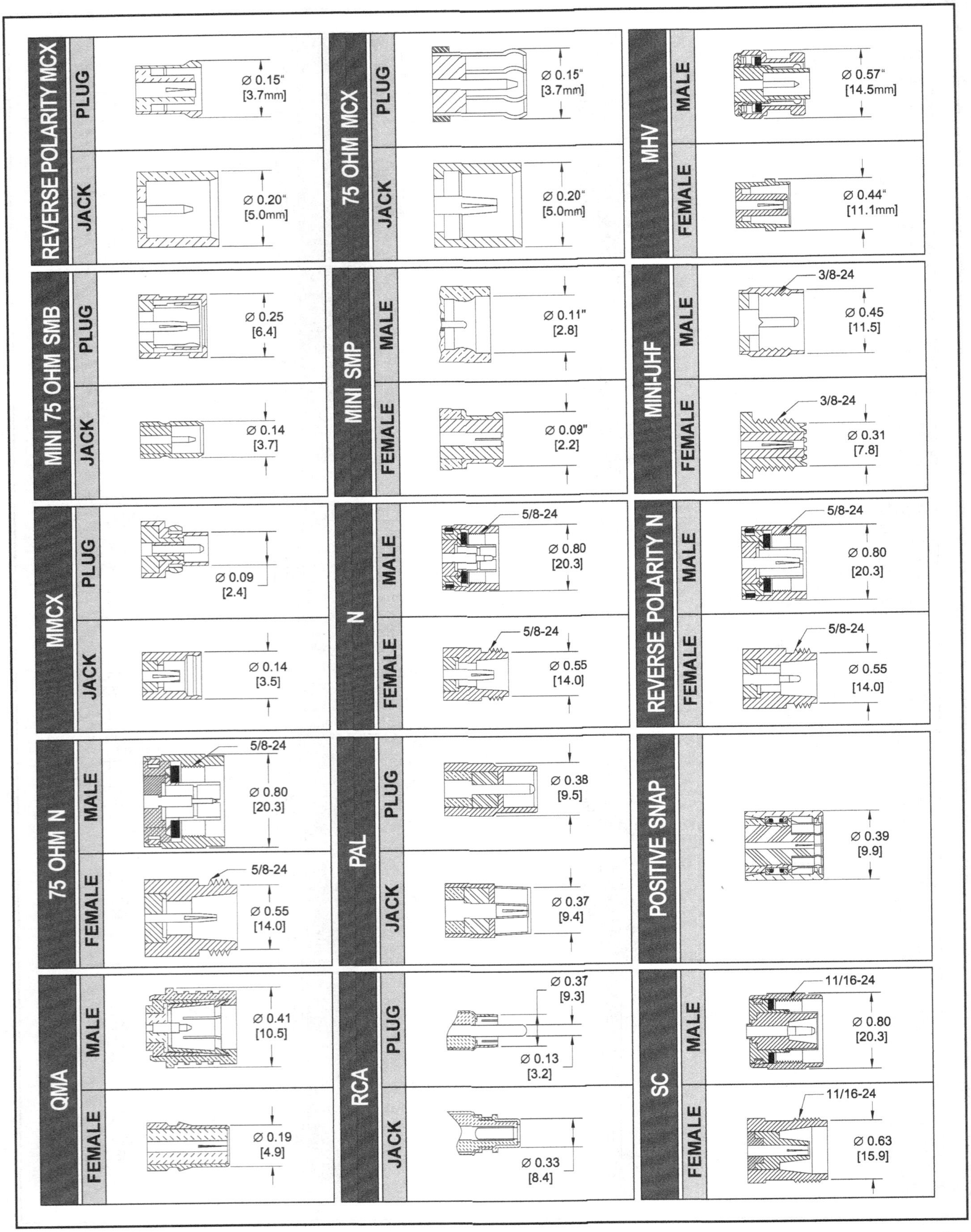

Figure 7C — Connector side views, set 3 of 4. [Courtesy of Pasternack]

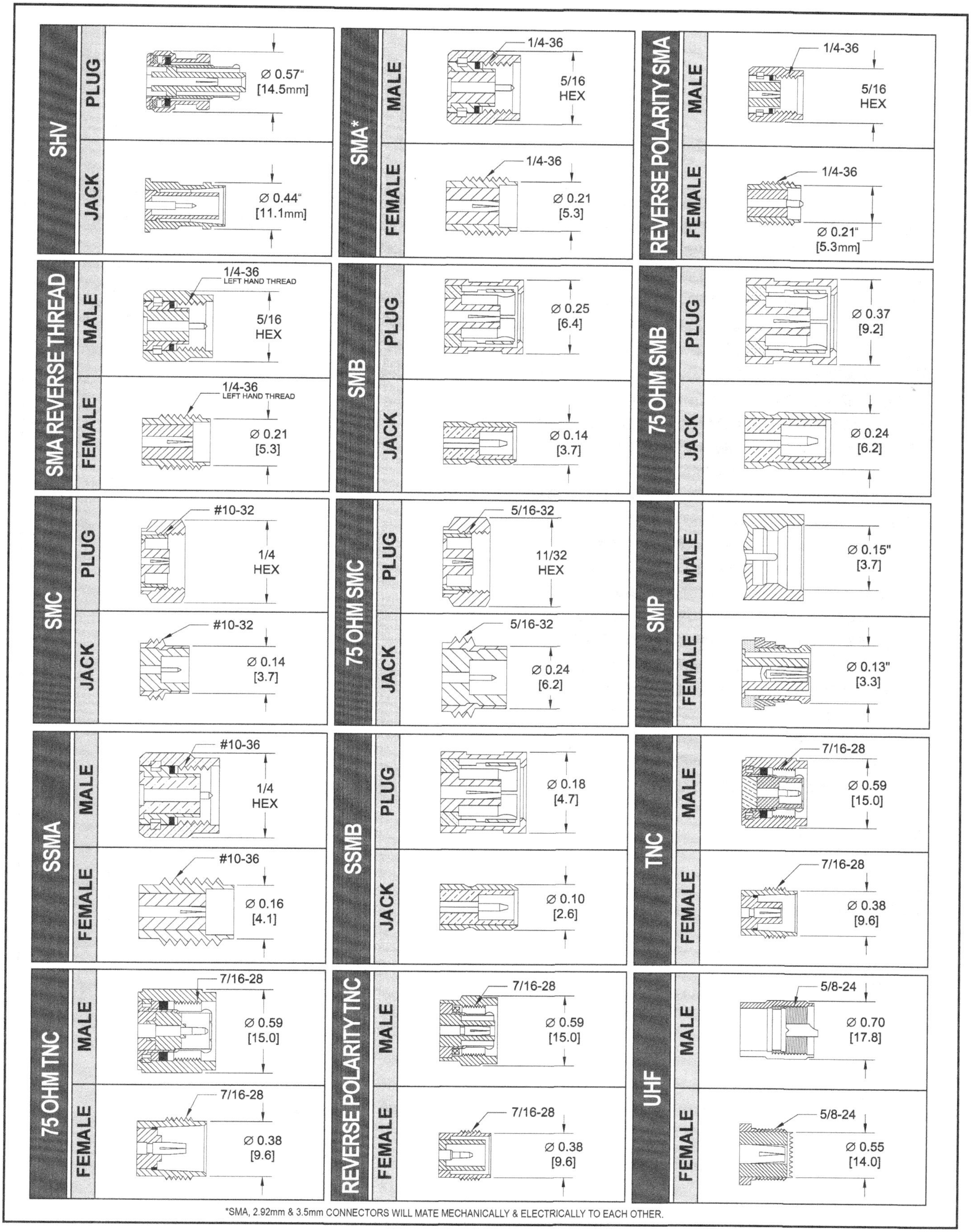

Figure 7D — Connector side views, set 4 of 4. [Courtesy of Pasternack]

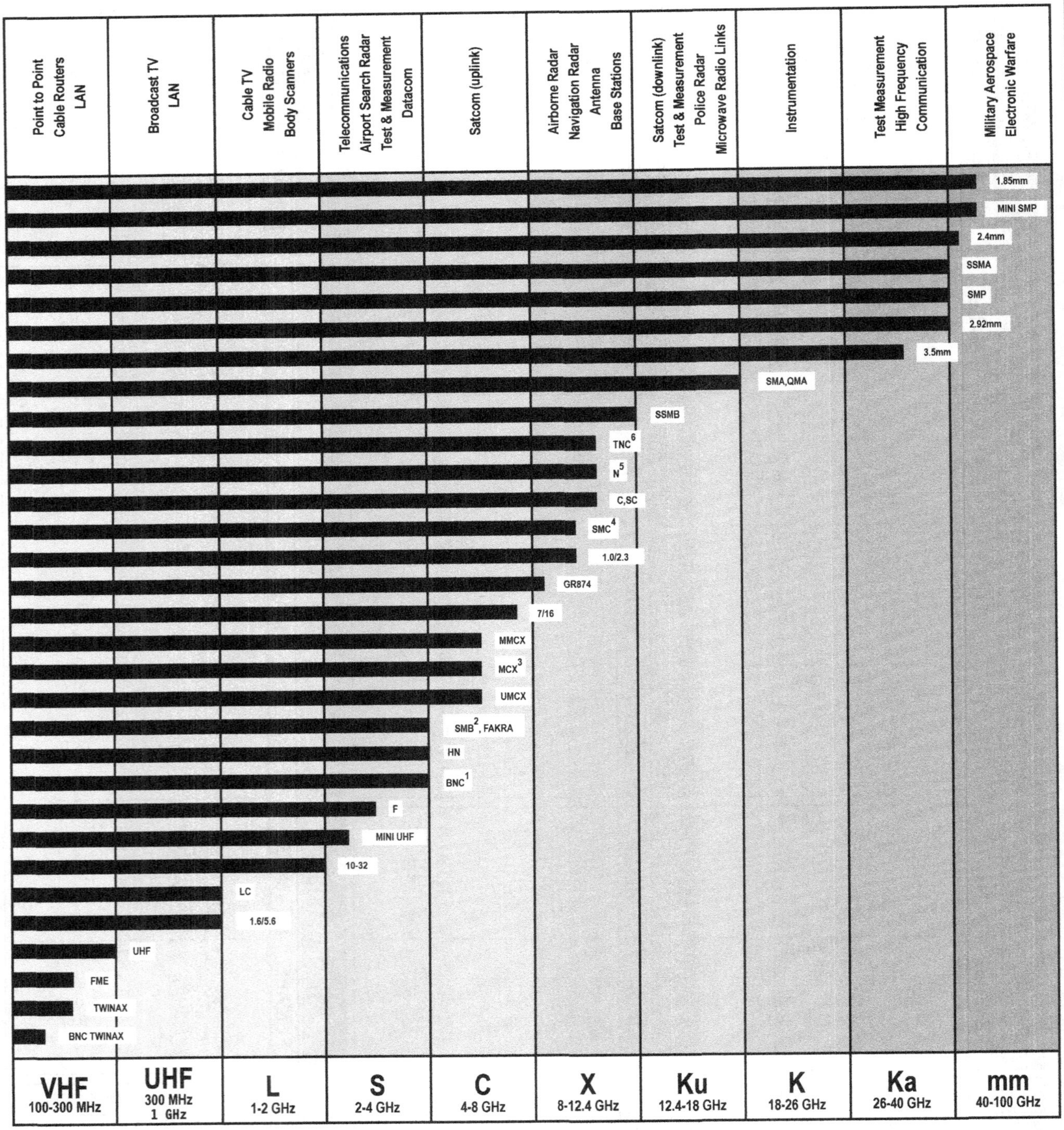

Notes:

1: BNC-75 Ohm connectors operate up to 1 GHz
2: SMB-75 Ohm & Mini SMB-75 Ohm connectors operate up to 4 GHz
3: MCX-75 Ohm connectors operate up to 6 GHz
4: SMC-75 Ohm connectors operate up to 10 GHz
5: N-75 Ohm connectors operate up to 1.5 GHz
6: TNC-75 Ohm connectors operate up to 1 GHz

Figure 8 — Recommended frequency ranges by connector type. [Courtesy of Pasternack]

Avoiding Poor Quality UHF Connectors

[This guidance is provided by Hal Kennedy, N4GG, and is repeated in the ***Station Construction*** *chapter. Many easily-avoided station problems are caused by poor-quality or poorly attached connectors or adapters — Ed.]*

PL-259s have four parts: the outer sleeve called the "knurled nut," the connector body, the insulator/dielectric and the center pin. All four components can be compromised to the point of making a bargain connector useless.

Problems frequently encountered:

- **Finish:** Bargain connectors sometimes have a finish you can't solder to! They may have a chrome-like appearance, but the plating may not take solder well and has to be filed down for a good connection.
- **Threading:** The internal threads at the rear of the body are there to accept a UG-style insert that narrows the connector barrel to accept smaller diameter coax such as RG-8X or RG-58. The threads may be metric! UG inserts also sometimes appear in the US market with metric threads. Either way, the insert will not screw into the body.
- **Dielectric:** Good connectors use quality phenolic or Teflon insulation between the center pin and the body. Bargain connectors might use anything, including materials such as polystyrene, which will melt when the center pin is soldered.
- **Center pin diameter:** This is one of the most common and insidious problems in mystery PL-259s. The center pin outer diameter (OD) is almost always slightly smaller than it should be and it's hard to notice. The center pin connection between a PL-259 and an SO-239 or barrel connector depends on the male side pin OD being correct and the matching fingers on the female side being the correct diameter and made of the proper spring material.
- **Center socket spring tension:** If the SO-239 socket metal relaxes over time and/or temperature, an intermittent connection will be created that can be very hard to track down.
- **Mating indentions:** The indentations on the end of the SO-239 that mate to a PL-259 (the annulus flange) may only have four indentations to match up with the short prongs on the body of the male connector. A quality SO-239 or barrel connector has indentations all the way around. If the PL-259 and SO-239 don't seat completely, an intermittent connection is likely to develop.
- **Tee and right-angle (elbow) UHF adapters:** The center conductor has to make a right-angle turn inside the shell. In poor-quality adapters the right-angle connection is done with a spring contact — these do not hold up. Quality tee and right-angle adapters are reliable because the internal conductors are tapped and threaded — the conductors are screwed together within the body at the right angle junction.

How can we tell the good connectors? If the price is too good to be true — well, it is. PL-259s with good silver plating have a dull appearance. Good connectors have a part number and manufacturer's name stamped into them. You can look up the connector's specifications if it's marked. An example is the connectors made by Amphenol — all of which have parts numbers such as 83-1SP (PL-259) or 83-1R (SO-239) stamped into or onto the connector body.

Table 3
Powdered-Iron Toroidal Cores: Magnetic Properties

There are differing conventions for referring to the type of core material: #, mix and type are all used. For example, all of the following designate the same material: #12, Mix 12, 12-Mix, Type 12 and 12-Type.

Inductance and Turns Formula

The turns required for a given inductance or inductance for a given number of turns can be calculated from:

$$N = 100\sqrt{\frac{L}{A_L}} \qquad L = A_L\left(\frac{N^2}{10{,}000}\right) \qquad \text{(Amidon cores)}$$

$$N = \sqrt{\frac{L}{A_L}} \qquad L = A_L \times N^2 \qquad \text{(Non-Amidon cores)}$$

where N = number of turns; L = desired inductance; A_L = inductance index (µH per 100 turns-squared for Amidon cores; nH or µH per turns-squared for non-Amidon cores; see core data sheet)

Amidon Associates literature gives the value of A_L as inductance per 100 turns but the correct units are inductance per 100 turns-squared. The units of inductance are generally in nH but may also be mH. Make sure you understand which units apply and use the A_L value and formula provided by the manufacturer of the core to calculate number of turns or inductance.

Toroid diameter is indicated by the number following "T" — T-200 is 2.00 in. dia; T-68 is 0.68 in. diameter, etc.

AL Values

	Mix										
Size	**26***	**3**	**15**	**1**	**2**	**7**	**6**	**10**	**12**	**17**	**0**
T-12	na	60	50	48	20	18	17	12	7.5	7.5	3.0
T-16	145	61	55	44	22	na	19	13	8.0	8.0	3.0
T-20	180	76	65	52	27	24	22	16	10.0	10.0	3.5
T-25	235	100	85	70	34	29	27	19	12.0	12.0	4.5
T-30	325	140	93	85	43	37	36	25	16.0	16.0	6.0
T-37	275	120	90	80	40	32	30	25	15.0	15.0	4.9
T-44	360	180	160	105	52	46	42	33	18.5	18.5	6.5
T-50	320	175	135	100	49	43	40	31	18.0	18.0	6.4
T-68	420	195	180	115	57	52	47	32	21.0	21.0	7.5
T-80	450	180	170	115	55	50	45	32	22.0	22.0	8.5
T-94	590	248	200	160	84	na	70	58	32.0	na	10.6
T-106	900	450	345	325	135	133	116	na	na	na	19.0
T-130	785	350	250	200	110	103	96	na	na	na	15.0
T-157	870	420	360	320	140	na	115	na	na	na	na
T-184	1640	720	na	500	240	na	195	na	na	na	na
T-200	895	425	na	250	120	105	100	na	na	na	na

*Mix-26 is similar to the older Mix-41, but can provide an extended frequency range.

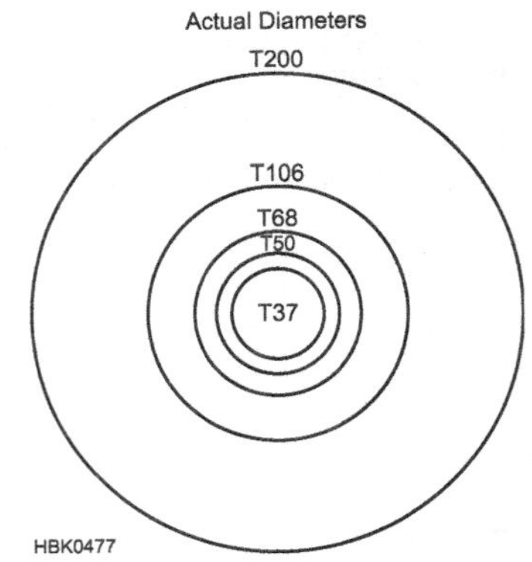

Magnetic Properties Iron Powder Cores

Mix	Color	Material	µ	Temp stability (ppm/°C)	f (MHz)	Notes
26	Yellow/white	Hydrogen reduced	75	825	dc - 1	Used for EMI filters and dc chokes
3	Gray	Carbonyl HP	35	370	0.05 - 0.50	Excellent stability, good Q for lower frequencies
15	Red/white	Carbonyl GS6	25	190	0.10 - 2	Excellent stability, good Q
1	Blue	Carbonyl C	20	280	0.50 - 5	Similar to Mix-3, but better stability
2	Red	Carbonyl E	10	95	2 - 30	High Q material
7	White	Carbonyl TH	9	30	3 - 35	Similar to Mix-2 and Mix-6, but better temperature stability
6	Yellow	Carbonyl SF	8	35	10 - 50	Very good Q and temperature stability for 20-50 MHz
10	Black	Powdered iron W	6	150	30 - 100	Good Q and stability for 40 - 100 MHz
12	Green/white	Synthetic oxide	4	170	50 - 200	Good Q, moderate temperature stability
17	Blue/yellow	Carbonyl	4	50	40 - 180	Similar to Mix-12, better temperature stability, Q drops about 10% above 50 MHz, 20% above 100 MHz
0	Tan	phenolic	1	0	100 - 300	Inductance may vary greatly with winding technique

Courtesy of Amidon Assoc and Micrometals
Note: Color codes hold only for cores manufactured by Micrometals, which makes the cores sold by most Amateur Radio distributors.

Table 4

Powdered-Iron Toroidal Cores: Dimensions

Toroid diameter is indicated by the number following "T" — T-200 is 2.00 in. dia; T-68 is 0.68 in. diameter, etc.
See Table 3 for a core sizing guide.

Red E Cores—500 kHz to 30 MHz (μ = 10)

No.	OD (in)	ID (in)	H (in)
T-200-2	2.00	1.25	0.55
T-94-2	0.94	0.56	0.31
T-80-2	0.80	0.50	0.25
T-68-2	0.68	0.37	0.19
T-50-2	0.50	0.30	0.19
T-37-2	0.37	0.21	0.12
T-25-2	0.25	0.12	0.09
T-12-2	0.125	0.06	0.05

Black W Cores—30 MHz to 200 MHz (μ=6)

No.	OD (In)	ID (In)	H (In)
T-50-10	0.50	0.30	0.19
T-37-10	0.37	0.21	0.12
T-25-10	0.25	0.12	0.09
T-12-10	0.125	0.06	0.05

Yellow SF Cores—10 MHz to 90 MHz (μ=8)

No.	OD (In)	ID (In)	H (In)
T-94-6	0.94	0.56	0.31
T-80-6	0.80	0.50	0.25
T-68-6	0.68	0.37	0.19
T-50-6	0.50	0.30	0.19
T-26-6	0.25	0.12	0.09
T-12-6	0.125	0.06	0.05

Number of Turns vs Wire Size and Core Size

Approximate maximum number of turns—single layer wound—enameled wire.

Wire Size	T-200	T-130	T-106	T-94	T-80	T-68	T-50	T-37	T-25	T-12
10	33	20	12	12	10	6	4	1		
12	43	25	16	16	14	9	6	3		
14	54	32	21	21	18	13	8	5	1	
16	69	41	28	28	24	17	13	7	2	
18	88	53	37	37	32	23	18	10	4	1
20	111	67	47	47	41	29	23	14	6	1
22	140	86	60	60	53	38	30	19	9	2
24	177	109	77	77	67	49	39	25	13	4
26	223	137	97	97	85	63	50	33	17	7
28	281	173	123	123	108	80	64	42	23	9
30	355	217	154	154	136	101	81	54	29	13
32	439	272	194	194	171	127	103	68	38	17
34	557	346	247	247	218	162	132	88	49	23
36	683	424	304	304	268	199	162	108	62	30
38	875	544	389	389	344	256	209	140	80	39
40	1103	687	492	492	434	324	264	178	102	51

Actual number of turns may differ from above figures according to winding techniques, especially when using the larger size wires. Chart prepared by Michel J. Gordon, Jr, WB9FHC.
Courtesy of Amidon Assoc.

NOTES

NOTES

NOTES